# Probabilistic Analysis
and Related Topics
Volume 1

# Contributors

S. D. CHATTERJI
PAO-LIU CHOW
RUTH F. CURTAIN
D. KANNAN
V. MANDREKAR
MARK A. PINSKY

# Probabilistic Analysis and Related Topics

*Edited by* A. T. BHARUCHA-REID

DEPARTMENT OF MATHEMATICS
WAYNE STATE UNIVERSITY
DETROIT, MICHIGAN

**Volume 1**

ACADEMIC PRESS    New York  San Francisco  London    1978
*A Subsidiary of Harcourt Brace Jovanovich, Publishers*

COPYRIGHT © 1978, BY ACADEMIC PRESS, INC.
ALL RIGHTS RESERVED.
NO PART OF THIS PUBLICATION MAY BE REPRODUCED OR
TRANSMITTED IN ANY FORM OR BY ANY MEANS, ELECTRONIC
OR MECHANICAL, INCLUDING PHOTOCOPY, RECORDING, OR ANY
INFORMATION STORAGE AND RETRIEVAL SYSTEM, WITHOUT
PERMISSION IN WRITING FROM THE PUBLISHER.

ACADEMIC PRESS, INC.
111 Fifth Avenue, New York, New York 10003

*United Kingdom Edition published by*
ACADEMIC PRESS, INC. (LONDON) LTD.
24/28 Oval Road, London NW1 7DX

LIBRARY OF CONGRESS CATALOG CARD NUMBER: 77–0531

ISBN 0–12–095601–2

AMS (MOS) 1970 Subject Classifications: 60–06,
60H10, 60H15, 60H20, 60H99

PRINTED IN THE UNITED STATES OF AMERICA

# Contents

*List of Contributors* vii
*Preface* ix

## Stochastic Partial Differential Equations in Turbulence Related Problems

PAO-LIU CHOW

| | | |
|---|---|---|
| I. | Introduction | 2 |
| II. | Theory of Stochastic Partial Differential Equations | 3 |
| III. | Linear Stochastic Partial Differential Equations in Weak Turbulence | 18 |
| IV. | Partial Differential Equations in Stochastic Wave Propagation | 22 |
| V. | Stochastic Equations in Turbulent Transport Theory | 28 |
| VI. | Markovian Model Equations in Turbulence | 33 |
| | References | 40 |

## Estimation and Stochastic Control for Linear Infinite-Dimensional Systems

RUTH F. CURTAIN

| | | |
|---|---|---|
| I. | Introduction | 45 |
| II. | The Semigroup Description of Linear Autonomous Systems | 46 |
| III. | Stochastic Evolution Equations | 51 |
| IV. | Deterministic Quadratic Cost-Control Problem | 60 |
| V. | State Estimation | 64 |
| VI. | The Separation Principle for Stochastic Optimal Control | 69 |
| VII. | Extensions | 73 |
| | References | 84 |

## Random Integrodifferential Equations

D. KANNAN

| | | |
|---|---|---|
| I. | Introduction and Preliminaries | 87 |

|      |                                              |     |
| ---- | -------------------------------------------- | --- |
| II.  | Existence and Uniqueness of Solution         | 103 |
| III. | Some Stochastic Properties of Solution Processes | 126 |
| IV.  | Small Perturbations                          | 148 |
| V.   | Vibrating String                             | 157 |
|      | References                                   | 165 |

## Equivalence and Singularity of Gaussian Measures and Applications

S. D. CHATTERJI AND V. MANDREKAR

|       |                                                       |     |
| ----- | ----------------------------------------------------- | --- |
| I.    | Introduction                                          | 169 |
| II.   | General Problem of Equivalence and Singularity        | 171 |
| III.  | Reproducing Kernel Hilbert Spaces and Gaussian Processes | 175 |
| IV.   | Equivalence and Singularity of Gaussian Processes     | 180 |
| V.    | Conditions for Equivalence: Special Cases             | 185 |
| VI.   | Applications                                          | 189 |
| VII.  | Concluding Remarks                                    | 194 |
|       | Appendix                                              | 194 |
|       | References                                            | 195 |

## Stochastic Riemannian Geometry

MARK A. PINSKY

|      |                                       |     |
| ---- | ------------------------------------- | --- |
| I.   | Introduction                          | 199 |
| II.  | Brownian Motion                       | 201 |
| III. | Semilocal Properties                  | 210 |
| IV.  | Asymptotic Properties, $t \to 0$      | 216 |
| V.   | Asymptotic Properties, $t \to \infty$ | 221 |
| VI.  | Bibliographical Remarks               | 234 |
|      | References                            | 234 |

*Index*  237

# List of Contributors

Numbers in parentheses indicate the pages on which the authors' contributions begin.

S. D. CHATTERJI (169), Département de Mathématiques, Ecole Polytechnique Federale de Lausanne, Lausanne, Switzerland

PAO-LIU CHOW (1), Department of Mathematics, Wayne State University, Detroit, Michigan

RUTH F. CURTAIN (45), Mathematics Institute, University of Groningen, The Netherlands

D. KANNAN (87), Department of Mathematics, University of Georgia, Athens, Georgia and Department of Mathematics and Statistics, University of Guelph, Guelph, Ontario, Canada

V. MANDREKAR (169), Department of Statistics and Probability, Michigan State University, East Lansing, Michigan

MARK A. PINSKY (199), Department of Mathematics, Northwestern University, Evanston, Illinois

# Preface

Probabilistic analysis is that branch of the general theory of random functions (or stochastic processes) that is primarily concerned with the analytical properties of random functions. Early research in the field was concerned with the continuity, differentiability, and integrability of random functions. In recent years probabilistic analysis has evolved into a very dynamic area of mathematical research that utilizes and extends concepts and results from functional analysis, operator theory, measure theory, and numerical analysis, as well as other branches of mathematics. The study of random equations is one of the most active areas of probabilistic analysis, and many recent results in the field are due to research on various classes of random equations.

"Probabilistic Analysis and Related Topics," which will be published in several volumes at irregular intervals, is devoted to current research in probabilistic analysis and its applications in the mathematical sciences. We propose to cover a rather wide range of general and special topics. Each volume will contain several articles, and each article will be by an expert in the subject area. Although these articles are reasonably self-contained and fully referenced, it is assumed that the reader is familiar with measure-theoretic probability, the basic classes of stochastic processes, functional analysis, and various classes of operator equations. The individual articles are not intended to be popular expositions of the survey type, but are to be regarded, in a sense, as brief monographs that can serve as introductions to specialized study and research.

In view of the above aims, the nature of the subject matter, and the manner in which the text is organized, these volumes will be addressed to a broad audience of mathematicians specializing in probability and stochastic processes, applied mathematical scientists working in those areas in which probabilistic methods are being employed, and other research workers interested in probabilistic analysis and its potential applicability in their respective fields.

# Stochastic Partial Differential Equations in Turbulence Related Problems[*]

*PAO-LIU CHOW*

DEPARTMENT OF MATHEMATICS
WAYNE STATE UNIVERSITY
DETROIT, MICHIGAN

| | | |
|---|---|---|
| I. | Introduction | 2 |
| II. | Theory of Stochastic Partial Differential Equations | 3 |
| | A. Introduction | 3 |
| | B. Probabilistic Background | 4 |
| | C. Stochastic Partial Differential Equations | 9 |
| | D. Functional Differential Equations for Solution Processes | 13 |
| III. | Linear Stochastic Partial Differential Equations in Weak Turbulence | 18 |
| | A. Introduction | 18 |
| | B. Linearized Navier–Stokes Equations with Random Forcing | 19 |
| | C. Computations of Solutions | 20 |
| IV. | Partial Differential Equations in Stochastic Wave Propagation | 22 |
| | A. Introduction | 22 |
| | B. Parabolic Approximation in Random Wave Propagation | 24 |
| | C. Method of Functional Integration | 25 |
| | D. Some Related Problems | 26 |
| V. | Stochastic Equations in Turbulent Transport Theory | 28 |
| | A. Introduction | 28 |
| | B. A Random Evolution Model | 29 |
| | C. Turbulent Transport with Molecular Diffusion | 30 |
| VI. | Markovian Model Equations in Turbulence | 33 |
| | A. Introduction | 33 |
| | B. A Langevin Equation Model | 33 |
| | C. Random Coupling Model | 34 |
| | D. Existence of Solutions | 37 |
| | References | 40 |

[*] The preparation of this article was supported by the US Army Research Office at Research Triangle Park, North Carolina, under Grant No. DAAG29-76-G-0141, and by NASA Langley Research Center, Hampton, Virginia, under Grant NSG1330.

## I. Introduction

The theory of stochastic differential equations has become increasingly popular because of its importance in applications. After the first appearance of Uhlenbeck and Ornstein's paper [82] on a dynamical theory of Brownian motion, the question on the mathematical meaning of the stochastic equation and its solution led Doob [28] and K. Itô [47] to introduce the notion of stochastic integrals, generalizing Wiener's result [85]. In particular, the work of Itô has been monumental in the subsequent development of this subject. His definition of stochastic integrals, generally known as *Itô integrals*, has been adopted by many workers as the basis for studying stochastic differential or integral equations and their ramifications. For complete theory and extensive references, one is referred to the books by Bharucha-Reid [7], Friedman [34], Gikhman and Skorokhod [39], and McKean [65].

Until 1960, most works on stochastic equations had been confined to theory of stochastic *ordinary* differential equations (stochastic ODEs). Since then, spurred by the demand from modern technology, stochastic *partial* differential equations (PDEs) have started to attract the attention of many researchers, who are mainly interested in their applications to physical and biological sciences. Notable examples are turbulent fluid dynamics [4], stochastic wave propagation [51], turbulent transport theory [18], quantum mechanics [67], and population biology [25]. The progress in this field has been slow for several reasons. The main reason may be attributed to the difficulty in dealing with multiparameter (space-time) stochastic processes and the greater variability of solutions to PDEs. Another reason is the lack of concerted efforts among theorists and practitioners in this area. In the early days, the classical theory of PDEs was advanced rapidly by physicists and mathematicians in close collaboration. However, this has not yet happened to stochastic PDEs. Most of the probabilistic analysts working on this subject are preoccupied with existence, uniqueness, and measure-theoretical questions on solutions of abstract equations, ignoring specific questions such as how to solve a concrete problem. On the other hand, the applied scientists tend to seek their solutions in the classical domain and are skeptical about the potential usefulness of certain abstract results put forth by the analysts.

This chapter constitutes an attempt to bridge the distance between divergent ways of thinking reflected in theory and practice. It will be done from the vantage point of an applied mathematician. To appeal to analysts who might be induced to work in this area, we shall present a few loosely assembled topics biased toward applications. At the same time, it will be shown that certain abstract results from probabilistic functional analysis may be useful in the study of stochastic PDEs.

The remaining portion consists of five sections. In Section II we present several relevant results in probabilistic analysis, especially Gaussian measures in function spaces and the theory of stochastic PDEs of Itô type. Since problems in turbulence provide the major source of stochastic PDEs, Sections III–VI will be devoted to these problems. To be specific, Section III is concerned with the analysis of linearized Navier–Stokes equations with a random forcing. Stochastic equations for waves in random media will be discussed in Section IV. The next section pertains to certain model equations in turbulent transport theory. Finally, in Section VI, some Markovian model equations in turbulence are treated from the viewpoint of stochastic differential equations.

Most results given in this chapter are not new. However, our approach to stochastic PDEs by employing the theories of Brownian motion and diffusion in infinite dimensions, functional differential equations, and functional integration is not conventional. Furthermore, certain minor and preliminary results included in this chapter have not been published elsewhere. Also, some specific open questions will be pointed out with the hope that they may challenge some capable analysts to get these problems solved.

## II. Theory of Stochastic Partial Differential Equations

### A. Introduction

Let $D$ be a domain in $\mathbb{R}^N$ with a smooth boundary $\partial D$. We consider the system of $M$ partial differential equations of evolutional type

$$\partial_t u_j = \sum_{k=1}^{M} \sum_{l,m=1}^{N} a_{jk}^{lm}(t,x)\, \partial_{lm}^2 u_k + \sum_{k=1}^{M} \sum_{l=1}^{N} b_{jk}^{l}(t,x,u)\, \partial_l u_k + f_j(t,x,u),$$

$$0 < t < T, \quad x \in D, \quad j = 1, 2, \ldots, M, \tag{2.1}$$

where $u = (u_1, u_2, \ldots, u_M)$, $\partial_t = \partial/\partial t$, $\partial_j = \partial/\partial x_j$, and $\partial_{lm}^2 = \partial_l \partial_m$; the functions $a_{jk}^{lm}$ may be complex-valued and are defined on $D_T = (0,T) \times D$; $b_{jk}^l$ and $f_j$ are real or complex-valued functions on $D_T \times \mathbb{R}^M$.

For each $y \in \mathbb{R}^N$, define the $M \times M$ matrix-valued functions

$$A(t,x,iy) = \left\{ \sum_{l,m=1}^{N} a_{jk}^{lm}(t,x)(iy_l)(iy_m) \right\}_{M \times M}, \tag{2.2}$$

$$B(t,x,u,iy) = \left\{ \sum_{l=1}^{N} b_{jk}^{l}(t,x,u)(iy_l) \right\}_{M \times M}, \tag{2.3}$$

where $i = \sqrt{-1}$ and $(t,x) \in D_T$.

If $a_{jk}^{lm} \not\equiv 0$, let $\lambda_j(t,x,y)$, $j = 1, 2, \ldots, M$, be the eigenvalues of the principal matrix $A(t,x,iy)$, and let

$$\lambda_0(y) = \max_{1 \leq j \leq M} \sup_{(t,x) \in D_T} \operatorname{Re}\{\lambda_j(t,x,y)\}. \tag{2.4}$$

Then the system (2.1) is said to be (uniformly and strongly) *parabolic* in $D_T$ if $A$ is *elliptic*, that is, if there exists a constant $\delta > 0$ such that

$$\sup_{|y|=1} \lambda_0(y) \leq -\delta, \quad \text{where} \quad |y| = \left(\sum_{j=1}^{N} y_j^2\right)^{\frac{1}{2}}. \tag{2.5}$$

The system is *hyperbolic* in $D_T$ if $\lambda_0 \equiv 0$.

If $a_{jk}^{lm} \equiv 0$, we let $\sigma_j(t, x, u, y)$, $j = 1, 2, \ldots, M$, denote the eigenvalues of the secondary matrix $B(t, x, u, y)$. Then the reduced equations of (2.1) form a *hyperbolic system* in $D_T$ if $\text{Re}\{\sigma_j\} \equiv 0$, $j = 1, 2, \ldots, M$.

Suppose that any one of coefficients $a_{jk}^{lm}$, $b_{jk}^{l}$, or $f_j$ depends on a space-time random process. Then Eqs. (2.1) are termed *random* or *stochastic partial differential equations*. In what follows, we shall first collect a set of results pertinent to studying stochastic PDEs. Most results are taken from a recent book by Kuo [58] on Gaussian measures in Banach spaces. Then the existence theorems for stochastic PDEs of Itô type are discussed. But we shall be confined to linear and monotone, nonlinear cases. Finally, we shall employ the method of functional differential equations to study the solution processes.

### B. Probabilistic Background

Let $\{\Omega, \mathscr{S}, \mu\}$ be a fixed probability space, where $\Omega$ is a set with elements $\omega$; $\mathscr{S}$, the $\sigma$-field generated by Borel subsets of $\Omega$; and $\mu$, a probability measure. We denote by $\mathscr{H}$ a real, separable Hilbert space consisting of $\mathbb{R}^M$-valued functions on $D$ with the inner product $\langle \cdot, \cdot \rangle$, and norm $|\cdot|$. If $\mathscr{F}$ is the Borel field of $\mathscr{H}$, and $\theta: \Omega \to \mathscr{H}$ is a single-valued mapping, then $P$ is a probability measure on $(\mathscr{H}, \mathscr{F})$ induced by $\theta$ so that $\forall F \in \mathscr{F}$, $P\{F\} = \mu\{\theta(\omega) \in F\}$. These two probability spaces will be used interchangeably for convenience.

Given the probability space $(\mathscr{H}, \mathscr{F}, P)$, the *characteristic functional* of $P$ is defined by [83]

$$Q(\lambda) = E\{e^{i\langle \lambda, h \rangle}\} = \int e^{i\langle \lambda, h \rangle} P(dh), \quad \lambda \in \mathscr{H}, \tag{2.6}$$

where $E\{\cdot\}$ signifies the mathematical expectation and the integration is over $\mathscr{H}$. This functional is of basic importance in stochastic processes [75, 78]. It is positive definite with $|Q(\lambda)| \leq 1$. For every $f, g \in \mathscr{H}$, the *mean vector* $a$ and the *covariance operator* $R$ in $\mathscr{H}$ are defined as

$$\langle f, a \rangle = E\langle f, h \rangle = \int \langle f, h \rangle P(dh), \tag{2.7}$$

$$\langle Rf, g \rangle = E\{\langle f, h-a \rangle \langle g, h-a \rangle\} = \int \langle f, h-a \rangle \langle g, h-a \rangle P(dh). \tag{2.8}$$

If they exist, $a$ is an element of $\mathscr{H}$ and $R$ is a positive semidefinite self-adjoint operator of trace class, i.e.,

$$\operatorname{tr} R = \int |h-a|^2 P(dh) < \infty. \tag{2.9}$$

In $\mathscr{H}$, let $\|\cdot\|$ denote another norm, and $\{\pi_n\}$ be the set of finite-dimensional projections. Then $\|\cdot\|$ is called a *P-measurable norm* if, for every $\varepsilon > 0$, there exists a finite-dimensional projection $\pi_0$ such that

$$P\{\|\pi h\| > \varepsilon\} < \varepsilon$$

whenever $\pi$ is a finite-dimensional projection of $\mathscr{H}$ orthogonal to $\pi_0$. Let $\mathscr{B}$ denote the completion of $\mathscr{H}$ with respect to the norm $\|\cdot\|$, and let $i$ be the inclusion map of $\mathscr{H}$ into $\mathscr{B}$. The triplet $(i, \mathscr{H}, \mathscr{B})$ is called the *abstract Wiener space* [42, 43], which will be abbreviated as *Wiener space* hereafter. Throughout, we shall assume that the Banach space $\mathscr{B}$ has a Schauder basis.

For each $t \geq 0$, let $P_t$ be the *Gaussian measure* on $\mathscr{H}$ of the parameter $t$. Then its characteristic functional takes the form

$$Q_t(\lambda) = e^{-t\langle \lambda, \lambda \rangle/2}. \tag{2.10}$$

A *Wiener process* or a *Brownian motion* $\{W(t), t \geq 0\}$ in $(\mathscr{H}, \mathscr{B})$ is a $\mathscr{B}$-valued stochastic process of independent increments so that for every $F \in \mathscr{F}$ and $f, g \in \mathscr{B}^*$, the dual of $\mathscr{B}$, we have

$$\operatorname{Prob}\{W(t) \in F\} = P_t\{F\}, \tag{2.11}$$

$$E(f, W(t)) = 0, \tag{2.12}$$

$$E(f, W(t))(g, W(s)) = \min(t, s)\langle f, g \rangle, \tag{2.13}$$

where $(\cdot, \cdot)$ denotes the duality between $\mathscr{B}$ and $\mathscr{B}^*$.

Suppose that $\mathscr{X}$ is a Banach space with norm $\|\cdot\|$ and $\mathscr{F}_t$ designates the complete $\sigma$-field generated by $\{W(s), 0 \leq s \leq t\}$. An $\mathscr{X}$-valued stochastic process $\xi(t, \omega)$ is called *nonanticipating* if $\xi$ is jointly measurable in $(t, \omega)$ and is $\mathscr{F}_t$-measurable for each $t \geq 0$. We denote by $\mathscr{M}[\mathscr{X}]$ the space of nonanticipating processes $\xi(t)$ in $\mathscr{X}$ such that $E \int_0^s \|\xi(t)\| dt < \infty$, $\forall s \geq 0$. A process $\xi_0(t) \in \mathscr{M}[\mathscr{X}]$ is *simple* if there exists a finite set of points $0 = t_0 < t_1 < \cdots < t_n$, and $\xi_0(t) = \xi(t_j)$, for $t_j \leq t < t_{j+1}$, $j = 0, 1, \ldots, n-1$, $\xi(t) = \xi(t_n)$, $t \geq t_n$. We let $\mathscr{K}$ be a Hilbert space and set $\mathscr{X} = \mathscr{L}(\mathscr{B}, \mathscr{K})$, the space of bounded linear operators from $\mathscr{B}$ into $\mathscr{K}$. Also, we denote by $\mathscr{L}_{(2)}(\mathscr{H}, \mathscr{K})$ the space of *Hilbert-Schmidt* operators $A: \mathscr{H} \to \mathscr{K}$, for which the Hilbert-Schmidt norm of $A$, $\|A\|_{(2)} = \{\sum_{j=1}^{\infty} \|Ae_j\|_{\mathscr{K}}^2\}^{\frac{1}{2}} < \infty$, for some orthonormal basis $\{e_j\}$ of $\mathscr{H}$. Now we introduce a stochastic integral $J_\xi$ of $\xi$ over the Wiener space $(\mathscr{H}, \mathscr{B})$ in the form

$$J_\xi(t, \omega) = \int_0^t \xi(s, \omega) dW(s, \omega), \tag{2.14}$$

which takes values in $\mathscr{M}[\mathscr{X}]$. For a simple process, $\xi_0(t) \in \mathscr{L}(\mathscr{B}, \mathscr{K})$, the integral is defined as

$$J_{\xi_0}(t) = \sum_{j=0}^{n-1} \xi(t_j)[W(t_{j+1}) - W(t_j)] + \xi(t_n)[W(t) - W(t_n)]. \quad (2.15)$$

It is easy to check that $J_{\xi_0}(t)$ is a continuous martingale with

$$EJ_{\xi_0}(t) = 0, \quad (2.16)$$

$$E\|J_{\xi_0}(t)\|_{\mathcal{K}}^2 = E\int_0^t \|\tilde{\xi}_0(s)\|_{(2)}^2 ds, \quad (2.17)$$

where $\tilde{\xi}$ means the restriction of $\xi$ to $\mathcal{L}_{(2)}(\mathcal{H}, \mathcal{K})$, hence, a Hilbert–Schmidt operator. By means of (2.17), it can be shown [58] that for every $\xi \in \mathcal{M}[\mathcal{L}_{(2)}(\mathcal{H}, \mathcal{K})]$ there exists a sequence of simple processes $\{\xi_n\}$ in $\mathcal{M}[\mathcal{L}(\mathcal{B}, \mathcal{K})]$ so that $\{J_{\xi_n}\}$ converges to $J_\xi$ in the mean square, i.e.,

$$\lim_{n\to\infty} E\|J_{\xi_n}(t) - J_\xi(t)\|_{\mathcal{K}}^2 = \lim_{n\to\infty} E\int_0^t \|\tilde{\xi}_n(s)\|_{(2)}^2 ds = 0. \quad (2.18)$$

In fact, the following theorem holds.

**Theorem 2.1** *To each $\xi \in \mathcal{M}[\mathcal{L}_{(2)}(\mathcal{H}, \mathcal{K})]$, there corresponds a unique stochastic integral* (2.14) *such that*

(a) $J_\xi(t)$ *is a martingale with continuous sample paths,*
(b) $\text{Prob}\{\sup_{0 \leq t \leq s} \|J_\xi(t)\|_{\mathcal{K}} > \delta\} \leq \delta^{-2} E\|J_\xi(s)\|_{\mathcal{K}},$ (2.19)
(c) $EJ_\xi(t) = 0$ *and* $E\|J_\xi(t)\|_{\mathcal{K}}^2 = E\int_0^t \|\xi(s)\|_{(2)}^2 ds.$ (2.20)

*Remarks* (a) By convention, set

$$\mathcal{L}_{(2)}(\mathcal{H}, \mathbb{R}) = \mathcal{H}, \quad \text{and} \quad J_\xi(t) = \int_0^t \langle \xi(s), dW(s) \rangle.$$

(b) Let $R = S^*S$ and $S^*$ be the adjoint of $S$. If $S \in \mathcal{L}(\mathcal{B}, \mathcal{K})$ and its restriction $\tilde{S}$ to $\mathcal{H}$ is Hilbert–Schmidt, then $J(t) = \int_0^t S\, dW(t) = SW(t)$ is a Gaussian process in $\mathcal{K}$, and its formal derivative $\dot{J}(t) = S\dot{W}(t)$ will be called a *Gaussian white noise*.

(c) Sometimes for physical reasons [80, 86] we need to define a stochastic integral as the limit of a sequence of sums similar to (2.15), where $\xi(t_j) = \xi[t_j, W(t_j)]$ is replaced by $\xi\{t_j, [W(t_j)+W(t_{j+1})]/2\}$. This integral $\hat{J}_\xi(t) = \int_0^t \xi(s)\, \hat{d}W(s)$ will be called a *central stochastic integral*.

The proof of this theorem was given by Kuo [58] and Remark (c) can be verified as in finite dimensions. Abstract stochastic integrals have also been defined, in different settings, by Cabaña [11], Daletskii [24], Kannan and Bharucha-Reid [50], and others. To proceed, we need a few definitions from functional calculus.

Let $\phi$ be a function defined in an open subset $V$ of $\mathscr{B}$ with values in a Banach space $\mathscr{X}$. The function $\phi$ is said to be *Fréchet differentiable* at $\eta \in V$ if there exists $\phi' \in \mathscr{L}(\mathscr{B}, \mathscr{X})$ such that $\|\phi(\eta+\xi) - \phi(\eta) - \phi'(\xi)\|_{\mathscr{X}} = o(\|\xi\|)$ for every $\xi \in \mathscr{B}$. The $k$th *Fréchet derivative* $\phi^{(k)}$, $k \geq 2$, is similarly defined as a $k$-linear operator on the product space

$$\mathscr{B}^k = \overbrace{\mathscr{B} \cdot \mathscr{B} \cdot \cdots \cdot \mathscr{B}}^{k}.$$

Since Fréchet differentiability is too restrictive, a weaker notion of differentiability in $\mathscr{H}$ is necessary. For $h \in \mathscr{H} \cap (V - \eta)$, let $\hat{f}(h) = \phi(\eta + h)$. The function $\phi$ is said to be $k$-times *differentiable in the direction of $\mathscr{H}$* or, simply, $\mathscr{H}$-*differentiable* at $\eta$ if $\hat{f}(h)$ is $k$-times Fréchet differentiable at 0. In this case we set $\delta^k \phi(\eta) = \hat{f}^{(k)}(0)$. As it turns out, any (bounded) uniformly continuous function $\phi$ on $\mathscr{B}$ can be approximated uniformly by $\mathscr{H}-C^\infty$ functions. In applications, of course, the approximation theory is important.

Now we turn to a fundamental result in stochastic calculus in Wiener space. The following theorem is known as Itô's lemma in infinite dimensions.

**Theorem 2.2** *Let $\mathscr{X}$ be the completion of the Hilbert space $\mathscr{H}$ with respect to a weaker norm in $\mathscr{H}$. Let $\phi(t, \eta)$ be a real-valued continuous function on $[0, \infty] \times \mathscr{X}$, satisfying the following:*

(a) *For each $\eta \in \mathscr{X}$, $\phi(\cdot, \eta)$ is continuously differentiable and $\partial \phi / \partial t$ is continuous on $[0, \infty) \times \mathscr{X}$.*

(b) *For each $t \geq 0$, $\phi(t, \cdot)$ is twice $\mathscr{H}$-differentiable with continuous first two derivatives from $[0, \infty) \times \mathscr{X}$ into $\mathscr{H}$ and the trace class $\mathscr{L}_{(1)}(\mathscr{H})$, respectively.*

*Suppose that for a given $\eta \in \mathscr{X}$,*

$$u(t) = \eta + \int_0^t a(s)\,ds + \int_0^t \xi(s)\,dW(s), \tag{2.21}$$

*where $a \in \mathscr{M}[\mathscr{H}]$ and $\xi \in \mathscr{M}[\mathscr{L}_{(2)}(\mathscr{H}, \mathscr{X})]$. Then*

$$\phi[t, u(t)] = \phi(0, \eta) + \int_0^t \langle \xi^*(s)\delta\phi[s, u(s)], dW(s) \rangle$$
$$+ \int_0^t \left\{ \frac{\partial \phi}{\partial s}[s, u(s)] + \langle \delta\phi[s, u(s)], a(s) \rangle \right.$$
$$\left. + \frac{1}{2} \operatorname{tr} \xi^*(s)\delta^2\phi[s, u(s)]\xi(s) \right\} ds. \tag{2.22}$$

*Remarks* (a) By convention, the integral formula (2.21) is taken to be equivalent to the differential form

$$du(t) = a(t)\,dt + \xi(t)\,dW(t), \quad u(0) = \eta. \tag{2.23}$$

(b) If the stochastic integral in (2.21) is replaced by a *central* stochastic integral $\int_0^t \xi(s)\, \partial W(s)$ (see Remark (c) after Theorem 2.1). The last term $(\frac{1}{2} \operatorname{tr} \xi^* \delta^2 \phi \xi)$ should be replaced by $\frac{1}{2} \operatorname{tr}(\delta \xi^* \delta \phi) \xi$.

The proof of this theorem is omitted here. We wish to point out that there is no analog of Itô's formula (2.22) if the Wiener process $W(t)$ is replaced by a general Gaussian process. However, by an integration by parts in function space, formula (2.20) can be generalized to other Gaussian processes. This was first done by Cameron [13], for Wiener measure, and then extended to other Gaussian measures by Boylan [10], and Daletskii and Paramonov [24], and to the most general form by Kuo [57]. Let $P$ be a centered Gaussian measure on the Hilbert space $\mathcal{H}$ with the covariance operator $R$. Then the following theorem holds.

**Theorem 2.3** *Let $f: \mathcal{H} \to \mathbb{R}$ be a bounded measurable function and $\mathcal{H}$-differentiable. Then, for every $h, h' \in \mathcal{H}$, the following integral formulas hold:*

(a) $$\int_{\mathcal{H}} f(\eta) \langle h, \eta \rangle P(d\eta) = \int_{\mathcal{H}} \langle \delta f(\eta), Rh \rangle P(d\eta) \tag{2.24}$$

(b) $$\int_{\mathcal{H}} f(\eta) \langle h, \eta \rangle \langle h', \eta \rangle P(d\eta) = \int_{\mathcal{H}} f(\eta) \langle Rh, h' \rangle P(d\eta)$$
$$+ \int_{\mathcal{H}} \langle \delta^2 f(\eta) Rh, h' \rangle P(d\eta). \tag{2.25}$$

A short proof of this theorem is provided by a corresponding theorem on Wiener measure [57]. To do so, let us introduce the Hilbert space $\mathcal{H}_R = R^{1/2} \mathcal{H}$ with inner product $\langle h, h' \rangle_R = \langle R^{1/2} h, R^{1/2} h' \rangle$. Then $(i, \mathcal{H}_R, \mathcal{H})$ is a Wiener space [58] on which the Wiener measure $P_t$ is defined. We observe that the Gaussian measure $P$ on $\mathcal{H}$ is the same as the Wiener measure $P_1$ on $(\mathcal{H}_R, \mathcal{H})$. By Theorem 1 of Kuo [57],

$$\int f(\eta) \langle h^*, \eta \rangle_R P_1(d\eta) = \int f'(\eta)(h^*) P_1(d\eta), \quad h^* \in \mathcal{H}^*, \tag{2.26}$$

where $f'$ is defined as

$$f'(x)(h) = \lim_{\varepsilon \to 0} \frac{f(x + \varepsilon h) - f(x)}{\varepsilon}, \quad h \in \mathcal{H}_R. \tag{2.27}$$

But,

$$\int f(\eta) \langle h', \eta \rangle_R P_1(d\eta) = \int f(\eta) \langle R^{-1} h', \eta \rangle P_1(d\eta)$$
$$= \int f(\eta) \langle h, \eta \rangle P(d\eta), \quad \text{with} \quad h = R^{-1} h'.$$
$$\tag{2.28}$$

Noting (2.26) and the fact that $f'(\eta)(Rh) = \langle \delta f(\eta), Rh \rangle$, we get formula (2.24) from (2.28). Formula (2.25) can be verified by a repeated application of (2.24).

In passing, we give a linear transformation formula for Wiener measure. Suppose that $T$ is a linear operator of $\mathscr{B}$. Let $P_t \circ T$ denote the linear transformation of Wiener measure $P_t$ under $T$. Then we have

**Theorem 2.4** *Let $T = I + K$ be a linear transformation $\mathscr{B}$ into itself, where $I$ is the identity operator and $K$ satisfies*

(a) $K(\mathscr{B}) \subset \mathscr{H}$,
(b) $(I+K)^{-1}$ *exists in* $\mathscr{H}$,
(c) $K \in \mathscr{L}_{(1)}(\mathscr{H})$.

*Then $P_t \circ T$ and $P_t$ are equivalent and the Radon–Nikodym derivative is given by*

$$dP_t \circ T/dP_t = \det|T| \cdot \exp -\tfrac{1}{2}\{2(K\eta, \eta) + |K\eta|^2\}, \qquad \eta \in \mathscr{B}, \quad (2.29)$$

*where $\det|T|$ denotes the (Fredholm) determinant of $T$.*

This formula may be useful in studying the solution process of certain linear equations with a random forcing term.

### C. Stochastic Partial Differential Equations

For many statistical models in turbulence, the governing equations are of the form (2.1), where the coefficients $b_{jk}^l$ and $f_j$ and the initial states of $u_j$ may be random functions. Thus we have to consider the *stochastic initial-boundary value problem*

$$\partial_t u = A(t, x, \nabla)u + V(t, x, u, \nabla u, \omega) + f(t, x, u, \nabla u, \omega),$$
$$(t, x) \in D_T, \qquad (2.30)$$
$$u(0, x, \omega) = \eta(x, \omega), \qquad x \in D, \qquad (2.31)$$
$$\alpha u + \beta(\partial u/\partial n)|_{\partial D} = 0, \qquad 0 \leq t \leq T. \qquad (2.32)$$

Here $u = (u_1, u_2, \ldots, u_M)$, $\nabla = (\partial_1, \partial_2, \ldots, \partial_N)$; the definition of $A$ is given by (2.2); $V$ is a generalization of $B$; and $V(t, x, \cdot, \cdot, \omega)$ may be an integral operator; $\alpha, \beta$ are constant $M \times M$ matrices, and $\partial/\partial n$ denotes the directional derivative in the outward normal direction to $\partial D$. The random system (2.30) is classified to be *parabolic* or *hyperbolic* if the classification holds a.s. in $\omega$.

For convenience, we shall suppress the boundary condition (2.32) and consider the problem (2.30)–(2.32) as an initial value (Cauchy) problem for a vector $u$ in function space. To this end, let $\mathscr{H}$ be a Hilbert space (possibly complex) with inner product $\langle \cdot, \cdot \rangle$ and norm $|\cdot|$; $\mathscr{X}$, a Banach space contained in $\mathscr{H}$ with norm $\|\cdot\|$; $\mathscr{X}^*$, the dual of $\mathscr{X}$ with norm $\|\cdot\|_*$; and $(f, g)$, the natural pairing between $f \in \mathcal{X}^*$, $g \in \mathcal{X}$. We assume that $\mathcal{X}$ and $\mathcal{H}$ are separable, $\mathcal{X}$ is reflexive, and the injections

$$\mathcal{X} \subset \mathcal{H} \subset \mathcal{X}^*$$

are continuous, where each space is dense in the next. In the sequel, $\mathcal{H}$ usually denotes the space or subspace of $\mathbb{R}^M$-valued square-integrable functions on $D$ to be signified by $\mathbb{L}^2(D)$, in contrast with the scalar $L^2$-space, while $\mathcal{X}$ is normally taken to be a dense subspace of the Sobolev space [33]

$$W^{m,p}(D) = \{\mathbf{f} \in \mathbb{L}^p(D) \mid \partial_j^k f \in \mathbb{L}^p(D), j = 1, 2, \ldots, N; k = 1, 2, \ldots, m\},$$

$$p \geq 2. \tag{2.33}$$

In particular, we put

$$H^m(D) = W^{m,2}(D). \tag{2.34}$$

If $C_0^\infty(D)$ denotes the space of infinitely differentiable, $\mathbb{R}^M$-valued functions with compact support in $D$, we let $W_0^{m,p}(D)$ be the completion of $C_0^\infty(D)$ in $W^{m,p}(D)$.

Suppose that $\mathcal{X}$ and $\mathcal{H}$ are dense subspaces of $H_0^1(D)$ and $\mathbb{L}^2(D)$, respectively, with $\mathcal{X} \subset \mathcal{H}$. We consider $u(t) = u(t, \cdot)$ as a vector in $\mathcal{X}$. Similarly, by dropping the dependence on $x$, let $A(t) = A(t, \cdot, \nabla)$ be a linear operator from $\mathcal{X}$ into $\mathcal{X}^*$ defined, for every $u, v \in \mathcal{X}$, by

$$(A(t)u, v) = - \sum_{l,m=1}^{N} \sum_{j,k=1}^{M} \int_D \partial_l [a_{jk}^{lm}(t, x) v_j] \partial_m u_k \, dx, \tag{2.35}$$

where the coefficients $a_{jk}^{lm}(t, x)$ are assumed to be continuous and their partial derivatives are bounded, continuous on $\bar{D}_T$. Also, we regard $V(t, u, \omega) = V(t, \cdot, u, \nabla u, \omega)$, $f(t, u, \omega) = f(t, \cdot, u, \nabla u, \omega)$, and $\eta(\omega) = \eta(\cdot, \omega)$ as vectors in $\mathcal{H}$, where $u \in \mathcal{X}$. Thereby we can recast the concrete problem (2.30)–(2.32) as an abstract initial value (Cauchy) problem in $\mathcal{X}^*$:

$$du/dt = A(t)u + V(t, u, \omega) + f(t, u, \omega), \quad 0 < t < T, \tag{2.36}$$

$$u(0) = \eta(\omega). \tag{2.37}$$

Of particular interest is the abstract equation of Itô type. In this case, the forcing term $f(t, u, \omega)$ is a Gaussian white noise (see remark (b) after Theorem 2.1), and Eq. (2.36) becomes

$$du = A(t) \, dt + V(t, u, \omega) \, dt + \xi(t, u) \, dW(t), \tag{2.38}$$

where $W(t)$ is the Wiener process on $(\mathcal{H}, \mathcal{B})$ and $\xi(\cdot, u) \in \mathcal{M}[\mathcal{L}_{(2)}(\mathcal{H}, \mathcal{K})]$.

If $A$ is independent of $t$, and generates an equicontinuous semigroup of bounded strongly continuous linear operators [87] $U(t)$, $t \geq 0$, from $\mathcal{H}$ into $\mathcal{X}$, we shall interpret the initial value problem for (2.38) as the stochastic integral equation

$$u(t) = U(t)\eta + \int_0^t U(t-s)V(s,u(s))\,ds + \int_0^t U(t-s)\xi(s,u(s))\,dW(s).$$
(2.39)

In general, Eq. (2.38) and condition (2.37) have the usual interpretation:

$$u(t) = \eta + \int_0^t A(s)u(s)\,ds + \int_0^t V(s,u(s))\,ds + \int_0^t \xi(s,u(s))\,dW(s).$$
(2.40)

We shall present two existence theorems pertaining to the stochastic equations (2.39) and (2.40).

**Theorem 2.5** *Let the following hypotheses be satisfied*:

(a) $A$ *generates an equicontinuous semigroup* $\{U(t), t \geq 0\}$ *of bounded, strongly continuous linear operators from* $\mathcal{K}$ *into* $\mathcal{X}$ *so that*

$$\|U(t)\|_{\mathcal{K},\mathcal{X}} \leq Me^{rt}, \quad \text{for some constants } M, r > 0.$$
(2.41)

(b) For each $u \in \mathcal{X}$, the mapping $V(\cdot, u, \omega): [0, T] \to \mathcal{K}$ is continuous a.s. and $V(t, u, \omega)$ is $\mathcal{F}_t$ measurable.

(c) For each $u \in \mathcal{X}$, $\xi(\cdot, u): [0, T] \to \mathcal{M}[\mathcal{L}_{(2)}(\mathcal{H}, \mathcal{K})]$ is continuous.

(d) For every $u, v \in \mathcal{X}$ and $t \in [0, T]$, there exists a constant $K > 0$ such that

$$|V(t,u,\omega)|^2 + \|\xi(t,u)\|_{(2)}^2 \leq K(1 + \|u\|^2),$$
(2.42)

$$|V(t,u,\omega) - V(t,v,\omega)| + \|\xi(t,u) - \xi(t,v)\|_{(2)} \leq K\|u-v\|,$$
(2.43)

*where* $\|\cdot\|_{(2)}$ *denotes the Hilbert–Schmidt norm in* $\mathcal{L}_{(2)}(\mathcal{H}, \mathcal{K})$.

(e) $\eta$ *is* $\mathcal{F}_0$-*measurable and* $E|\eta|^2 < \infty$.

*Then there exists a unique solution of the stochastic integral equation* (2.39), *and the solution* $u(t)$ *is a homogeneous Markov process in* $\mathcal{X}$ *for* $0 \leq t \leq T$.

*Proof* The theorem can be proved easily by the well-known Picard iteration procedure. For brevity, this will be done by a probabilistic fixed-point theorem [8, 58].

Let $\mathscr{L} = \{u \in C\{[0,T], \mathcal{M}[\mathcal{X}]\} \mid \sup_{0 \leq t \leq T} \|u(t)\|^2 < \infty\}$. Then $\mathscr{L}$ is the Banach space of nonanticipating continuous processes in $\mathcal{X}$ with the norm $\|\|u\|\| = \{\sup_{0 \leq t \leq T} E\|u(t)\|^2\}^{1/2}$.

Define the operator $\mathscr{T}$ on $\mathscr{L}$ as follows:

$$\mathscr{T}u(t) = U(t)\eta + \int_0^t U(t-s)V(s,u(s))\,ds + \int_0^t U(t-s)\xi(s,u(s))\,dW(s).$$
(2.44)

Since by hypotheses and equality (2.20)

$$E\|\mathcal{T}u(t)\|^2 \leq 3\left\{E\|U(t)\eta\|^2 + E\left\|\int_0^t U(t-s)V(s,u(s))\,ds\right\|^2\right.$$
$$\left. + E\int_0^t \|U(t-s)\xi(s,u(s))\|_{(2)}^2\,ds\right\}$$
$$\leq 3M^2 e^{2rT}\{E\|\eta\|^2 + KT(T+1)\|\|u\|\|^2\}, \tag{2.45}$$

clearly, $\mathcal{T}u(t)$ is nonanticipating and continuous in $t$. Hence, in view of (2.45), $\mathcal{T}: \mathscr{L} \to \mathscr{L}$. In fact, $\mathcal{T}$ is continuous because for every $u, v \in \mathscr{L}$

$$E\|\mathcal{T}u(t) - \mathcal{T}v(t)\|^2 \leq 2\left\{E\left\|\int_0^t U(t-s)[V(s,u(s)) - V(s,v(s))]\,ds\right\|^2\right.$$
$$\left. + E\int_0^t \|U(t-s)[\xi(s,u(s)) - \xi(s,v(s))]\|_{(2)}^2\,ds\right\}$$
$$\leq 2KM^2(T+1)e^{2rT}E\int_0^t \|u(s) - v(s)\|^2\,ds$$
$$\leq 2KM^2 T(T+1)e^{2rT}\|\|u-v\|\|^2. \tag{2.46}$$

From (2.46) we can show by iterations that, for any integer $k \geq 1$,

$$E\|\mathcal{T}^k u(t) - \mathcal{T}^k v(t)\| \leq \frac{[2KM^2 T(T+1)e^{2rT}]^k}{k!}\|\|u-v\|\|. \tag{2.47}$$

Hence $\mathcal{T}^k: \mathscr{L} \to \mathscr{L}$ is a contraction map for sufficiently large $k \geq k_0$, where $(k_0!) > [2KM^2 T(T+1)e^{2rT}]^{k_0}$. By appealing to a fixed-point theorem [8, Theorem 3], there exists a fixed point $u \in \mathscr{L}$ of $\mathcal{T}$ such that $u(t) = \mathcal{T}u(t)$. Thus the stochastic integral equation (2.39) has a unique solution $u(t)$ in $\mathscr{X}$. The homogeneous Markov property of the solution process can be checked by direct computation, but is omitted here.

The next theorem is concerned with a monotone stochastic differential equation (2.36) and (2.37) in the sense of (2.40).

**Theorem 2.6** *In the integral equation* (2.40), *we assume the following conditions are satisfied* (a.s. *in* $\omega$ *if necessary*).

(a) *For each* $t \in [0,T]$, *the linear operator* $A(t): \mathscr{X} \to \mathscr{X}^*$, *defined by* (2.35), *is bounded and the function* $A(\cdot): [0,T] \to \mathscr{L}(\mathscr{X}, \mathscr{X}^*)$ *is continuous.*

(b) *For each* $u \in \mathscr{X}$, $V(\cdot, u) \in C([0,T], \mathscr{K})$ *and* $V(t,u)$ *is* $\mathscr{F}_t$-*measurable*, $t \in [0,T]$. *Furthermore, for every* $u, v \in \mathscr{X}$, *there exist constants* $K > 0$ *and* $p \geq 2$ *such that for each* $t \in [0,T]$

$$|V(t,u)| \leq K\|u\|^{p-1}, \tag{2.48}$$

$$|V(t,u) - V(t,v)| \leq K\|u-v\|, \tag{2.49}$$

(c) $\xi(\cdot, u) \in \mathcal{M}[\mathcal{L}_{(2)}(\mathcal{H}, \mathcal{K})]$, satisfying

$$\|\xi(t, u) - \xi(t, v)\|_{(2)} \leq K \|u - v\|. \tag{2.50}$$

(d) For every $u, v \in \mathcal{X}$, there exists a constant $\lambda, K > 0, p \geq 2$ such that

$$-2\langle A(t)u - A(t)v, u-v\rangle - 2\langle V(t,u) - V(t,v), u-v\rangle + \lambda \|u-v\|^2$$
$$\geq \|\xi(t, u) - \xi(t, v)\|_{(2)}, \tag{2.51}$$

and the following coercive condition holds:

$$2\langle A(t)u, u\rangle + 2\langle V(t,u), u\rangle + \|\xi(t,u)\|_{(2)}^2 - \lambda |u|^2 \leq K \|u\|^p. \tag{2.51'}$$

(e) The initial vector $\eta$ is $\mathcal{F}_0$-measurable, and $E|\eta|^2 < \infty$.

Then the stochastic integral equation (2.40) has a unique, nonanticipating solution $u(t)$, which is a Markov process with continuous sample paths in $\mathcal{H}$ for $0 \leq t \leq T$.

A theorem of this kind was first proved by Bensoussan and Teman [5], and later generalized by Pardoux [70], where the proof of this theorem can be found.

The problems of existence and uniqueness of stochastic differential equations in Hilbert space have been investigated by many authors, including Baklan [2], Daletskii [21], Curtain and Falb [20], Kuo [55], and others (see the references in [7] and [26]). Also, we mention that stochastic PDEs in population genetics were studied by Dawson [25], Fleming [31], and Viot [84]. An existence theorem for stochastic Navier–Stokes equations will be given in Section VI. Finally, we remark that as a special case $\xi \equiv 0$, Theorems 2.5 and 2.6 yield the existence statements for regular stochastic PDEs.

### D. Functional Differential Equations for Solution Processes

According to the existence theorems 2.5 and 2.6, the solution processes are Markov diffusions in $\mathcal{K}$ for $0 \leq t \leq T$. For each Borel subset $B$ of $\mathcal{K}$, define the *transition probability* as

$$q(s, t, \eta, B) = \text{Prob}\{u(t) \in B \mid u(s) = \eta\}, \quad 0 \leq s \leq t, \quad \eta \in \mathcal{K}. \tag{2.52}$$

Let $f$ be a real-valued function on $\mathcal{K}$, which is twice $\mathcal{K}$-differentiable so that the mappings $\delta f: \mathcal{K} \to \mathcal{K}$ and $\delta^2 f: \mathcal{K} \to \mathcal{L}_{(1)}(\mathcal{K})$ are continuous. We consider the *conditional expectation functional* of $f$:

$$F(s, t, \eta) = E_{s,\eta}\{f[u(t)]\} = E\{f[u(t)] \mid u(s) = \eta\}$$
$$= \int_{\mathcal{K}} f(h) q(s, t, \eta, dh). \tag{2.53}$$

As in finite dimensions, the functional $F$ will satisfy the Kolmogorov equation in $\mathscr{X}$.

**Theorem 2.7** *Let the expectational functional $F(s, t, \eta)$ be defined by (2.53). If $u(t)$ is a solution of the stochastic differential equation (2.38), whose existence is ensured by either Theorem 2.5 or 2.6, and $\delta F: \mathscr{X} \to \mathscr{X}$, $S^2 F: \mathscr{X} \to \mathscr{L}_{(1)}(\mathscr{K})$ are continuous, then $F$ satisfies the Kolmogorov equation in $\mathscr{X}$:*

$$\partial_s F(s, t, \eta) - (A(s)\eta, \delta F(s, t, \eta)) + \langle V(s, \eta), \delta F(s, t, \eta) \rangle$$
$$+ \tfrac{1}{2} \operatorname{tr} \{\xi^*(s, \eta) \delta^2 F(s, t, \eta) \xi(s, \eta)\} = 0, \qquad 0 \leq s \leq t \leq T \quad (2.54)$$

*and the condition*
$$F(0, t, \eta) = f(\eta). \quad (2.55)$$

*Proof* For brevity, we only sketch the proof. The major step is the following computation.

By the Markov property,

$$F(s, t, \eta) = E_{s,\eta}\{f[u(t)]\} = E_{s,\eta} E_{\sigma, u(\sigma)}\{f[u(t)]\}$$
$$= E_{s,\eta} F[\sigma, t, u(\sigma)], \qquad 0 \leq s \leq \sigma \leq t, \quad (2.56)$$

we get, granting that $F$ is as smooth as $f$,

$$F(\sigma, t, \eta) - F(s, t, \eta) = -E_{s,\eta}\{F[\sigma, t, u(\sigma)] - F(s, t, \eta)\}. \quad (2.57)$$

Noting Theorems 2.1 and 2.2 and Eq. (2.38),

$$E_{s,\eta}\{F[\sigma, t, u(\sigma)] - F(s, t, \eta)\}$$
$$= E_{s,\eta} \int_s^\sigma \langle \xi^*[\tau, u(\tau)] \delta F[\tau, t, u(\tau)] \xi[\tau, u(\tau)], dW(\tau) \rangle$$
$$+ E_{s,\eta} \int_s^\sigma \{(\partial/\partial \tau) F[\tau, t, u(\tau)] + (Au(\tau), \delta F[\tau, t, u(\tau)])$$
$$+ \langle V[\tau, u(\tau)], \delta F[\tau, t, u(\tau)] \rangle$$
$$+ \tfrac{1}{2} \operatorname{tr} \xi^*[\tau, u(\tau)] \delta^2 F[\tau, t, u(\tau)] \xi[\tau, u(\tau)]\} d\tau$$
$$= E_{s,\eta} \int_s^\sigma \{(A(\tau)u(\tau), \delta F[\tau, t, u(\tau)]) + \langle V[\tau, u(\tau)], \delta F[\tau, t, u(\tau)] \rangle$$
$$+ \tfrac{1}{2} \operatorname{tr} \xi^*[\tau, u(\tau)] \delta^2 F[\tau, t, u(\tau)] \xi[\tau, u(\tau)]\} d\tau$$
$$= (\sigma - s)\{(A(s)\eta, \delta F[s, t, u(s)]) + \langle V(s, \eta), \delta F(s, t, \eta) \rangle$$
$$+ \tfrac{1}{2} \operatorname{tr} \xi^*(s, \eta) \delta^2 F(s, t, \eta) \xi(s, \eta)\} + o(\sigma - s), \quad (2.58)$$

where the last equality holds because the integrands are continuous, as a consequence of the hypotheses of Theorems 2.5 and 2.6. Now the Kolmogorov equation follows after dividing (2.58) by $(\sigma-s)$ and invoking (2.57), then taking the limit as $\sigma \downarrow s$. Condition (2.55) is clearly satisfied.

We remark that, if $A$, $V$, and $\xi$ in (2.38) are independent of time $t$, the solution process is a homogeneous Markov diffusion. In this case, the transition probability

$$q(s,t,\eta,\beta) = q(t-s,0,\beta-\eta) \equiv q_{t-s}(\beta-\eta). \tag{2.59}$$

Set

$$F(t-s,\eta) = \int f(h+\eta) q_{t-s}(dh) = F(s,t,\eta), \quad 0 \leqslant s \leqslant t. \tag{2.60}$$

In view of (2.60), the following result is an immediate consequence of Theorem 2.7.

**Corollary 2.7** *If the solution process $u(t)$ is time-homogeneous, then the expectional functional (2.60) satisfies the initial-value problem in $\mathscr{X}$:*

$$\partial_t F(t,\eta) = (A\eta, \delta F(t,\eta)) + \langle V(\eta), \delta F(t,\eta) \rangle$$
$$+ \tfrac{1}{2} \mathrm{tr}[\xi^*(\eta) \delta^2 F(t,\eta) \xi(t,\eta)], \tag{2.61}$$

$$F(0,\eta) = f(\eta). \tag{2.62}$$

*Furthermore, if the stochastic differential in (2.38) is "central," the term $(\tfrac{1}{2} \mathrm{tr}\, \xi^* \delta^2 F \xi)$ should be replaced by $\tfrac{1}{2} \mathrm{tr}(\delta \xi^* \delta F) \xi$.*

Due to the lack of a Lebesgue integral in infinite dimensions [42], there is no analogy of the probability density function as in finite dimensions. Although it is possible to introduce a forward equation for the transition probability function [23], which is regarded as a generalized measure, the associated calculus is too complicated to be of immediate use. For this reason, we shall introduce the Hopf equation [46] for the *condition characteristic functional* of the solution process,

$$Q(t,\lambda) = E\{\exp i \langle u(t), \lambda \rangle \,|\, u(0) = \eta\} = E_\eta\{\exp i \langle u(t), \lambda \rangle\}, \quad \lambda \in \mathscr{K}, \tag{2.63}$$

to replace the Fokker–Planck equation. The former may be regarded as a formal Fourier transform of the latter. Therefore, the Hopf equation is well-defined if the $V(t,\eta)$ and $\xi(t,\eta)$ are polynomials in $\eta$. We say that $F: \mathscr{K} \to \mathscr{L}$ is a *polynomial function* of degree $k$ if there exist $j$-linear operators $F_j$, $j = 0, 1, 2, \ldots, k$, from

$$(\mathscr{K})^j = \overbrace{\mathscr{K} \times \cdots \times \mathscr{K}}^{j}$$

into $\mathscr{L}$, a Banach space, such that

$$F(\eta) = \sum_{j=0}^{k} F_j(\eta), \qquad (2.64)$$

where

$$F_j(\eta) = \overbrace{F_j(\eta, \ldots, \eta)}^{j},$$

and $F_j$ will be called the *coefficients* of $\eta^j$. In deriving the Hopf equation, we shall assume that $Q$ is sufficiently smooth and $V(t, -i\delta)$, $\xi(t, -i\delta)$ are $\mathscr{H}$-differential operators defined on $C_0^\infty(\mathscr{H})$.

**Theorem 2.8** *Let $u(t)$ be a solution of the stochastic equation (2.38) with values in $\mathscr{X}$. If $V(t, \eta)$ and $\xi(t, \eta)$, for each $t$, are polynomials in $\eta \in \mathscr{H}$ with coefficients $V_j(t, \cdot)$ and $\xi_j(t, \cdot)$ in $\mathscr{X}$ and $\mathscr{L}_{(2)}(\mathscr{H}, \mathscr{H})$, respectively, and the characteristic functional $Q$ is sufficiently smooth so that for each $t \in [0, T]$*

(a) $V(t, -i\delta) Q(t, \lambda)$ *and* $\xi(t, -i\delta) \xi^*(t, -i\delta) Q(t, \lambda)$ *exist for each* $\lambda \in \mathscr{H}$, *and*

(b) *for every bounded, closed subset $B$ of $\mathscr{H}$, $V(t, -i\delta) Q(t, \lambda) \in \mathscr{X}$ and $\xi(t, -i\delta) \xi^*(t, -i\delta) Q(t, \lambda) \in \mathscr{L}_{(2)}(\mathscr{H})$ are uniformly bounded and continuous on $[0, T] \times B$,*

*then the functional $Q$ satisfies the Hopf equation in $\mathscr{X}$:*

$$\partial_t Q(t, \lambda) = (A^*(t) \lambda, \delta Q(t, \lambda)) + i \langle V(t, -i\delta) Q(t, \lambda), \lambda \rangle$$
$$- \tfrac{1}{2} \langle [\xi(t, -i\delta) \xi^*(t, -i\delta) Q(t, \lambda)] \lambda, \lambda \rangle \qquad (2.65)$$

*and*

$$Q(0, \lambda) = \exp i \langle \lambda, \eta \rangle. \qquad (2.66)$$

*For a central stochastic equation, the last term on the right-hand side of (2.65) should be replaced by $-\tfrac{1}{2} \langle \xi(t, -i\delta) \xi^*(t, -i\delta) \lambda Q, \lambda \rangle$.*

*Proof* Equation (2.65) can be verified by computations similar to that in the previous theorem.

For $0 \leq t \leq \sigma \leq T$, we write

$$Q(\sigma, \lambda) = E_\eta [\exp i \langle u(t), \lambda \rangle] E_{t, u(t)} [\exp i \langle u(\sigma) - u(t), \lambda \rangle]. \qquad (2.67)$$

Setting $f(\eta) = \exp i \langle \eta - u(t), \lambda \rangle$, it follows from Theorem 2.2 that

$$E_{t, u(t)} [\exp i \langle u(\sigma) - u(t), \lambda \rangle]$$
$$= 1 + E_{t, u(t)} \int_t^\sigma \{i \langle A(\tau) u(\tau), \lambda \rangle + i \langle V(\tau, u(\tau)], \lambda \rangle$$
$$- \tfrac{1}{2} \langle \lambda, \xi[\tau, u(\tau)] \xi^*[\tau, u(\tau)] \lambda \rangle \} f[u(\tau)] \, d\tau$$

$$= 1 + (\sigma - t)\{i\langle A(t)u(t), \lambda\rangle + i\langle V[t, u(t)], \lambda\rangle$$
$$- \tfrac{1}{2}\langle \lambda, \xi[t, u(t)]\xi^*[t, u(t)]\lambda\rangle\}f[u(t)] + o(\sigma - t), \quad (2.68)$$

where the last step can be shown by a continuity argument. Suppose that $V[t, u(t)] = V_1(t)u(t)$ with $V_1(t) \in \mathscr{L}_{(2)}(\mathscr{K})$. Then

$$E_\eta \langle V[t, u(t)], \lambda\rangle f[u(t)] = E_\eta \langle V_1(t) i(t), \lambda\rangle f[u(t)]$$
$$= E_\eta \langle V_1^*(t)\lambda, -i\delta f[u(t)]\rangle = \langle \lambda, V_1(t)(-i\delta) Q(t, \lambda)\rangle$$
$$= \langle \lambda, V(t, -i\delta) Q\rangle. \quad (2.69)$$

In general, $V_\lambda(t, \eta) = \langle V(t, \eta), \lambda\rangle$ contains a $j$-linear form, which can be expressed as a linear combination of a $j$-tensor product of linear forms [79] $V_\lambda^{(1)} \otimes V_\lambda^{(2)} \otimes \cdots \otimes V_\lambda^{(j)}$ so that it suffices to consider the case

$$V_\lambda(t, \eta) = \langle V_\lambda^{(1)}, \eta\rangle \langle V_\lambda^{(2)}, \eta\rangle \cdots \langle V_\lambda^{(j)}, \eta\rangle. \quad (2.70)$$

Now, by a repeated application of (2.69), we get

$$E_\eta \langle V[t, u(t)], \lambda\rangle f[u(t)] = EV_\lambda[t, u(t)]f[u(t)]$$
$$= \langle V_\lambda^{(1)}, -i\delta\rangle \cdots \langle V_\lambda^{(j)}, -i\delta\rangle Q(t, \lambda)$$
$$= \langle \lambda, V(t, -i\delta) Q\rangle. \quad (2.71)$$

Similarly, if $\langle \lambda, \xi(t, \eta)\xi^*(t, \eta)\lambda\rangle$ is a multilinear form, the previous argument holds. Thus

$$E_\eta \langle \lambda, \xi[t, u(t)]\xi^*[t, u(t)]\lambda\rangle f[u(t)]$$
$$= \langle [\xi(t, -i\delta)\xi^*(t, -i\delta) Q(t, \lambda)]\lambda, \lambda\rangle. \quad (2.72)$$

In view of (2.67)–(2.72), we see that

$$\partial_t Q(t, \lambda) = \lim_{\sigma \downarrow t} \frac{Q(\sigma, \lambda) - Q(t, \lambda)}{\sigma - t}$$
$$= (A^*(t)\lambda, \delta Q(t, \lambda)) + i\langle V(t, -i\delta) Q(t, \lambda), \lambda\rangle$$
$$- \tfrac{1}{2}\langle [\xi(t, -i\delta)\xi^*(t, -i\delta) Q(t, \lambda)]\lambda, \lambda\rangle. \quad (2.73)$$

The remaining part of the proof is obvious.

We remark that, in general, the Hopf equation cannot be defined as a differential equation in function space. Then one is faced with the mathematical dilemma of introducing a Fokker–Planck equation or defining a "pseudo"-differential operator [35] in function space. In practice, however, most model equations are of polynomial type in nonlinearity, to which the above theorem applies. Parabolic equations in function space have been investigated by many authors, e.g., see [3, 22, 56, 71].

## III. Linear Stochastic Partial Differential Equations in Weak Turbulence

### A. Introduction

Let us consider an incompressible viscous fluid contained in a domain $D \subset \mathbb{R}^3$. If the domain $D$ is bounded, the boundary $\partial D$ will be assumed to be sufficiently smooth throughout this chapter to avoid unnecessary complications. If $u(t, x)$ denotes the velocity field, $p(t, x)$ the pressure, $f(t, x)$ the external force, and $\nu$ the kinematic viscosity, the Navier–Stokes equations and the continuity equation [4] read as

$$\partial_t u + (u \cdot \nabla) u = -\nabla p + \nu \Delta u + f(t, x), \qquad (3.1)$$

$$\nabla \cdot u = 0, \qquad (3.2)$$

where $\Delta = \nabla^2$ is the Laplacian operator, and the initial condition and the homogeneous boundary conditions $u|_{\partial D} = 0$ must be imposed.

The statistical theory of turbulence is usually concerned with the solution of (3.1) and (3.2) in an unbounded domain when the initial condition and the external force may be random processes. This problem will be discussed at the end of the chapter. To illustrate the basic ideas in solving a stochastic PDE, a simple exactly solvable model will be treated first. The model arises in hydrodynamic fluctuations at equilibrium. Let $v$ be a constant equilibrium solution of the homogeneous Navier–Stokes equations (3.1) and (3.2). We suppose that $u(t, x)$ is the fluctuating velocity about the equilibrium state $v$ due to a weak perturbation by the random force $f$. The resulting weak turbulence is governed by the linearized Navier–Stokes equations:

$$\partial_t u + (v \cdot \nabla) u = -\nabla p + \nu \Delta u + f(t, x, \omega). \qquad (3.3)$$

By invoking (3.2) and imposing boundedness condition at $|x| = \infty$, the pressure $p$ in (3.3) can be determined in terms of $f$:

$$p(t, x) = \frac{1}{4\pi} \int \frac{1}{|x-y|} \nabla \cdot f(t, y) \, dy. \qquad (3.4)$$

Let $K$ denote the singular integral operator [12]

$$(Kg)(x) = g(x) - \int \nabla_x \frac{1}{4\pi |x-y|} \nabla_y \cdot g(y) \, dy$$

$$= \int \left[ \nabla_x \cdot \nabla_y \frac{1}{4\pi |x-y|} \right] g(y) \, dy$$

$$= \lim_{\varepsilon \downarrow 0} \int_{|x-y| > \varepsilon} \left[ \nabla_x \cdot \nabla_y \frac{1}{4\pi |x-y|} \right] g(y) \, dy, \qquad (3.5)$$

which is defined as a principal-value integral and is bounded on $\mathbb{L}^2$. It is also easy to verify that $K$ is formally self-adjoint and idempotent so that $K^* = K$ and $K^2 = K$. By virtue of (3.4) and (3.5), the pressure term in (3.3) may be eliminated to give

$$\partial_t u + (v \cdot \nabla) u = v \Delta u + Kf(t, x). \tag{3.6}$$

In the remaining section, we shall analyze the above equation when $f(t, \cdot) = S\dot{W}(t, \cdot)$ is a Gaussian white noise.

### B. Linearized Navier–Stokes Equations with a Random Forcing

Let $\mathcal{V} = \{g \in C_0^\infty(D) | \nabla \cdot g = 0\}$ be the admissible velocity field. We take $\mathcal{H}$ to be the completion of $\mathcal{V}$ in $\mathbb{L}^2(D)$, and $\mathcal{X}$ the completion of $\mathcal{V}$ in $H_0^1$. Let $S$ be a bounded linear operator from $\mathcal{H}$ into $\mathcal{X}$ so that $KS \in \mathcal{L}_{(2)}(\mathcal{H}, \mathcal{X})$. We consider (3.6) as the stochastic system in $\mathcal{X}^*$:

$$du = Au\, dt - (v \cdot \nabla) u\, dt + KS\, dW(t), \qquad u(0) = \eta, \tag{3.7}$$

where $A = v\Delta$, $W(t)$ is the Wiener process on $(\mathcal{H}, \mathcal{X}^*)$, and $\eta(\omega) \in \mathcal{X}$. Let $U(t)$, $t \geq 0$, denote the semigroup on $\mathcal{X}$ generated by $A$. For every $g \in \mathcal{X}$,

$$U(t)g(x) = \int h(t, x-y) g(y)\, dy, \tag{3.8}$$

where $h(t, x)$ is the well-known heat kernel. Now we write (3.7) as integral equation in $\mathcal{X}$:

$$u(t) = U(t)\eta - \int_0^t U(t-s)(v \cdot \nabla) u(s)\, ds + \int_0^t U(t-s) KS\, dW(s). \tag{3.9}$$

In comparison with (2.39), we see that $V(u) = -(v \cdot \nabla) u$ and $\xi(u) = KS$. Thus, by applying Theorems 2.5, 2.7, and 2.8 we have

**Theorem 3.1** *If the force field $f(t, x, \omega) = S\dot{W}(t, x)$ is a Gaussian white noise with covariance $(SS^*) \in \mathcal{L}_{(1)}(\mathcal{H})$, the initial state $\eta(x, \omega)$ is $\mathcal{F}_0$-measurable with $E|\eta|^2 < \infty$ and $R_1 = KS \in \mathcal{L}_{(2)}(\mathcal{H}, \mathcal{X})$, then the initial-boundary value problem for Eqs. (3.1) and (3.2), or (3.7) in abstract form, has a unique solution $u(t, \cdot, \omega)$, which is a homogeneous Markov diffusion in $\mathcal{X}$ with continuous sample paths. Furthermore, the Kolmogorov and Hopf equations read*

$$\partial_t F = v(\Delta \eta, \delta F) - \langle (v \cdot \nabla)\eta, \delta F \rangle + \tfrac{1}{2} \operatorname{tr}(R_1^* \delta^2 F R_1),$$
$$F(0, \eta) = f(\eta) \tag{3.10}$$

and
$$\partial_t Q = v(\Delta\lambda, \delta Q) + \langle (v \cdot \nabla)\lambda, \delta Q \rangle + \tfrac{1}{2}\langle R\lambda, \lambda \rangle Q, \qquad (3.11)$$
$$Q(0,\lambda) = \exp i\langle \eta, \lambda \rangle,$$
where $R = R_1 R_1^*$.

A proof of the theorem follows from the invoked theorems by showing that their hypotheses are satisfied. We note that the above equations may be put in concrete form. For example, the Hopf equation (3.11) takes the form

$$\partial_t Q(t,\lambda) = -v \sum_{j,k=1}^{3} \int \frac{\partial \lambda_j(x)}{\partial x_k} \frac{\partial}{\partial x_k} \frac{\delta Q(t,\lambda)}{\delta \lambda_j(x)} dx + \sum_{j,k=1}^{3} \int v_k \frac{\partial \lambda_j(x)}{\partial x_k} \frac{\delta Q(t,\lambda)}{\delta \lambda_j(x)} dx$$
$$+ \frac{1}{2} \sum_{j,k=1}^{3} \iint R_{jk}(x,y) \lambda_j(x) \lambda_k(y) \, dx \, dy, \qquad (3.12)$$

where $R_{jk}(x,y)$ are components of $R(x,y)$, the kernel of $R$, defined in terms of the kernels $K(x-y)$ and $S_2(x,y)$ of $K$ and $(SS^*)$, respectively, as follows:

$$R(x,y) = \iint K(x-x') S_2(x',y') K(y'-y) \, dx' \, dy'. \qquad (3.13)$$

For conciseness, we shall not write down the concrete forms, such as the above, unless it is necessary.

## C. Computation of Solutions

It is instructive to work out the solution of this problem. To do so we let

$$B = v\Delta - (v \cdot \nabla), \qquad (3.14)$$

and denote by $e^{tB}$, $t \geq 0$, the semigroup generated by $B$ in $\mathcal{H}$. Then the system (3.7) can be integrated to yield

$$u(t) = e^{tB}\eta + \int_0^t e^{(t-s)B} KS \, dW(s), \qquad (3.15)$$

from which we obtain the mean vector

$$m(t) = E u(t) = e^{tB} a, \qquad \text{with} \quad a = E\eta, \qquad (3.16)$$

and the covariance operator

$$\Gamma(t) = \text{Cov}\{u(t)\} = E\{[u(t)-m] \otimes [u(t)-m]\}$$
$$= e^{tB} \Gamma_0 e^{tB^*} + \int_0^t e^{(t-s)B} R e^{(t-s)B^*} \, ds, \qquad (3.17)$$

where $\otimes$ signifies the tensor product and $R = (KSS^*K)$.

If $h_1(t, x, y)$ denotes the fundamental solution of the convective heat equation associated with $B$ and the homogeneous boundary condition, then the mean and covariance functions take the following form:

$$m(t, x) = \int h_1(t, x, y) a(y) \, dy, \tag{3.18}$$

$$\Gamma(t, x, y) = \int\int h_1(t, x, x') \Gamma_0(x', y') h_1^*(t, y', y) \, dx' \, dy'$$
$$+ \int_0^t \int\int h_1(t-s, x, x') R(x', y') h_1^*(t-s, y', y) \, ds \, dx' \, dy', \tag{3.19}$$

where $R(x, y)$ was defined in (3.13).

It is interesting to obtain the above results by solving the Hopf equation (3.31), which may be regarded as a linear "hyperbolic" system in $\mathscr{X}$. To solve it, we try the exponential solution

$$Q(t, \lambda) = \exp\{i\langle \beta(t), \lambda \rangle + \langle \Sigma(t) \lambda, \lambda \rangle\}, \tag{3.20}$$

where $\beta(t)$ is a vector and $\Sigma(t)$ is a linear operator in $\mathscr{H}$, as yet to be determined. A substitution of (3.20) into (3.31) yields

$$\partial_t \beta = B\beta, \quad \beta(0) = \eta, \tag{3.21}$$

and

$$\partial_t \Sigma = B\Sigma + \Sigma B^* - \tfrac{1}{2} R, \quad \Sigma(0) = 0. \tag{3.22}$$

The above linear systems have the solutions

$$\beta(t) = e^{tB} \eta \tag{3.23}$$

and

$$\Sigma(t) = -\tfrac{1}{2} \int_0^t e^{(t-s)B} R e^{(t-s)B^*} \, ds. \tag{3.24}$$

By comparing (3.23) and (3.24) with (3.16) and (3.17) at a fixed $\eta$, it is evident that $\beta = m$, $\Sigma = -\tfrac{1}{2}\Gamma$, so that (3.20) becomes

$$Q(t, \lambda) = \exp\{i\langle m(t), \lambda \rangle - \tfrac{1}{2} \langle \Gamma(t) \lambda, \lambda \rangle\}. \tag{3.25}$$

Therefore, $Q(t, \lambda)$ is the characteristic functional for a Gauss–Markov measure $q_t$ of $(m, \Gamma)$, or a generalized Ornstein–Uhlenbeck process in $\mathscr{H}$.

Since, for each $g \in \mathscr{K}$,

$$\lim_{t \to \infty} e^{tB} g = 0, \tag{3.26}$$

one expects that the stochastic equation (3.7) has an equilibrium solution $u_\infty$, and the corresponding transition probability measure $q_t$ tends to an equilibrium (invariance) measure $q_\infty$, which may be obtained by seeking a time-independent solution $Q_\infty$ of the Hopf equation (3.11) in the form

(3.20). This procedure gives $\beta_\infty \equiv 0$ and an operator equation for $\Sigma_\infty$ in $\mathscr{L}_{(2)}(\mathscr{H})$:

$$B\Sigma_\infty + \Sigma_\infty B^* = \tfrac{1}{2}R. \tag{3.27}$$

If the above problem is well-posed, the equilibrium measure $q_\infty$ is Gaussian $(0, -2\Sigma_\infty)$. Let us summarize the results.

**Theorem 3.2** *Let $u(t, x)$ be the fluctuating velocity field about the mean velocity $v$ due to a Gaussian white noise forcing with covariance $(SS^*)$. Then the random velocity $u(t, x)$ is a generalized Gauss–Markov process with mean vector and covariance operator given by (3.18) and (3.19), with $\Gamma_0 \equiv 0$. Furthermore, if the problem (3.27) is well-posed, the equilibrium velocity $u_\infty(x)$ is a Gaussian field $(0, -2\Sigma_\infty)$.*

We remark that this problem can be solved for a general Gaussian force field $f(t, x, \omega)$ in (3.3). In particular, the Hopf equation is obtainable by applying the integration-by-parts formula (2.24). Though unrelated to this problem, the Ornstein–Uhlenbeck process in Weiner space was studied in [72] and [59].

## IV. Partial Differential Equations in Stochastic Wave Propagation

### A. Introduction

The problems of wave propagation in turbulent media are often governed by hyperbolic systems with random coefficients. Familiar examples are acoustic, electromagnetic waves propagating in turbulent atmosphere, ocean, or plasma [36, 51]. When $a_{jk}^{lm} \equiv 0$ in (2.1), it reduces to a first-order quasilinear random system

$$\partial_t u = B(t, x, u, \nabla u, \omega) + f(t, x, u, \omega), \tag{4.1}$$

where $B(t, x, u, iy)$ is defined in (2.3). If the matrix $B$ has only imaginary eigenvalues a.s. in $\omega$, the system (4.1) is hyperbolic. Such a random system has not been investigated. For the linear case, Eq. (4.1) becomes

$$\partial_t u_j = \sum_{k=1}^{M} \sum_{l=1}^{N} b_{jk}^l(t, x, \omega)\, \partial_l u_k + \sum_{k=1}^{M} c_{jk}(t, x, \omega) u_k + f_j(t, x, \omega),$$
$$j = 1, 2, \ldots, M. \tag{4.2}$$

The above system may be treated by the *method of characteristics*. For example, in one space variable $N = 1$, a linear transformation of (4.2) yields the canonical form [19]

$$\partial_t v_j - \lambda_j(t, x, \omega)\, \partial_x v_j = \sum_{k=1}^{M} c_{jk}(t, x, \omega) v_k + f_j(t, x, \omega), \tag{4.3}$$

where $\lambda_j(t, x, \omega)$ is the $j$th eigenvalue of the random matrix $\{b_{jk}(t, x, \omega)\}$.

Let $\sigma_j(t, \omega)$ denote the arc length along the $j$th random characteristic (curve) $l_j(\omega)$ through $(t, x)$ defined by

$$d\sigma_j = \{1 + \lambda_j^2(s, \sigma_j)\}^{\frac{1}{2}} ds, \qquad \sigma_j(t) = x, \qquad t \geq s \geq 0. \tag{4.4}$$

Let $y = \sigma_j(s, t, x, \omega)$ denote the random solution of (4.4). Then (4.3) can be integrated along the characteristic $l_j(\omega)$ to give

$$v_j(t, x) = v_j(0, \sigma_j(0, t, x)) + \int_0^t \left\{ \sum_{k=1}^M c_{jk}[s, \sigma_j(s, t, x)] v_k[s, \sigma_j(s, t, x)] \right.$$
$$\left. + f_j[s, \sigma_j(s, t, x)] \right\} ds. \tag{4.5}$$

For many space variables, the same method applied to (4.2) leads to integral equations similar to (4.5). We observe that the integral equation (4.5) is unusual in the sense that the integration is along a random curve $l_j(\omega)$. For example, if $f_j$ is a Gaussian white noise, we have to define a stochastic integral along a random curve. These problems have not yet been resolved. Further, it is necessary to introduce a theory of weak random solution for a hyperbolic system, which is not covered by the theorems in Section II.

Most work in random wave propagation is concerned with random scattering problems governed by the scalar wave equation

$$\partial_t^2 v = c^2(x, \omega) \Delta v + f(t, x, \omega), \tag{4.6}$$

where $c$ is the local speed of propagation characterizing the random medium and the source $f$ may be randomly distributed. Of course, if necessary, the wave equation can be rewritten as a linear hyperbolic system. By a Laplace transform or otherwise, the wave equation is reduced to a *random Helmholtz equation* [51]

$$\Delta u + k^2 n^2(x, \omega) u = q(x, \omega), \tag{4.7}$$

which must satisfy the physical radiation condition

$$\lim_{|x| \to \infty} |x| \left( \frac{\partial u}{\partial |x|} - iku \right) = 0. \tag{4.8}$$

Here the wave number $k$ is complex with $\operatorname{Im} k > 0$, and $n(x, \omega)$, known as the refractive index, is inversely proportional to $c(x, \omega)$. If the random function is sufficiently smooth, the analytical questions, such as existence and uniqueness, can be settled as in the nonrandom case. Therefore the central problem lies in the construction of solutions. Another problem that is closely related to (4.7) is the random parabolic equation

$$\partial_t u = (\alpha/2) \Delta u + \beta \mu(t, x, \omega) u, \qquad u(0, x) = \eta(x), \tag{4.9}$$

where $\alpha$, $\beta$ are complex constants and $\operatorname{Re} \alpha \geq 0$. If $\mu$ is independent of $t$, a Laplace transform of (4.9) yields (4.7) after adjusting the constants properly.

The time-dependent case occurs in the so-called "parabolic approximation" of (4.7), where $t = x_1$ corresponds to the first space coordinate, and $x = (x_2, x_3)$ is the transverse vector. The rest of the section will be concerned with problems related to the random parabolic equation (4.9), where $\mu$ need not be a white noise.

### B. Parabolic Approximation in Random Wave Propagation

The approximation of the reduced wave equation (4.7) by a parabolic equation of the form (4.9) with a white noise coefficient has received a great deal of attention in recent years due to its tractability. It will be shown that the functional formalism [15, 52] in solving the equation can be justified in our abstract setting presented in Section II.

Let $\mathscr{K} = L^2(\mathbb{R}^2)$ of complex-valued functions on $R^2$, $\mathscr{X} = H^1(\mathbb{R}^2)$, $A = (\alpha/2)\Delta: \mathscr{X} \to \mathscr{X}^*$. For $S \in \mathscr{L}_{(2)}(\mathscr{H}, \mathscr{K})$, we assume that $\mu = S\dot{W}$ is a central Gaussian white noise with covariance $R = SS^*$. We may regard (4.9) as a stochastic equation in $\mathscr{X}^*$:

$$du = Au\, dt + \beta uS\, \partial W(t), \quad u(0) = \eta. \tag{4.10}$$

By applying Theorem 2.5, system (4.10) has a unique (complex) solution $u(t)$ in $\mathscr{X}$. However, before invoking Theorem 2.8, there is a need for redefining the characteristic functional $Q(t, \lambda)$ for a complex-valued process $u(t)$:

$$Q(t,\lambda,\bar{\lambda}) = E\{\exp i[\langle u(t),\lambda\rangle + \langle \bar{u}(t),\bar{\lambda}\rangle] \mid u(0) = \eta\}, \tag{4.11}$$

where the overbar denotes the complex conjugate. By considering $u(t)$ and $\bar{u}(t)$ as the two components of a vector, a complex version of Theorem 2.8 implies that $Q$ satisfies

$$\partial_t Q = \tfrac{1}{2}\{\beta^2 \langle \delta_\lambda R \delta_\lambda Q, \lambda\rangle + \bar{\beta}^2 \langle \delta_{\bar{\lambda}} R \delta_{\bar{\lambda}} Q, \bar{\lambda}\rangle\}$$
$$+ \tfrac{1}{2}\alpha(\Delta\lambda, \delta_\lambda Q) + \tfrac{1}{2}\bar{\alpha}(\Delta\bar{\lambda}, \delta_{\bar{\lambda}} Q), \quad \lambda \in \mathscr{X}, \tag{4.12}$$

$$Q(0,\lambda,\bar{\lambda}) = \exp i\{\langle \eta, \lambda\rangle + \langle \bar{\eta}, \bar{\lambda}\rangle\}. \tag{4.13}$$

Equation (4.12) can be solved by a formal power series solution of the form

$$Q(t,\lambda,\bar{\lambda}) = 1 + \sum_{n=1}^{\infty} \sum_{j+k=n} \frac{i^n}{n!}\binom{n}{j} \Gamma_{jk}(t,\lambda,\bar{\lambda}), \tag{4.14}$$

where $\Gamma_{jk}$ with $n = j+k$ is an $n$-linear form on $\mathscr{X}^n$ and

$$\Gamma_{jk}(t,\lambda,\bar{\lambda}) = \Gamma_{jk}(t,\overbrace{\lambda,\lambda,\ldots,\lambda}^{j},\overbrace{\bar{\lambda},\bar{\lambda},\ldots,\bar{\lambda}}^{k})$$

for each $t \geq 0$. $\Gamma_{jk}$ has the interpretation as the $n$th moment-functional of the solution $u(t)$. A direct substitution of (4.14) yields closed equations for moments. This procedure will be illustrated for a turbulent diffusion problem in Section V.

### C. Method of Functional Integration

As mentioned before, the random radiation problem (4.7) can be imbedded in the parabolic initial-value problem (4.9). Here we shall give a brief discussion of the method based on a functional integration of (4.9). Though formal in nature, the method has proven to be powerful in application. Therefore its justification poses a significant problem in analysis.

In (4.9), let $\mu(t, \cdot, \omega)$ be a real, continuous Gaussian process in $C_0^\delta(\mathbb{R}^n)$, the space of Hölder continuous functions of exponent $\delta > 0$ with compact support in $\mathbb{R}^n$, $n = 2$ or 3. For simplicity we assume that $\mu$ is stationary and homogeneous with mean $m$ and covariance function $R(t, x)$, which is continuous and bounded above on $[0, \infty) \times \mathbb{R}^n$. Let us define the moment-generating functional

$$G(\lambda) = E \exp[\lambda, \mu] = \exp\{[\lambda, m] + \tfrac{1}{2}[R\lambda, \lambda]\}, \tag{4.15}$$

where $[\lambda, \cdot]$ stands for a continuous linear functional $\lambda$ on $C_0(\mathbb{R}^{n+1})$. Regarding $\lambda$ as a generalized function, we may write

$$[\lambda, \mu] = \int_{-\infty}^{\infty} \int_{\mathbb{R}^n} \mu(t, x) \lambda(t, x) \, dt \, dx. \tag{4.16}$$

To construct the moments of solution to (4.9), it is advantageous to represent the random solution, for each fixed $\omega$, as a functional integral by the Feynman–Kac formula [30, 48]

$$u(t, x, \omega) = M_x^\alpha \exp\left\{\beta \int_0^t \mu[\tau, z_x(\tau), \omega] \, d\tau\right\} f[z_x(t)]. \tag{4.17}$$

Here $z_x(t)$ with $z_x(0) = x$, is the $n$-dimensional (pseudo) Brownian motion with complex parameter $\alpha$, and $M_x^\alpha g_t(z_x)$ denotes the *sequential Wiener integral* over $z_x(\tau)$, $0 \leq t \leq \tau$, in the sense of Cameron [14]. Formally interchanging the expectation $E$ and the operation $M_x^\alpha$, the mean solution of (4.17) can be expressed in terms of $G(\lambda)$ as follows

$$\Gamma_1(t, x) = Eu(t, x) = M_x^\alpha\{G(\lambda_1) f[z_x(t)]\}, \tag{4.18}$$

where $\lambda_1$ is a generalized function,

$$\lambda_1(\tau, y) = \beta h(t-\tau) \delta[y - z_x(\tau)], \tag{4.19}$$

and $h(t)$, $\delta(y)$ denote the Heaviside and Dirac delta functions, respectively. By the same token, moments of higher order can also be written. For instance, the second moment $\Gamma_{11}$ reads

$$\Gamma_{11}(t, x, x') = Eu(t, x) u(t, x') = M_x^\alpha M_{x'}^{\bar{\alpha}} \{G(\lambda_{11}) f[z_x(t)] \bar{f}[z_{x'}(t)]\}, \quad (4.20)$$

$$\lambda_{11}(\tau, y) = h(t-\tau) \{\beta \delta[y - z_x(t)] + \bar{\beta} \delta[y - z_{x'}(t)]\}. \quad (4.21)$$

In view of (4.15), Eqs. (4.18) and (4.20) yield

$$\Gamma_1(t, x)$$
$$= e^{\beta m t} M_x^\alpha \exp\left\{\tfrac{1}{2} \int_0^t \int_0^t R[\tau_1 - \tau_2, z_x(\tau_1) - z_x(\tau_2)] \, d\tau_1 \, d\tau_2\right\} f[z_x(t)], \quad (4.22)$$

$$\Gamma_{11}(t, x, x')$$
$$= e^{(\beta + \bar{\beta})m t} M_x^\alpha M_{x'}^{\bar{\alpha}} \exp\left\{\tfrac{1}{2} \int_0^t \int_0^t [\beta^2 R(\tau_1 - \tau_2, z_x(\tau_1) - z_x(\tau_2))\right.$$
$$\left. + 2\beta\bar{\beta} R(\tau_1 - \tau_2, z_x(\tau_1) - z_{x'}(\tau_2)) + \bar{\beta}^2 R(\tau_1 - \tau_2, z_{x'}(\tau_1) - z_{x'}(\tau_2))]\right\}$$
$$\times f[z_x(t)] \bar{f}[z_{x'}(t)]. \quad (4.23)$$

Sometimes it is possible to evaluate the above Wiener integrals asymptotically, as $|\alpha| \to \infty$ or $t \to \infty$. Results of these kinds can be found in the references [15, 17, 27]. Also, the method of differential equations may be adapted to their evaluation as will be shown.

The method of constructing solutions given here resembles what Feynman [30], Kac [48], Gelfand and Yaglom [38], and many others did in quantum physics. The difficulty in justifying this method lies in defining the Wiener integral with the complex parameter $\alpha$ in a measure-theoretical sense. However, we believe that the method is justifiable if the operator $M_x^\alpha$ is regarded as a random contraction operator in a suitable function space.

### D. Some Related Problems

The real parabolic equation (4.9) ($\alpha$, $\beta$ being real) has no physical significance in the context of wave propagation. But it may serve as a model equation for testing the validity of certain analytic techniques for constructing approximate solutions. This parabolic equation is connected to turbulence in a different aspect, namely, the randomly forced Burgers equation

$$\partial_t v + (v \cdot \nabla) v = \nu \Delta v + f(t, x, \omega), \quad (4.24)$$

which has often been adopted as a turbulence model in place of the Navier–Stokes equations for simplicity. If the force field is derivable from a random potential $\mu$ so that $f = 2\beta \nu \nabla \mu$, then a substitution of $v = -2\nu \nabla \ln u$ and

$\alpha = 2\nu$ in (4.24) produces the parabolic equation (4.9), as asserted. For $\alpha$ being real and positive, the process $z_x(t)$ is a genuine Brownian motion, and the results obtained previously by functional integration can be readily verified, simply by appealing to the Fubini theorem [76]. In fact one can prove the theorem concerning existence, uniqueness, and so on (for details see [16]). Here we shall state two major theorems without proof.

**Theorem 4.1** *Let $\mu(t, \cdot, \omega)$, $t \geq 0$, be a continuous Gaussian process in $C_0^\delta(\mathbb{R}^n)$ for some $\delta > 0$. Then for each $f \in C_0^\delta$, the initial-value problem (4.9) for the parabolic equation has a unique random solution $u(t, x, \omega)$ in $C_0^\delta$. Furthermore, the random solution has a Wiener integral representation (4.17). For $0 \leq t \leq T$, all moments of the solution*

$$\Gamma_m(t, x', x'', \ldots, x^{(m)}) = E\{u(t, x')u(t, x') \cdots u(t, x^{(m)})\}$$

*exist, are continuous on $[0, \infty) \times \mathbb{R}^m$, and vanish at $|x^{(j)}| = \infty$ for each $j$ with $1 \leq j \leq m$, $m = 1, 2, \ldots$.*

The proof is essentially the same as the deterministic case, which may be verified by direct computation from the solution formula (4.17). A more interesting result is concerned with certain closure approximations and their errors. By a *closure approximation of order m*, we mean a set of approximating moment-functions $\tilde{\Gamma}_j$, $j = 1, 2, \ldots, m$, which are determined by a system of $m$ differential equations. Let $\gamma_m(t)$ denote the *closure error of order m* defined by

$$\gamma_m(t) = \sup_{x \in \mathbb{R}^n} \left| \frac{\Gamma_m(t, x, \ldots, x) - \tilde{\Gamma}_m(t, x, \ldots, x)}{\tilde{\Gamma}_m(t, x, \ldots, x)} \right|. \tag{4.25}$$

A well-known closure scheme in turbulence is the direct interaction approximation due to Kraichnan [53, 54] based on physical grounds. This formalism, when applied to the fundamental solution of (4.9), gives the following closure approximations of second order:

$$\partial_t \tilde{\Gamma}_1(t, x) = \tfrac{1}{2}\alpha \Delta \tilde{\Gamma}_1(t, x) + \tfrac{1}{2}\beta^2 \int_0^t \int R(t-s, x-y) \tilde{\Gamma}_1(t-s, x-y) \tilde{\Gamma}_1(s, y) \, ds \, dy,$$

$$\tilde{\Gamma}_1(0, x) = \delta(x), \tag{4.26}$$

and

$$\partial_t \tilde{\Gamma}_2(t, x, x') = \tfrac{1}{2}\alpha(\Delta + \Delta') \tilde{\Gamma}_2(t, x, x') + \beta^2 \int_0^t \int\int [R(t-s, x-y)$$

$$+ R(t-s, x'-y') + R(t-s, x-y') + R(t-s, x'-y)]$$

$$\times \tilde{\Gamma}_2(t-s, x-y, x'-y') \tilde{\Gamma}_2(s, y, y') \, ds \, dy \, dy',$$

$$\tilde{\Gamma}_2(0, x, x') = \delta(x)\delta(x'). \tag{4.27}$$

In the above equations, we assumed $E\mu = 0$. As it turns out, this approximation can be justified by our Wiener integral approach. To this end, we make the following assumptions on the covariance function $R$:

(a) $R(t, x)$ is nonnegative and continuous for $t \geq 0$, $x \in \mathbb{R}^n$.

(b) There exists an even function $R_0(t)$ on $\mathbb{R}$ such that
$$R(t, x) \leq R_0(t), \quad \text{uniformly in} \quad x \in \mathbb{R}^n.$$

(c) $\int_0^t \tau R_0(\tau) \, d\tau \leq M < \infty, \quad \forall t \geq 0.$ \hfill (4.28)

**Theorem 4.2** *Let $\mu(t, x, \omega)$, $(t, x) \in [0, \infty) \times \mathbb{R}^n$ be a space–time homogeneous Gaussian field with a zero mean and the covariance function $R(t, x)$ satisfying hypotheses (a)–(c) in (4.28) immediately above. If $\{\tilde{\Gamma}_1, \tilde{\Gamma}_2\}$ is the closure approximation determined by the nonlinear integrodifferential equations (4.27) and (4.28), then the corresponding closure error has the following bound:*

$$0 \leq \gamma_2(t) \leq \left\{1 - \exp\left[-4\beta^2 \int_0^t s R_0(s) \, ds\right]\right\}, \quad \forall t \geq 0.$$

This theorem shows the critical dependence of closure error on the time-dependence behavior of the covariance function.

## V. Stochastic Equations in Turbulent Transport Theory

### A. Introduction

Analytic study of dispersion of suspended particles in turbulent fluid was initiated by Taylor [81]. The physical model consists of a particle moving along a line with constant speed and reversing its direction at random times. It was then shown by Goldstein [40] that, under a certain limit, the mean concentration satisfies the Telegrapher's equation. Many years later, Kac [49] recovered this result by introducing a random walk model.

Let $v(t)$ denote the velocity of a marked particle moving along a line at time $t$, and let $|v(t)| = v_0$, a constant. If the number of time reversals $N(t)$ is the Poisson process with the parameter $a$, the particle trajectory $\xi(t)$ satisfies the kinematic relation

$$d\xi/dt = \xi_0(-1)^{N(t)}, \quad (5.1)$$

where $\xi_0 = \pm v_0$ with equal probability. Let $\phi(x)$ be a smooth real function on $\mathbb{R}$. We define the expectational function

$$u(t, x) = E\{\phi[\xi(t)] \mid \xi(0) = x\}. \quad (5.2)$$

It was shown by Kac [49] that $u$ satisfies the dissipative wave equation

$$\partial_t^2 u + 2a\,\partial_t u = v_0^2\,\partial_x^2 u, \tag{5.3}$$

and the initial conditions

$$u(0,x) = \phi(x), \qquad \partial_t u(0,x) = 0. \tag{5.4}$$

In view of (5.3), only when the observation time is long or both the intensity and speed are high, do we have the "turbulent diffusion" approximation. This probabilistic model has been generalized to include dispersion of particles with many speeds. More realistically, a particle moves with the local fluid velocity and is perturbed by random impacts from the surrounding fluid molecules. These models will be discussed separately in what follows.

### B. A Random Evolution Model

In a more general situation, we consider a particle cloud transported by the turbulent motion of a fluid. Let $u(t,x,\omega)$ denote the particle concentration at time $t$ and position $x \in \mathbb{R}^3$. In the absence of sinks and sources, the function $u$ satisfies the *random transport equation*

$$\partial_t u + \nabla \cdot (vu) = 0, \qquad u(0,x) = \phi(x), \tag{5.5}$$

where $v(t,x,\omega)$ is the local fluid velocity and $\phi$ the initial distribution.

In the random evolution model, the velocity $v$ is assumed to be independent of $x$, so that $v = \xi(t,\omega)$ is a stationary Markov process with state space $\mathbb{R}^3$ and infinitesimal generator $A$. For an $n$-state Markov chain, $A$ is an $n \times n$ matrix, while $A$ is a second-order elliptic operator if $\xi(t,\omega)$ is a diffusion process. For each $y \in \mathbb{R}^3$, let $V(y) = -(y \cdot \nabla)$ be a bounded operator in a Banach space $\mathscr{X}$, in which $A$ is also defined. Then the system takes the simple abstract form

$$du/dt = V[\xi(t)]u, \qquad u(0) = \phi, \tag{5.6}$$

which is an initial-value problem in $\mathscr{X}$.

Let $M(s,t)$, $0 \leqslant s \leqslant t$, denote the random fundamental solution of (5.6). Since $M(s,t)M(t,\tau) = M(s,\tau)$ for $s < t < \tau$, $M$ is an operator-valued *multiplicative functional* in the sense of Pinsky [73]. Suppose that $\xi(t,\omega)$ is an $n$-state Markov chain with jump times $\tau_j(\omega)$ and $N(t,\omega)$ is the number of jumps up to time $t$; then

$$M(t) = M(0,t) = M(0,\tau_1)M(\tau_1,\tau_2)\cdots M(\tau_{N(t)},t) \tag{5.7}$$

was termed a *random evolution* (actually the adjoint of $M(t)$) by Greigo and Hersh [41]. Let us consider the $\mathscr{X}$-valued functions

$$F_j(t) = E\{M(t) \,|\, \xi(0) = y_j\}, \qquad j = 1, 2, \ldots, n. \tag{5.8}$$

It can be shown by a direct computation [41, 69] that $F_j$ satisfies

$$\frac{dF_j}{dt} = V(y_j)F_j + \sum_{k=1}^{n} a_{jk}F_k, \qquad F_j(0) = I, \qquad j = 1, 2, \ldots, n, \qquad (5.9)$$

where $a_{jk}$ are elements of the matrix $A$ and $I$ is the identity operator on $\mathscr{X}$. If we apply the result (5.9) to (5.5), the mean concentration vector $F_j(t, x) = E\{u(t, x) | \xi(0) = y_j\}$ is determined by

$$\partial_t F_j(t, x) = (y_j \cdot \nabla) F_j(t, x) + \sum_{k=1}^{n} a_{jk} F_k(t, x),$$

$$F_j(0, x) = \phi(x), \qquad j = 1, 2, \ldots, n, \qquad (5.10)$$

which generalizes Kac's result (5.3) and (5.4). It is interesting to see how the turbulent diffusion approximation is obtained in this case. The main idea involves the balance between small stochastic perturbation and large observation time.

Let $\varepsilon > 0$ be a small parameter and set $\xi = \varepsilon \tilde{\xi}$. We rewrite Eq. (5.6) as

$$du^\varepsilon/dt = \varepsilon V[\tilde{\xi}(t)]u^\varepsilon. \qquad (5.11)$$

The desired result is a special case of asymptotic limit theorem for stochastic differential equations [69]. The theorem says that if $\tilde{\xi}(t)$ is a stationary process with $E\tilde{\xi} = 0$ and satisfies a strong mixing condition, among other technical conditions, then as $\varepsilon \to 0$ and $t \to \infty$ with $\tau = \varepsilon^2 t$ held fixed, the mean concentration $\Gamma(t, x) = Eu(t, x)$ approaches the solution of the diffusion equation

$$\partial_\tau \tilde{\Gamma}(\tau, x) = \sum_{j,k=1}^{n} v_{jk} \partial_{jk}^2 \tilde{\Gamma}(\tau, x), \qquad \tilde{\Gamma}(0, x) = \phi(x), \qquad (5.12)$$

where

$$v_{jk} = \lim_{t \to \infty} \frac{1}{t} \int_0^t \int_s^t E\{\tilde{\xi}(t)\tilde{\xi}^*(s)\} \, ds \, dt. \qquad (5.13)$$

The above results, though interesting, have no physical significance because the velocity field is spatial-dependent and the random fluctuations are not small. A realistic solution seems difficult. As a first step, one may seek a generalization of the asymptotic limit theorem for (5.11) where $\tilde{\xi}$ is a space–time process. For a complete list of references on multiplicative functionals and random evolution, one should consult the interesting articles by Pinsky [74] and Hersh [45].

### C. Turbulent Transport with Molecular Diffusion

We consider a marked Brownian particle drifting with the local turbulent fluid velocity $v(t, x, \omega)$. Let $\sigma(x)$ denote the diffusion matrix for the

Brownian motion $z(t)$ in $\mathbb{R}^3$. Then the trajectory $\xi(t, \omega)$ of the particle must obey the randomized (central) stochastic equation,

$$d\xi(t) = v[t, \xi(t), \omega] \, dt + \sigma[\xi(t)] \, \hat{d}z(t). \tag{5.14}$$

Let $\mathscr{F}_t$ denote the complete $\sigma$-field generated by the turbulent velocity $v(t, \cdot, \omega)$ up to time $t$. Suppose that the fluid is contained in a domain $D \subset \mathbb{R}^3$, and, for every set $B \subset D$, the transition probability

$$P(s, y, t, B, \omega) = E\{\xi(t) \in B \mid \mathscr{F}_t, \xi(s) = y\}$$

has a smooth probability density $p(s, y, t, x, \omega)$ as a function of $(t, x) \in D_T$ a.s. in $\omega$. Then the particle concentration function

$$u(t, x, \omega) = \int_D \eta(y) \, p(0, y, t, x, \omega) \, dy \tag{5.15}$$

satisfies the stochastic Fokker–Planck equation

$$\partial_t u + v \cdot \nabla u = \tfrac{1}{2} \nabla \cdot [\sigma(x) \nabla \sigma(x) \mathbf{u}], \tag{5.16}$$

and the initial-boundary conditions

$$u(0, x) = \eta(x), \qquad \left. \frac{\partial u}{\partial n} \right|_{\partial D} = 0. \tag{5.17}$$

Here we tacitly assumed that $\sigma$ is twice differentiable and that the fluid is incompressible. The system (5.16) and (5.17) will be the starting point of our analysis. This problem was treated from a different viewpoint by Chow [18], where $v(t, x)$ was taken as a generalized Gaussian process in the sense of Daletskii and Paramonova [24].

In our abstract setting, let $\mathscr{X}$ be the space $H_0^1(D)$ of real-valued functions $\eta$ in $H^1(D)$ with $(\partial \eta / \partial n)|_{\partial D} = 0$, and let $\mathscr{K} = L^2(D)$. As in Section II, the velocity field $v$ belongs to the admissible set $\mathscr{V}$, whose completions in $\mathbb{L}^2(D)$ and $H_0^1(D)$ are denoted here by $\mathscr{B}$ and $\mathscr{H}$, respectively. We consider $W(t)$ as a Wiener process over $(\mathscr{H}, \mathscr{B})$, and assume that $dv(t, \cdot) = S \, \hat{d}W(t)$, i.e., $v(t, \cdot)$ is a central Gaussian white noise with covariance $R = SS^*$ in $\mathscr{B}$. If we put $A = \tfrac{1}{2} \nabla \cdot (\sigma \nabla \sigma) \colon \mathscr{X} \to \mathscr{X}^*$, Eqs. (5.15) and (5.16) may be regarded as a system in $\mathscr{X}^*$:

$$du = Au \, dt - (\nabla u) \cdot S \, \hat{d}W(t), \qquad u(0) = \eta, \tag{5.18}$$

which may be written as an integral equation in an obvious way:

$$u(t) = e^{tA} \eta - \int_0^t e^{(t-s)A} [\nabla u(s)] \cdot S \, \hat{d}W(s). \tag{5.19}$$

By Theorems 2.5 and 2.8, we have

**Theorem 5.1** *Assume that the diffusion matrix* $\sigma(x) \in C^{2+\delta}(D)$, *and the* covariance operator $R = SS^*$ of the velocity field $v(t, x)$ is so smooth that its kernel $R(x, y)$ has components $\rho_{jk}(x, y) \in C^\alpha(D^2) \cap H^1(D^2)$, $j, k = 1, 2, 3$. Then if $\eta \in L^2(D)$, the turbulent transport equation (5.16) subject to condition (5.17) has a unique solution $u(t, x)$. Moreover, the solution is a Markov diffusion in $H_0^1(D)$ and its conditional characteristic functional $Q(t, \lambda)$ satisfies

$$\partial_t Q = \tfrac{1}{2}(\sigma \nabla \cdot [\sigma \nabla \lambda], \delta Q) + \tfrac{1}{2}\langle \nabla \cdot (\delta R \delta) \cdot \nabla \lambda \, Q, \lambda \rangle,$$

$$Q(0, \lambda) = \exp i \langle \eta, \lambda \rangle, \qquad \lambda \in H_0^1. \tag{5.20}$$

As mentioned in Section IV.B, Eq. (5.20), similar to Eq. (4.12), admits a power series solution of the form

$$Q(t, \lambda) = 1 + \sum_{k=1}^{\infty} \frac{(i)^k}{k!} \Gamma_k(t, \lambda), \tag{5.21}$$

where $\Gamma_k(t, \lambda)$ is a $k$-linear functional on $\mathscr{X}^k$. In terms of its kernel function, $\Gamma_k$ takes the form

$$\Gamma_k(t, \lambda) = \int_{D^k} \Gamma_k(t, x', x'', \ldots, x^{(k)}) \lambda(x') \cdots \lambda(x^{(k)}) \, dx' \cdots dx^{(k)}. \tag{5.22}$$

Substituting (5.21) into (5.20) yields the main result:

**Theorem 5.2** *The Cauchy problem* (5.20) *admits a series solution of the form* (5.21), *where the kernels of the coefficients* $\Gamma_k(t, \cdot)$ *in* (5.22), $k = 1, 2, \ldots,$ *are determined as follows. For* $k = 1$,

$$\partial_t \Gamma_1(t, x) = \tfrac{1}{2} \nabla \cdot (\sigma \nabla \sigma) \Gamma_1(t, x) + \tfrac{1}{2} \sum_{j,k=1}^{3} \rho_{jk}(x, x) \partial_{jk}^2 \Gamma_1(t, x),$$

$$\Gamma_1(0, x) = \eta(x), \qquad \left. \frac{\partial \Gamma_1}{\partial n} \right|_{\partial D} = 0, \tag{5.23}$$

*and for* $k \geq 2$,

$$\partial_t \Gamma_k(t, x', x'', \ldots, x^{(k)})$$

$$= \frac{1}{2} \{[\nabla \cdot (\sigma \nabla \sigma)]' + \cdots + [\nabla \cdot (\sigma \nabla \sigma)]^{(k)}\} \Gamma_k(t, x', \ldots, x^{(k)})$$

$$+ \frac{1}{2} \sum_{l,m=1}^{3} \sum_{p,q=1}^{k} \rho_{lm}(x^{(p)}, x^{(q)}) \frac{\partial^2}{\partial x_l^{(p)} \partial x_m^{(q)}} \Gamma_k(t, x', \ldots, x^{(k)}), \tag{5.24}$$

$$\Gamma_k(0, x', \ldots, x^{(k)}) = \prod_{j=1}^{k} \eta(x^{(j)}),$$

$$\frac{\partial}{\partial n_j} \Gamma_k(t, x', \ldots, x^{(k)}) = 0, \qquad \text{for } x^{(j)} \in \partial D, \quad j = 1, 2, \ldots, k, \tag{5.25}$$

*where* $\partial/\partial n_j$ *designates the normal derivative of* $\Gamma_k$ *in the* $j$-*variable.*

A proof of the above theorems in concrete form and other results can be found in [18]. Also, precise conditions on the covariance functions $\rho_{jk}(x, y)$ were given to ensure that the sequence of diffusion equations (5.23)–(5.25) has unique bounded solutions.

## VI. Markovian Model Equations in Turbulence

### A. Introduction

In this last section, we shall give a brief account of a difficult subject, the statistical theory of turbulence. A very limited part of the theory, namely, certain Markovian models in fully developed turbulence, will be treated from a stochastic equation viewpoint. At an earlier stage, the mathematical theory of turbulence had been formulated as a stochastic initial value problem for the Navier–Stokes equation [9], in a manner similar to the classical statistical mechanics. However, due to dissipation, the system does not seem to possess an invariant distribution. An alternative idea of introducing random forcing was proposed by Landau and Lipschitz [61], Novikov [68], Edwards [29], and others. In this case, one regards the randomly forced Navier–Stokes equation as Langevin equations in the classical theory of noises [82]. The goal is to seek the invariant measure, which depends on the random force and the boundary condition, but not on the initial state. The existence of such a measure is suggested by a formal analogy to the dynamical theory of Brownian motion [82], but has not been proved. So far no equilibrium solutions to the Langevin–Navier–Stokes equations have been found due to computational difficulty. Consequently, all practical analyses have been concerned with the derivation of moment equations under all kinds of closure assumptions. The most interesting closure assumption is the Markovian random coupling. The Markovian force alone is not an unreasonable assumption. It gives rise to a diffusion in function space, which does not yield closed moment equations. The random coupling model, though, giving the closed moment equation, is difficult to justify.

Here we shall attempt to give a unified approach to some Markovian model equations in turbulence based on our abstract stochastic equation formulation. An existence theorem, due to Bensoussan and Temam [6], will be presented at the end. For the theory and literature of turbulence, we cite the books by Batchelor [4], Monin and Yaglom [66]. More recent work may be found in the expository articles in the volume edited by Lumley and Rosenblatt [64], and the lecture note by Ruelle [77].

### B. A Langevin Equation Model

We consider the turbulence problem for an incompressible fluid in an unbounded domain. If the energy flow is induced by the action of a random force $f(t, x, \omega)$, the turbulent velocity $u(t, x)$ satisfies Eqs. (3.1) and (3.2) of Section III. As before, it is convenient to eliminate the pressure in (3.1) by the continuity equation. With the operator $K$ defined as in (3.5), we get

$$\partial_t u + K(u \cdot \nabla) u = \nu \Delta u + K f(t, x, \omega), \qquad u(0, x) = \eta(x, \omega). \tag{6.1}$$

Recall that $\mathscr{V} = \{u \in C_0^\infty \mid \nabla \cdot u = 0\}$ denotes the admissible velocity field. Let $p \geq 2$ and $W^{1,p}$ be the Sobolev space of the $\mathbb{L}^2$-function whose first derivatives also belong to $\mathbb{L}^2$. Then we take $\mathscr{H}$, $\mathscr{X}$, $\mathscr{Y}$ to be the completion of $\mathscr{V}$ in $\mathbb{L}^2$, $W^{1,2}$, and $W^{1,4}$, respectively. The inclusions $\mathscr{Y} \subset \mathscr{X} \subset \mathscr{H}$ are dense. Letting $A = \nu \Delta$ and $V(u) = -K(u \cdot \nabla) u$, we consider $A$, $B$ as bounded operators from $\mathscr{X}$, $\mathscr{Y}$ into $\mathscr{X}^*$, $\mathscr{H}$, respectively. Again, if $f$ is a Gaussian white noise with covariance $(SS^*)$, system (6.1) can be put in the abstract form in $\mathscr{X}^*$:

$$du = Au\, dt + V(u)\, dt + KS\, dW(t), \qquad u(0) = \eta(\omega), \tag{6.2}$$

where the initial state $\eta$ is independent of the random force $f$. We note that since in (6.2) the operator $V$ is neither monotone nor Lipschitzian, the existence theorems given in Section II fail to apply. This question will be discussed later. Let us assume that system (6.2) has a solution $u(t)$, which is a homogeneous Markov diffusion in $\mathscr{H}$. Then Theorem 2.7 says that the conditional functional $F(t, \eta)$ satisfies the Kolmogorov equation in $\mathscr{X}$:

$$\partial_t F = \nu(\Delta \eta, \delta F) - \langle K(\eta \cdot \nabla)\eta, \delta F \rangle - \tfrac{1}{2} \operatorname{tr}(S^* K \delta^2 F K S). \tag{6.3}$$

According to Theorem 2.8, we get the Hopf equation for $Q(t, \lambda)$:

$$\partial_t Q = \nu(\Delta \lambda, \delta Q) + \tfrac{1}{2} \langle R\lambda, \lambda \rangle Q - (1/i)\langle K(\delta \cdot \nabla)\delta Q, \lambda \rangle, \qquad \lambda \in \mathscr{X}, \tag{6.4}$$

where $R = SKS^*$. Unlike the "hyperbolic" system (3.11), the "Schrödinger-type" equation (6.4) is difficult to solve. An attempt to use a power series solution as in the previous section will lead to a hierarchy of moment equations. The diffusion equation (6.3) appears simpler, but it is not known how to construct its solution by a formal functional calculus. Therefore, an approximate solution by using certain closure schemes is of practical interest.

### C. Random Coupling Model

For closure approximations, various random coupling models were introduced by Kraichnan [53, 44] and others. We only present a slightly modified version due to Frisch et al. [37]. However, it will be done from the stochastic equation point of view in more generality. Further, to avoid the difficulty of defining the probability density in infinite dimensions, we shall work with the characteristic functional.

Instead of (6.1), we consider an ensemble of $N$ turbulent flows

$$\{v_j(t,x), j = 1, 2, ..., N\}$$

that are randomly coupled and independently forced by Gaussian white noises. The main assumption here is to replace the single set of Navier–Stokes equations by a family of $N$ sets of equations that are coupled through the nonlinear terms $(v_j \cdot \nabla) v_j$. It is proposed that the nonlinear terms fluctuate about their means $\mu_j = \mu(v_j) = E\{(v_j \cdot \nabla) v_j\}$ as follows:

$$(v_j \cdot \nabla) v_j = \mu_j + (1/N) \sum_{j=1}^{N} \gamma_{jkl}(t) (v_k \cdot \nabla) v_l. \tag{6.5}$$

Roughly speaking, the fluctuations are proportional to the average sizes of the quadratic terms. The real processes $\gamma_{jkl}(t)$ are independent, identical, central white noises and symmetric in all the indices $j, k, l$. In view of (6.5) and the above assumptions, we replace system (6.2) by the following supersystem in $\{\mathscr{X}^*\}^N$:

$$dv_j = Av_j\, dt - K\mu(v_j)\, dt - (1/N) \sum_{k,l=1}^{N} K(v_k \cdot \nabla) v_l\, dz_{jkl}(t) + KS\, dW_j(t)$$
$$v_j(0) = \eta_j, \quad j = 1, 2, ..., N, \tag{6.6}$$

where the $z_{jkl}(t)$ are independent, identical Brownian motions of variance one. This system is similar to (6.2) in structure except that the equations are doubly stochastic in $z(t)$ and $W(t)$. However, if we consider the joint process $(z(t), W(t))$, Eqs. (6.6) fit in our general setting.

Let

$$Q_N(t, \lambda_1, \lambda_2, ..., \lambda_N) = E\left\{\exp i \sum_{j=1}^{N} \langle v_j(t), \lambda_j \rangle \mid v_j(0) = \eta_j, j = 1, 2, ..., N\right\}. \tag{6.7}$$

Then, referring to Theorem 2.8, it is not difficult to show that $Q_N$ satisfies the following equation in $\mathscr{X}^N$:

$$\partial_t Q_N - v \sum_{j=1}^{N} (\Delta \lambda_j, \delta_j Q_N) + \tfrac{1}{2} \sum_{j=1}^{N} \langle R\lambda_j, \lambda_j \rangle Q_N - i \sum_{j=1}^{N} \langle K\mu_j, \lambda_j \rangle Q_N$$
$$+ (1/2N^2) \sum_{j,k,l=1}^{N} \langle \lambda_j, K(\delta_k \cdot \nabla) \delta_l \lambda_j K[(\delta_k \cdot \nabla) \delta_l Q_N]^+ \rangle = 0, \tag{6.8}$$

$$Q_N(0, \lambda_1, ..., \lambda_N) = \exp\left\{i \sum_{j=1}^{N} \langle \eta_j, \lambda_j \rangle\right\}, \tag{6.9}$$

where $\delta_j = (\delta/\delta \lambda_j)$ and the superscript plus means the transpose of a matrix.

Equation (6.8) seems much more formidable than the original equation (6.4). The remarkable thing is that as $N \to \infty$ Eq. (6.8) can be drastically simplified to yield the closed equation for moments. In particular, we shall show that the equations for the first two moments,

$$\Gamma_1(t, x) = \lim_{N \to \infty} (1/N) \sum_{j=1}^{N} E v_j(t, x), \tag{6.10}$$

$$\Gamma_2(t, x, y) = \lim_{N \to \infty} (1/N) \sum_{j=1}^{N} E v_j(t, x) v_j^+(t, y), \tag{6.11}$$

are closed in the limit. To this end let us assume that the limit $Q_\infty = \lim_{N \to \infty} Q_N$ exists. For every integer $m > 0$, we denote the $m$th marginal characteristic functional of $Q_\infty$ by

$$Q^{(m)}(t, \lambda_1, \ldots, \lambda_m) = \lim_{N \to \infty} Q_N(t, \lambda_1, \lambda_2, \ldots, \lambda_m, \ldots, \lambda_N)|_{\lambda_j = 0, j > m}. \tag{6.12}$$

Taking the above limit in (6.8) and making use of a simple combinatoric argument and symmetry in $\lambda_j$, we obtain the following equation for $Q^{(m)}$:

$$\partial_t Q^{(m)} - v \sum_{j=1}^{m} (\Delta \lambda_j, \delta_j Q^{(m)}) + \tfrac{1}{2} \sum_{j=1}^{m} \langle R\lambda_j, \lambda_j \rangle Q^{(m)} - i \sum_{j=1}^{m} \langle K\mu_j, \lambda_j \rangle Q^{(m)}$$

$$+ \tfrac{1}{2} \sum_{j=1}^{m} \langle K(\delta_{m+1} \cdot \nabla) \delta_{m+2}$$

$$\times K[(\delta_{m+1} \cdot \nabla) \delta_{m+2} Q^{(m+2)}]^+ \lambda_j, \lambda_j \rangle \Big|_{\lambda_{m+1} = \lambda_{m+2} = 0}$$

$$= 0. \tag{6.13}$$

To solve the above equation, we try a product solution of the form

$$Q^{(m)}(t, \lambda_1, \ldots, \lambda_m) = Q(t, \lambda_1), \ldots, Q(t, \lambda_m). \tag{6.14}$$

This amounts to assuming stochastic independence or complete chaos. After substituting (6.14) into (6.13) for $m = 1$, setting $\mu_j = \mu(v_j) = \nabla \operatorname{tr} \Gamma_2(t)$ for each $j$, we get

$$\partial_t Q - v(\Delta\lambda, \delta Q) + \tfrac{1}{2}\langle R\lambda, \lambda \rangle Q - (i/2)\langle K\nabla \operatorname{tr} \Gamma_2(t), \lambda \rangle Q$$

$$+ \tfrac{1}{2}\langle (K \otimes K) \Gamma_2(t)[(e \cdot \nabla) \otimes (e \cdot \nabla)] \Gamma_2(t)\lambda, \lambda \rangle Q = 0, \tag{6.15}$$

where $\Gamma_2(t) = \Gamma_2(t, \cdot, \cdot)$; $\operatorname{tr} \Gamma_2(t, x, y) = \operatorname{Diag}\{\Gamma_2(t, x, x)\}$; $e$ denotes a unit tensor (or dyadic) of rank two, and $\otimes$ stands for the tensor (or Kronecker) product so that

$$(K \otimes K)\{(u \otimes v)[(e \cdot \nabla) \otimes (e \cdot \nabla)](u' \otimes v')\}$$

$$= (K \otimes K)\{[(u \cdot \nabla) u'] \otimes [(v \cdot \nabla) v']\} = [K(u \cdot \nabla) u'] \otimes [K(v \cdot \nabla) v']. \tag{6.16}$$

In obtaining the last term in (6.15), use was made of the relation between the $k$th moment $\Gamma_k$ and $Q$:

$$\Gamma_k(t) = (i)^{-k} \left.\frac{\delta^k Q(t,\lambda)}{\delta \lambda^k}\right|_{\lambda=0}, \qquad k = 1, 2, \ldots. \tag{6.17}$$

Now if we differentiate (6.15) with respect to $\lambda$, and then set $\lambda = 0$, we obtain the first moment equation

$$\partial_t \Gamma_1 = v \Delta \Gamma_1 + \tfrac{1}{2} K \nabla \operatorname{tr} \Gamma_2. \tag{6.18}$$

Similarly, the second moment equation can be derived by differentiating (6.15) twice with respect to $\lambda$ and setting $\lambda = 0$,

$$\partial_t \Gamma_2 = v(\Delta \oplus \Delta) \Gamma_2 + \tfrac{1}{2}(K \nabla \operatorname{tr} \Gamma_2) \otimes \Gamma_1 + \tfrac{1}{2}\Gamma_1 \otimes (K \nabla \operatorname{tr} \Gamma_2)$$
$$+ (K \otimes K)\{\Gamma_2[(e \cdot \nabla) \otimes (e \cdot \nabla)]\Gamma_2\} + R, \tag{6.19}$$

where $(\Delta \oplus \Delta) \Gamma_2(t, x, x') = (\Delta + \Delta') \Gamma_2(t, x, x')$.

In view of (6.18) and (6.19), the first two moment equations, similar to Kraichnan's direct interaction equation, are closed. Once the correlation tensor $\Gamma_2$ is determined from the system (6.18) and (6.19), the Hopf equation (6.15) becomes a linear hyperbolic equation of first order, similar to Eq. (3.11), which can be solved exactly. As shown in Section III.B, the solution of (6.15) is the Gaussian characteristic functional with mean $\Gamma_1$ and covariance $(\Gamma_2 - \Gamma_1 \otimes \Gamma_1)$. If (6.18) and (6.19) have equilibrium solutions, then they determine the Gaussian invariance measure.

We remark that the prediction of normal law in turbulence from this model is certainly inconsistent with the experimental observations [4]. Perhaps the closure approximations for moments obtained in this way have a certain range of validity. However, it is not clear in what sense the moments derived from the limiting Hopf equation of the supersystem approximate the corresponding moments of the original system.

### D. Existence of Solutions

For the deterministic case, Leray [62] proved that the Navier–Stokes equations have a generalized solution for all time, but it is not known whether the solution is actually smooth and unique (for details and further references, see Ladyzhenskaya's book [60]). For its stochastic counterpart, Bensoussan and Teman [6] proved that there exists a random (weak) solution to the randomly forced Navier–Stokes equations (3.1) and (3.2) with a random initial condition, where $f$ is a Gaussian white noise. We shall summarize the relevant results without giving a proof.

Let $\mathcal{K}$, $\mathcal{X}$, $\mathcal{Y}$ by given as in Section VI.B so that the inclusions $\mathcal{Y} \subset \mathcal{X} \subset \mathcal{K} \subset \mathcal{X}^* \subset \mathcal{Y}^*$ are dense. In this Hilbert space setting, the existence theorem 2.6 was generalized to the following equation with a quadratic nonlinear term $B(u,u)$:

$$(du/dt) + Au + B(u,u) = f(t,\omega), \qquad u(0) = \eta(\omega) \qquad (6.20)$$

where $A$ is a monotone operator (independent of time) from $\mathcal{X}$ into $\mathcal{X}^*$, as defined in Theorem 2.6, and $B(u,v)$ is a bilinear operator from $\mathcal{X} \times \mathcal{X} \to \mathcal{Y}^*$ satisfying

(a) $((B(u,v),v)) = 0 \quad \forall u \in \mathcal{X}, \; v \in \mathcal{Y}$,
(b) $|||B(u,v)|||_* \leq c\{|u||v|\,\|u\|\,\|v\|\}^{\frac{1}{2}}$, for some constant $c > 0$,
(c) $B(u,v) \in \mathcal{K}$ if $u \in \mathcal{Y}, \; v \in \mathcal{X}$, or $u \in \mathcal{X}, \; v \in \mathcal{Y}$,
(d) $|B(u,v)| < c|||u|||\,\|v\|$, $u \in \mathcal{X}, \; v \in \mathcal{Y}$;
$\quad\; |B(u,v)| < c\|u\|\,|||v|||$, $u \in \mathcal{Y}, \; v \in \mathcal{X}$. $\qquad (6.21)$

In (6.21) $(\cdot,\cdot)$ denotes the natural pairing between $\mathcal{Y}$ and $\mathcal{Y}^*$; $|\cdot|$, $\|\cdot\|$, and $|||\cdot|||$, the norms in $\mathcal{K}$, $\mathcal{X}$, and $\mathcal{Y}$, respectively. Suppose that $f(t) = S\dot{W}(t)$ is a Gaussian white noise in $(S\mathcal{K},\mathcal{K})$ with covariance $R = SS^*$. Then it was shown [6, Theorem 3.3] by the method of multiple-valued random functions that the initial value problem (6.20) in $\mathcal{Y}^*$ has a random solution $u(t,\omega)$.

**Theorem 6.1** *Let $A: \mathcal{X} \to \mathcal{X}^*$ be a monotone operator, corresponding to $(A+V)$ in Theorem 2.6; and $B: \mathcal{X} \times \mathcal{X} \to \mathcal{Y}^*$, a bilinear operator satisfying hypotheses (a)–(d) in (6.21). If $f(t) = S\dot{W}(t)$ is a Gaussian white noise in $(S\mathcal{K},\mathcal{K})$ with the covariance operator $R = (SS^*) \in \mathcal{L}_{(1)}(\mathcal{K})$, and $\eta(\omega) \in \mathcal{K}$ with $E|\eta|^2 < \infty$, then the stochastic system (6.20) has a random solution $u(t,\omega) \in \mathbb{L}^2\{[0,T],\mathcal{X}\} \cap \mathbb{L}^\infty\{[0,T],\mathcal{K}\}$ so that*

$$E\|u\|^2_{\mathbb{L}^2\{[0,T],\mathcal{X}\}} + E\|u\|^2_{\mathbb{L}^\infty\{[0,T],\mathcal{K}\}} < \infty. \qquad (6.22)$$

This theorem is applicable to the Burgers equation (4.24). In this case $A = -\nu\Delta$ is defined by $(Au,v) = \nu\sum_{j=1}^{3}\langle\partial_j u, \partial_j v\rangle$ for $u,v \in \mathcal{X}$, while $B(u,v) = (u\cdot\nabla)v$ is defined by

$$((B(u,v),w)) = \int [(u\cdot\nabla)w]\cdot v\,dx, \qquad \text{for} \quad u,v,w \in \mathcal{X}. \qquad (6.23)$$

Since by (2.22)

$$((B(u,v),v)) = \tfrac{1}{2}\int (u\cdot\nabla)(u\cdot v)\,dx = -\tfrac{1}{2}\int (\nabla\cdot u)(v\cdot v)\,dx = 0, \qquad (6.24)$$

condition (6.21a) is satisfied. Conditions (b)–(d) are also fulfilled. For instance, to verify conditions (c) and (d), we apply the Hölder inequality to get

$$|((B(u,v), w))| \leq \sum_{j,k=1}^{3} \|u_j\|_{L^\infty} \|\partial_j v_k\|_{L^2} \|w_k\|_{L^2}$$
$$\leq C \|\|u\|\| \|v\| |w|, \quad \forall u \in \mathcal{Y}, \ v \in \mathcal{X}, \ w \in \mathcal{K}, \quad (6.25)$$

where the last step is obtained by making use of the fact that $\mathcal{Y} \subset \mathbb{L}^\infty$. This inequality implies condition (c) and the first part of (d). The second part of (d) follows from the fact that $\mathcal{Y} \subset W^{1,\infty}$ so that

$$|((B(u,v), w))| \leq \sum_{j,k=1}^{3} \|u_j\|_{L^2} \|\partial_j v_k\|_{L^\infty} \|w_k\|_{L^2} \leq C|u| \|\|v\|\| |v|. \quad (6.26)$$

Therefore, by invoking Theorem 6.1, we have the following lemma.

**Lemma 6.2** *We consider the initial value problem for the Burgers equations (3.1) and (3.2) with $p \equiv 0$ and $u(0, x) = \eta(x)$. If $f$ is a Gaussian white noise, as stated previously, and $\eta \in \mathbb{L}^2$, then the initial value problem has a random solution with the smooth properties depicted in Theorem 6.1.*

Finally, employing Leray's method for deterministic case, it can be shown that, if there is a random solution of the Burgers equation given by Lemma 6.2, there exists a pressure distribution $p(t, x)$ such that $u(t, x)$ satisfies the forced Navier–Stokes equations, i.e., there exists a solution to the randomly forced Navier–Stokes equations.

**Theorem 6.2** *Let the hypotheses of Theorem 6.1 be satisfied. Then the initial value problem for the randomly forced Navier–Stokes equations (3.1) and (3.2) has a random solution with the smooth properties indicated in Theorem 6.1.*

In conclusion, we remark that since the above pathwise solution inherits the same difficulty in uniqueness and smoothness questions from its deterministic counterpart, it seems more desirable to deal with equations in function space for the solution measure or its characteristic functional, which are stressed throughout this chapter. There has been work done in this direction. For example, Foias and Prodi [32] studied the stochastic initial-value problem for Navier–Stokes equations and showed that the solution measure becomes singular in finite time with respect to the initial turbulence measure. Their idea was extended to a generalized Hopf equation for the space-time characteristic functional [63] by Arsénev [1], who announced the existence of such a turbulence measure, or the solution of the generalized Hopf equation. However, no proof was provided there.

## ACKNOWLEDGMENTS

The author wishes to express his sincere thanks and gratitude to Professor Albert T. Bharucha-Reid for his unbound patience and kind understanding during the preparation of this article. To Professor Hui-Hsiung Kuo, the author owes a great deal for many helpful conversations that clarified a number of fine points in the theory of abstract Weiner space.

## REFERENCES

1. Arsenév, A. A., Construction of a turbulent measure for the system of Navier-Stokes equations, *Soviet Math. Dokl.* **16** (1975), 1422–1424.
2. Baklan, V. V., On the existence of solutions of stochastic equations in Hilbert space (Ukrainian), *Dopovidi Akad. Nauk Ukrain. RSR* (1963), 1299–1303.
3. Baklan, V. V., Equations in variational derivatives and Markov processes in Hilbert spaces, *Dokl. Akad. Nauk. SSSR* **159** (1964), 707–710.
4. Batchelor, G. K., "The Theory of Homogeneous Turbulence." Cambridge Univ. Press, Cambridge, England, 1959.
5. Bensoussan, A., and Temam, R., Équations aux dérivées partielles stochastiques non linéaires I, *Israel J. Math.* **11** (1972), 95–129.
6. Bensoussan, A., and Temam, R., Équations stochastiques du type Navier Stokes, *J. Functional Anal.* **13** (1973), 195–222.
7. Bharucha-Reid, A. T., "Random Integral Equations." Academic Press, New York, 1972.
8. Bharucha-Reid, A. T., Fixed point theorems in probabilistic analysis, *Bull. Amer. Math. Soc.* **82** (1976), 641–657.
9. Birkhoff, G., Bona, J., and Kampé De Fériet, J., Statistically well-set Cauchy problems, *in* "Probablistic Methods in Applied Mathematics" A. T. Bharucha-Reid, ed. Vol. 3. Academic Press, New York, 1973.
10. Boylan, S. L., Uniqueness of solutions to function-space differential equations using infinite-series and integral methods, *J. Functional Anal.* **13** (1973), 77–96.
11. Cabaña, E. M., On stochastic differentials in Hilbert spaces, *Proc. Amer. Math. Soc.* **20** (1969), 259–265.
12. Calderón, A. P., "Singular Integrals," Colloq. Lectures. *Amer. Math. Soc.*, Providence, Rhode Island 1965.
13. Cameron, R. H., The first variation of an indefinite Wiener integral, *Proc. Amer. Math. Soc.* **2** (1951), 914–924.
14. Cameron, R. H., A family of integrals serving to connect the Weiner and Feynman integrals, *J. Mathematics and Physics* **39** (1960), 126–140.
15. Chow, P. L., Applications of function space integrals to problems in wave propagation in random media, *J. Mathematical Phys.* **13** (1972), 1224–1236.
16. Chow, P. L., On the exact and approximate solutions of random parabolic equation, *SIAM J. Appl. Math.* **27** (1974), 376–397.
17. Chow, P. L., A functional phase integral method and applications to the laser beam propagation in random media, *J. Statist. Phys.* **12** (1975), 93–109.
18. Chow, P. L., Function-space differential equations associated with a stochastic partial differential equation, *Indiana Univ. Math. J.* **25** (1976), 609–627.
19. Courant, R., and Hilbert D., "Methods of Mathematical Physics," Vol. II. Wiley (Interscience), New York, 1962.
20. Curtain, R. F., and Falb, P. L., Stochastic differential equations in Hilbert space, *J. Differential Equations* **10** (1971), 412–430.

21. Daletskii, Ju. L., Differential equations with functional derivatives and stochastic equations for a generalized process, *Dokl. Akad. Nauk. SSSR* **166** (1966), 220–223.
22. Daletskii, Yu. L., Infinite dimensional elliptic operators and parabolic equations connected with them, *Uspekhi Mat. Nauk.* **22** (1967); Translation, *Russian Math. Surveys* **22** 1–53.
23. Daletskii, Yu. L., and Fomin, S. V., Generalized measures in Hilbert space and Kolmogorov's forward equation, *Soviet Math. Dokl.* **13** (1972), 993–997.
24. Daletskii, Ju. L., and Paramonova, S. N., Stochastic integrals with respect to a normally distributed additive set function, *Soviet Math. Dokl.* **14** (1973), 96–100.
25. Dawson, D. A., Stochastic evolution equations, *Math. Biosci.* **15** (1972), 287–316.
26. Dawson, D. A., Stochastic evolution equations and related measure processes, *J. Multivari. Anal.* **5** (1975), 1–52.
27. Donsker, M. D., and Varadhan, S. R. S., Asymptotic evaluation of certain Wiener integrals for large time, *in* "Functional Integration and its Applications." Oxford Univ. Press (Clarendon) London and New York, 1975, pp. 13–33.
28. Doob, J. L., The Brownian movement and stochastic equations, *Ann. of Math.* **43** (1942), 251–369.
29. Edwards, S. F., The statistical dynamics of homogeneous turbulence, *J. Fluid Mech.* **18** (1964), 239–273.
30. Feynman, R. P., Space-time approach to nonrelativistic quantum mechanics, *Rev. Modern Phys.* **20** (1948), 367–387.
31. Fleming, W. H., Distributed parameter stochastic systems in population biology, *Proc. IRIA Symp. on Control Theory, Numerical Meth. and Computer Systems Mod.*, (June, 1974).
32. Foias, C., and Prodi, G., Statistical study of Navier–Stokes equations, *Rend. Sem. Math. Univ. Padova* **48** (1972), 219–348.
33. Friedman, A., "Partial Differential Equations." Holt, New York, 1969.
34. Friedman, A., "Stochastic Differential Equations and Applications." Vols. I and II, Academic Press, New York, 1975.
35. Friedricks, K. O., "Pseudodifferential Operators, Lecture Notes." Courant Institute, New York Univ., New York 1970.
36. Frisch, U., Wave propagation in random media, *in* "Probabilistic Methods in Applied Mathematics" (A. T. Bharucha-Reid, ed). Academic Press, New York, 1968.
37. Frisch, U., Lesieur, M., and Brissaud, A., A Markovian random coupling model for turbulence, *J. Fluid Mech.* **65** (1974), 145–152.
38. Gelfand, I. M., and Yaglom, A. M., Integration in function spaces and its applications in quantum physics, *J. Mathematical Phys.* **1** (1960), 48–69.
39. Gihman, I. I., and Skorokhod, A. V., "Stochastic Differential Equations," Springer-Verlag, Berlin and New York, 1972.
40. Goldstein, S., On diffusion by discontinuous movements, and on the telegraph equation, *Quart. J. Mech. Appl. Math.* **4** (1951), 129–156.
41. Griego, R., and Hersh, R., Theory of random evolutions with applications to partial differential equations, *Trans. Amer. Math. Soc.* **56** (1971), 405–418.
42. Gross, L., Classical analysis on a Hilbert space, *in* "Analysis in Function Space." MIT Press, Cambridge, Massachusetts, 1964, pp. 51–68.
43. Gross, L., Abstract Wiener spaces, *Proc. 5th Berkeley Symp. Math. Statistic and Prob.*, Vol. II, 1967, 31–42.
44. Herring, J. R., and Kraichnan, R. H., Comparison of some approximations for isotropic turbulence, *in* "Statistical Models and Turbulence," Springer Publ. New York, 1972, pp. 148–177.

45. Hersh, R., "Random evolutions: A survey of results and problems, *Rocky Mountain J. Math.* **4** (1974), 443–496.
46. Hopf, E., Statistical hydromechanics and functional calculus, *J. Rat. Mech. Anal.* **1** (1952), 87–123.
47. Itô, K., Stochastic integral, *Proc. Imp. Acad. (Tokyo)* **20** (1944), 519–524.
48. Kac, M., On distribution of certain Wiener functionals, *Trans. Amer. Math. Soc.* **65** (1949), 1–13.
49. Kac, M., Some stochastic problems in physics and mathematics, Magnolia Petroleum Co., Lectures in Pure and Applied Science, No. 2, 1956.
50. Kannan, D., and Bharucha-Reid, A. T., Operator valued stochastic integral, *Proc. Jap. Acad.* **47** (1971), 472–476.
51. Keller, J. B., Stochastic equations and wave propagation in random media, *Proc. Symp. Appl. Math.* **16** (1964), 145–170.
52. Klyatskin, V. I., and Tatarski, V. I., The parabolic equation approximation for propagation of waves in a medium with random inhomogenities, *Soviet Phys. JETP* **31** (1970), 335–339.
53. Kraichnan, R. H., The structure of isotropic turbulence at very high Reynolds numbers, *J. Fluid Mech.* **5** (1959), 497–543.
54. Kraichnan, R., Dynamics of nonlinear stochastic systems, *J. Mathematical Phys.* **2** (1961), 124–148.
55. Kuo, H. H., Differential and stochastic equations in abstract Wiener space, *J. Functional Analysis* **12** (1973), 246–256.
56. Kuo, H. H., and Peich, M. A., Stochastic integrals and parabolic equations in abstract Wiener space, *Bull. Amer. Math. Soc.* **79** (1973), 478–482.
57. Kuo, H. H., Integration by parts for abstract Wiener measures, *Duke Math. J.* **41** (1974), 373–379.
58. Kuo, H. H., "Gaussian Measures in Banach spaces, Lecture Notes." Springer–Verlag, Berlin and New York, 1975.
59. **Kuo, H. H., Potential theory associated with Uhlenbeck-Ornstein process,** *J. Functional Analysis* **21** (1976), 63–75.
60. Ladyzhenskaya, O. A., "The Mathematical Theory of Viscous Incompressible Flow." Gordon & Breach, New York, 1963.
61. Landau, L., and Lifschitz, E., "Fluid Mechanics." Pergamon, Oxford, 1959.
62. Leray, J., Sur le mouvement din liquide visqueux emplisant l'espace, *Acta Math.* **63** (1934), 193–248.
63. Lewis, R. M., and Kraichnan, R. H., A space–time functional formalism for turbulence, *Comm. Pure Appl. Math.* **15** (1962), 397–411.
64. Lumley, J. L., and Rosenblatt, M., "Statistical Models in Turbulence." Springer–Verlag, Berlin and New York, 1972.
65. McKean, H. P., "Stochastic Integrals." Academic Press, New York, 1969.
66. Monin, A. S., and Yaglom, A. M., "Statistical Fluid Mechanics." MIT Press, Cambridge, Massachusetts, 1971.
67. Nelson, E., Quantum fields and Markov fields, *Proc. Symp. Pure Math.* **23** (1973), 413–421.
68. Novikov, E. A., Functionals and the random-force method in turbulence theory, *Sov. Phys. JETP* **20** (1965), 1290–1294.
69. Papanicolaou, G., and Varadham, S.R.S., A limit theorem with strong mixing in Banach space and two applications to stochastic differential equations, *Comm. Pure Appl. Math.* **26** (1973), 497–524.
70. Pardoux, E., Equations aux dérivées partielles stochastiques monotones, *CR. Acad. Sci.*, Paris, Ser. A, **275** (1972), 101–103.

71. Piech, M. A., Diffusion semigroups on abstract Wiener space, *Trans. Amer. Math. Soc.* **166** (1972), 411–430.
72. Piech, A. M., The Ornstein–Uhlenbeck semigroup in an infinite dimensional $L^2$ setting, *J. Functional Analysis* **18** (1975), 271–285.
73. Pinsky, M., Differential equations with a small parameter and central limit theorem for functions defined on a Markov chain, *Z. Wahrscheinlichkeitstheorie und Verw. Gebiete* **9** (1968), 101–111.
74. Pinsky, M., Multiplicative operator functionals of a Markov process, *Bull. Amer. Math. Soc.* **77** (1971), 377–380.
75. Prohorov, Yu. V., The methods of characteristic functionals, *Proc. 4th Berkeley Symp. Math. Statist. and Prob.*, Vol. II, 403–319, (1961).
76. Royden, H. L., "Real Analysis." Macmillan, New York, 1963.
77. Ruelle, D., "Statistical Mechanics and Dynamical Systems, Duke Univ. Math. Series III." Duke Univ., Durham, North Carolina, 1977.
78. Skorokhod, A. V., Constructive methods of specifying stochastic processes, *Russian Math. Surveys* **20** (1965), 63–83.
79. Spivak, M., "Calculus on Manifolds." Benjamin, New York, 1965.
80. Stratonovich, R. L., A new representation for stochastic integrals and equations, *J. SIAM Control* **4** (1966), 362–371.
81. Taylor, G. I., Diffusion by continuous movements, *Proc. London Math. Soc.*, Ser. 2, **20** (1921), 196–212.
82. Uhlenbeck, G. E., and Ornstein. L. S., On the theory of Brownian motion I, *Phys.* **38** (1930), 823–841.
83. Varadhan, S.R.S., "Stochastic Processes, Lecture Notes." Courant Institute, New York University, New York 1968.
84. Viot, M., A stochastic partial differential equation arising in population genetic theory, Tech. Rep. 75-3, Lefschetz Centre for Dynamical Systems, Brown University, 1975.
85. Wiener, N., "Nonlinear problems in Random Theory." MIT Press, Cambridge, Massachusetts, 1958.
86. Wong, E., and Zakai, M., On the convergence of ordinary integrals to stochastic integrals, *Ann. Math. Statist.* **36** (1965), 1560–1564.
87. Yosida, K., "Functional Analysis." Springer–Verlag, Berlin and New York, rev. ed., 1970.

# Estimation and Stochastic Control for Linear Infinite-Dimensional Systems

*RUTH F. CURTAIN*

MATHEMATICS INSTITUTE
UNIVERSITY OF GRONINGEN
THE NETHERLANDS

|      |                                                              |     |
|------|--------------------------------------------------------------|-----|
| I.   | Introduction                                                 | 45  |
| II.  | The Semigroup Description of Linear Autonomous Systems       | 46  |
| III. | Stochastic Evolution Equations                               | 51  |
| IV.  | Deterministic Quadratic Cost-Control Problem                 | 60  |
| V.   | State Estimation                                             | 64  |
| VI.  | The Separation Principle for Stochastic Optimal Control      | 69  |
| VII. | Extensions                                                   | 73  |
|      | A. Related Control and Estimation Problems                   | 73  |
|      | B. Time-Dependent System Operators                           | 74  |
|      | C. Nongaussian Noise Processes                               | 76  |
|      | D. Limited Sensing and Control                               | 77  |
|      | E. Delayed Observation and Delayed Control Action            | 80  |
|      | F. Systems with State-Dependent Noise                        | 82  |
|      | References                                                   | 84  |

## I. Introduction

Over the past few years there has been interest in estimation and control problems for systems described by linear delay equations and by partial differential equations (distributed systems). Initially, results were developed using methods specifically for distributed systems [3–5, 43] or for delay systems [24, 25, 39]. Several authors studying distributed systems used an abstract semigroup formulation [1, 7, 14, 22, 26] and using the Hilbert space setting for delay equations introduced in [24], it was realized that delay equations can also be treated using a semigroup approach [9, 10, 40]. More recently, this semigroup approach has been used by several authors to consider the major systems-theory concepts of controllability, observability, stability, optimal control, and estimation for linear infinite dimensional systems on Hilbert spaces [1, 11, 12, 13, 14, 16, 28, 37, 44, 45].

The semigroup formulation is an appealing one for an expository article because the abstract notation looks exactly like the finite dimensional notation with operators instead of matrices and because it provides a unified treatment of lumped, distributed, and delay systems. Moreover, it is the approach with which the author is familiar, and will be used here to present the theory of stochastic linear infinite-dimensional systems, comprising abstract stochastic evolution equations, estimation for linear systems, and the separation principle for the stochastic linear quadratic cost-control problem. To make the presentation self contained, there is also an introduction on semigroups and a brief treatment of the deterministic quadratic cost-control problem. For simplicity, the first sections treat gaussian time-invariant systems with bounded system operators and the extensions to nongaussian time-varying systems with unbounded system operators and other extensions are discussed in the final section. Although the theory may seem superficially similar to the finite-dimensional estimation and control, the examples of particular delay and distributed systems illustrating the theory will clearly indicate the complexities and subtleties of infinite-dimensional systems.

## II. The Semigroup Description of Linear Autonomous Systems

In this section we consider linear autonomous dynamical systems as they evolve in their free state, that is, not subject to any external inputs or random disturbances. Suppose the system's initial state at time zero is denoted by $z_0$, an element in a Banach space $Z$, and that its state at a subsequent time $t$ is denoted by $z(t)$. From our assumption that the dynamics governing the evolution from $z_0$ to $z(t)$ are linear and autonomous, we can define an operator $T_t$; $t \geq 0$ such that

$$z(t) = T_t z_0 \tag{2.1a}$$

$$T_t: Z \to Z; \quad T_0 = I, \quad \text{the identity on } Z. \tag{2.1b}$$

If we assume that the solution varies continuously with the initial state, then $T_t$ is bounded. Furthermore, if we assume that $z(t+s)$ is the same state as that reached by going from the state $z(s)$ at time $s$ for a further time $t$, for all initial states $z_0 \in Z$, then we have the so-called "semigroup" property of dynamical systems.

$$z(t+s) = T_{t+s}z_0 = T_t z(s) = T_t T_s z_0 \tag{2.2}$$

or

$$T_{t+s} = T_t T_s \quad \text{for all} \quad t, s \geq 0. \tag{2.3}$$

Finally, we impose some smoothness on the trajectory and suppose that $z(t) \to z_0$ as $t \to 0+$ for all $z_0 \in Z$, where the convergence is the usual norm convergence in $Z$. All of these physically motivated properties of linear autonomous dynamical systems are expressed very simply by requiring that $T_t$ be a strongly continuous semigroup.

**Definition 2.1** *Strongly continuous semigroups.* A strongly continuous semigroup is a map $T_t$ from $R^+$ to $\mathscr{L}(Z)$, which satisfies

$$T_{t+s} = T_t T_s, \quad 0 \leq s \leq t, \tag{2.4a}$$

$$T_0 = I, \tag{2.4b}$$

$$\|T_t z_0 - z_0\| \to 0 \quad \text{as} \quad t \to 0+ \ \forall z_0 \in Z. \tag{2.4c}$$

Useful consequences of (2.4a)–(2.4c) are that $T_t$ is in fact strongly continuous on $[0, \infty)$ and has the estimate

$$\|T_t\| \leq Me^{\omega t} \tag{2.5}$$

for constants $\omega$ and $M \geq 0$. Furthermore, for all $z_0$ from a dense subset of $Z$, $T_t z_0$ is differentiable and uniquely defines a linear closed operator $A$ on $Z$ by

$$Az_0 = \lim_{h \to 0+} [(T_t z_0 - z_0)/h], \tag{2.6}$$

and $\mathscr{D}(A)$ is the set of $z_0 \in Z$ such that (2.6) holds. $A$ is a closed, densely defined operator and is called the *infinitesimal generator of $T_t$* and

$$d/dt(T_t z_0) = AT_t z_0 = T_t A z_0 \quad \text{for all} \quad z_0 \in \mathscr{D}(A). \tag{2.7}$$

This provides the link between our description of a dynamical system by $z(t)$, $T_t z_0$ to the more usual differential one,

$$\dot{z} = Az, \quad z(0) = z_0 \in \mathscr{D}(A). \tag{2.8}$$

The unique solution of (2.8) is then $z(t) = T_t z_0$. There is a well-known theory of strongly continuous semigroups [27, 31] and the Hille–Yoshida theorem provides a complete characterization of those operators generating strongly continuous semigroups.

**Theorem 2.1** *Hille–Yoshida theorem.* *If $A$ is a closed linear operator with $\mathscr{D}(A) = Z$ and $(\lambda I - A)^{-1}$ exists for $\lambda > \omega$, with*

$$\|(\lambda I - A)^{-m}\| \leq [M/(\lambda - \omega)^m], \quad m = 1, 2, \ldots,$$

then $A$ generates a strongly continuous semigroup with
$$\|T_t\| \leq Me^{\omega t}.$$
As we shall only consider strongly continuous semigroups here, we shall use the abbreviation: *$T_t$ is a semigroup with generator $A$.*

**Example 2.1** For systems of ordinary differential equations
$$\dot{z} = Az; \quad z(0) = z_0,$$
where $Z = R^n$ and $A \in \mathcal{L}(R^n)$ is represented as a matrix, then
$$T_t = e^{At}.$$
So our semigroup is a generalization of $e^{At}$ to infinite dimensions.

**Example 2.2** *Delay Equations* We consider the linear system on $[0, t_1]$ described by
$$\dot{x}(t) = A_0 x(t) + \sum_{i=1}^{N} A_i \begin{cases} x(t+\theta_i); & t+\theta_i \geq 0 \\ h(t+\theta_i); & t+\theta_i < 0 \end{cases}$$
$$+ \int_{-b}^{0} A_{01}(\theta) \begin{cases} x(t+\theta) \\ h(t+\theta) \end{cases} d\theta; \quad \begin{matrix} t+\theta \geq 0 \\ t+\theta < 0 \end{matrix}$$
$$x(0) = h(0), \tag{2.9}$$

where $-b \leq \theta_N \leq \theta_{N-1} \leq \cdots \leq \theta_1 = 0$ and $b$ is a positive real number. $A_0, A_i; (i = 1, \ldots, N) \in \mathcal{L}(R^n)$ and $A_{01} \in C(-b, 0; \mathcal{L}(R^n))$.

It is possible to define this system on $C(-b, 0; R^n)$, but for control applications it is more convenient to use the Hilbert space, $\mathcal{M}^2 = \mathcal{M}^2(-b, 0; R^n)$, introduced by Delfour and Mitter in [24]. $\mathcal{M}^2$ is the quotient space of $L_2(-b, 0; R^n)$ under the norm,
$$\|h\|_{\mathcal{M}^2}^2 = \|h(0)\|_{R^n}^2 + \int_{-b}^{0} \|h(\theta)\|_{R^n}^2 d\theta,$$
and it is isometrically isomorphic to $R^n \times L_2(-b, 0; R^n)$. In [24] it is shown that (2.9) has a unique solution $x(t)$ for all initial states $h \in \mathcal{M}^2$ and gives rise to a strongly continuous semigroup $\tilde{T}_t$ on $\mathcal{M}^2$ via
$$(\tilde{T}_t h)(\theta) = \begin{cases} x(t+\theta); & t+\theta \geq 0 \\ h(t+\theta); & t+\theta < 0 \end{cases}.$$
The appropriate abstract evolution equation on $\mathcal{M}^2$ is
$$\dot{z} = \tilde{A}z, \quad z(0) = h, \quad h \in \mathcal{D}(\tilde{A}), \tag{2.10}$$
where $\tilde{A}$ is a closed linear operator on $\mathcal{M}^2$ with domain $W^{1,2}(-b, 0; R^n)$.
$$W^{1,2}(-b, 0; R^n) = \{x \in L_2(-b, 0; R^n) : Dx \in L_2(-b, 0; R^n)\}, \tag{2.11}$$

where $Dx$ denotes the distributed derivative of $x$. Furthermore,

$$(\tilde{A}h)(\theta) = \begin{cases} A_0 h(0) + \sum_{i=1}^{N} A_i h(\theta_i) + \int_{-b}^{0} A_{01}(\theta) h(\theta) \, d\theta; & \theta = 0 \\ \dfrac{dh}{d\theta}; & \theta \neq 0. \end{cases} \quad (2.12)$$

$\tilde{A}$ is the generator of $\tilde{T}_t$ and so (2.10) has the unique solution $\tilde{T}_t h$, and $x(t) = (\tilde{T}_t h)(0)$ provides the relationship with the original delay equation (2.9).

**Example 2.3** *Heat Equation*

$$\frac{\partial z}{\partial t} = \frac{\partial^2 z}{\partial x^2}$$

$$z(0, t) = 0 = z(1, t), \qquad z(0, x) = z_0(x). \quad (2.13)$$

If we take $H = L_2(0, 1)$ and introduce the operator $A$ on $H$ given by

$$Ah = \frac{\partial^2 h}{\partial x^2} \quad \text{for} \quad h \in \mathcal{D}(A), \quad (2.14)$$

where $\mathcal{D}(A) = \{h \in H : h_x, h_{xx} \in H; h(0) = 0 = h(1)\}$ then $A$ generates the strongly continuous semigroup given by

$$T_t h(x) = \sum_{n=1}^{\infty} 2e^{-n^2\pi^2 t} \sin n\pi x \int_0^1 \sin \pi n r h(r) \, dr. \quad (2.15)$$

Using separation of variables it is easily verified that the solution of (2.13) is just $(T_t z_0)(x)$ and so (2.13) is equivalent to the abstract evolution equation

$$\dot{z} = Az, \qquad z(0) = z_0,$$

where $A$ is given by (2.14).

Similarly, other parabolic partial differential equations can be expressed as abstract evolution equations on an appropriate Hilbert space. It is also possible to formulate hyperbolic systems in this way, both symmetric and nonsymmetric (see Curtain and Pritchard [21] for details). Thus this semigroup description of linear systems includes a wide variety of systems of quite different types.

For controllable dynamical systems, we are led to considering inhomogeneous abstract equations of the type

$$\dot{z} = Az + Bu, \qquad z(0) = z_0, \quad (2.16)$$

where $u \in L_2(0, t_1; U)$ represents some control input and $B \in \mathcal{L}(U, Z)$ describes the effect of the control on the system. ($U$ is another Banach space.)

If $Z$ is finite dimensional, it is an elementary fact that (2.10) has the unique

solution
$$z(t) = T_t z_0 + \int_0^t T_{t-s} Bu(s)\, ds \qquad (2.17)$$

However, if $Z$ is infinite dimensional this is not always true as (2.17) is not in $\mathscr{D}(A)$ in general. Sufficient conditions are that $Bu$ is continuously differentiable or that (2.17) represent a delay system as in Example 2.2 (see Curtain and Pritchard [21] for other conditions.)

These difficulties are neatly avoided by considering (2.17) to be the description of the system and defining it to be the *mild solution* of (2.16).

One can generalize all the finite-dimensional system theory concepts of controllability, observability, stability, and stabilizibility to describe abstract system behavior, though of course they are more complex (see [28, 44, 21]). Here we shall only be concerned with stability and stabilizibility.

**Definition 2.2** *Stable Semigroups* If $T_t$ is a semigroup that satisfies $\|T_t\| \leqslant Me^{-\omega t}$ where $\omega > 0$, then we say that $T_t$ or equivalently its generator $A$ is stable. The notion of stabilizibility arises from considering (2.16), when the control is a feedback one, $u(t) = -Fz(t)$, where $F \in \mathscr{L}(Z, U)$. Equation (2.16) then becomes a homogeneous system of the form of (2.8) with the new operator $A - BF$. It can be shown that $A - BF$ is the generator of a strongly continuous semigroup and so if $A$ is not itself stable, by suitably choosing $F \in \mathscr{L}(Z, U)$ it may be possible to "stabilize" the system.

**Definition 2.3** *Stabilizibility* We say $(A, B)$ is stabilizable if there exists an $F \in \mathscr{L}(Z, U)$ such that $A - BF$ is a stable generator.

Often $u(t)$ is a time-dependent feedback control, that is $u(t) = -F(t)z(t)$ for some $F(t) \in \mathscr{L}(Z, U)$ and then (2.16) becomes the time-dependent homogeneous equation
$$\dot{z} = (A - BF(t))z(t), \qquad z(0) = z_0. \qquad (2.18)$$

The relationship of (2.18) to (2.16) is clearly seen from the following perturbation result.

**Theorem 2.2** *Let*
$$\mathscr{B}_\infty(0, t_1; \mathscr{L}(Z)) = \left\{ \begin{array}{l} D : (0, t_1) \to \mathscr{L}(z) \text{ such that } D(\cdot)z \text{ is} \\ \text{strongly measurable for each } z \in Z \text{ and} \\ \quad \operatorname*{ess\,sup}_{0 \leqslant t \leqslant t_1} \|D(t)\|_{\mathscr{L}(Z)} < \infty \end{array} \right\}.$$

*Then for each $D \in \mathscr{B}_\infty(0, t_1; \mathscr{L}(Z))$, the following integral equation has a unique solution $U(t, s)$ for every $0 \leqslant s \leqslant t \leqslant t_1 =$*
$$U(t, s)z = T_{t-s}z + \int_s^t T_{t-r} D(r) U(r, s)z\, dr. \qquad (2.19)$$

$U(t, s)$ *is called the perturbation of $T_t$ by $D$ and is a mild evolution operator.*

ESTIMATION AND STOCHASTIC CONTROL FOR LINEAR SYSTEMS

**Definition 2.4** *Mild Evolution Operator* Let

$$\Delta(t_1) = \{(t,s); 0 \leq s \leq t \leq t_1\},$$

then $U(t,s) : \Delta(t_1) \to \mathscr{L}(Z)$ is a mild evolution operator if

$$U(t,t) = I, \quad t \in [0, t_1], \tag{2.20a}$$

$$U(t,r)U(r,s) = U(t,s), \quad 0 \leq s \leq r \leq t \leq t_1, \tag{2.20b}$$

and

$$U(\cdot, s) \text{ is strongly continuous on } [s, t_1]$$

$$\text{and } U(t, \cdot) \text{ is strongly continuous on } [0, t]. \tag{2.20c}$$

Clearly, $T_{t-s}$; $t \geq s \geq 0$ is a special case of a mild evolution operator and if $D$ is time invariant, the perturbation of $T_t$ by $D$ is a semigroup with generator $A+D$. We can also think of $A+D(t)$ as a generator of $U(t,s)$, although in general $U(t,s)z$ is *not* differentiable in $t$, but surprisingly, it is differentiable in $s$, with

$$\frac{\partial}{\partial s} U(t,s)z = -U(t,s)(A+D(s))z \quad \text{for} \quad z \in \mathscr{D}(A). \tag{2.21}$$

This is called the "quasi" property in the literature and if it is also differentiable in $t$ it has the "strong" property. These properties turn out to have special significance for control and filtering problems and we shall discuss this further in a later section.

## III. Stochastic Evolution Equations

The prototype stochastic differential equation in finite-dimensional stochastic control is

$$dz(t) = Az(t)\,dt + D\,dw(t), \quad z(0) = z_0, \tag{3.1}$$

where $A$ and $D$ are $n \times n$ and $n \times r$ matrices, respectively, $z_0$ is an $n$-dimensional vector random variable and $w(t)$ is an $r$-dimensional Wiener process; that is, the components $w_i(t)$ are mutually independent real Wiener processes for $i = 1, \ldots, r$.

Equation (3.1) has the unique solution

$$z(t) = e^{At}z_0 + \int_0^t e^{A(t-s)} D\,dw(s), \tag{3.2}$$

which is Gaussian and has a finite variance matrix if $z_0$ has these properties.

We wish to consider (3.1) on an abstract Hilbert space $H$, assuming that $A$ generates a strongly continuous semigroup $T_t$, $D \in \mathscr{L}(H, K)$, where $K$ is another Hilbert space and $w(t)$ is a $K$-valued Wiener process. Before presenting the theory of such abstract evolution equations, let us consider two typical examples of stochastic differential equations, which can be formulated as abstract equations on a Hilbert space.

**Example 3.1** *Noisy Heat Equation* This is essentially a stochastic version of Example 2.3, for we take $H = L_2(0, 1)$ and

$$Ah = \frac{\partial^2 h}{\partial x^2} \quad \text{for} \quad h \in \mathcal{D}(A) = \left\{ \begin{array}{l} h \in H : h_x, h_{xx} \in H \\ h(0) = 0 = h(1) \end{array} \right\}. \quad (3.3)$$

as $A$ generates the semigroup $T_t$ given by (2.15).

A naive attempt at modeling a distributed Wiener process leads us to suppose that

$$w(t) = \sum_{i=1}^{\infty} \beta_i(t) e_i, \quad (3.4)$$

where $\beta_i(t)$ are mutually independent real Wiener processes with incremental variance $\lambda_i$ and $e_i(x) = r_2 \sin \pi i x$ form an orthonormal basis for $H$.

Similarly, we suppose $z_0$ has expectation $a \in H$ and covariance operator $P_0 \in \mathcal{L}(H)$ such that

$$P_0 e_i = \alpha_i e_i; \quad i = 1, \ldots. \quad (3.5)$$

By analogy with the finite-dimensional case and the deterministic Example 2.3, we would expect the solution of (3.1), if it exists, to be given by

$$z(t) = \sum_{n=1}^{\infty} 2e^{-n^2 \pi^2 t} \sin n\pi x \int_0^1 \sin n\pi r z_0(r) \, dr$$

$$+ \sum_{n=1}^{\infty} \int_0^t e^{-n^2 \pi^2 (t-s)} \, d\beta_n(s) \sin n\pi x. \quad (3.6)$$

We shall see that this is the case under certain assumptions on the noise parameters $\lambda_i$ and $\alpha_i$.

**Example 3.2** *A Stochastic Delay Equation* Consider the stochastic version of Example 2.2.

$$dx(t) = A_0 x(t) \, dt + \sum_{i=1}^{N} A_i \left\{ \begin{array}{l} x(t + \theta_i); \ t + \theta_i \geq 0 \\ h(t + \theta_i); \ t + \theta_i < 0 \end{array} \right\} dt$$

$$+ \int_{-b}^{0} A_{01}(\theta) \left\{ \begin{array}{l} x(t + \theta) \\ h(t + \theta) \end{array} \right\} d\theta \, dt; \ \begin{array}{l} t + \theta \geq 0 \\ t + \theta < 0 \end{array} + D \, dw(t)$$

$$x(0) = h(0), \quad (3.7)$$

where $A_0$, $A_i$, and $A_{01}$ satisfy the assumptions of Example 2.2, $D \in \mathcal{L}(R^p, R^n)$, $h$ is a vector stochastic process on $(-b, 0)$ and $w(t)$ is a $p$-dimensional Wiener process, with matrix incremental variance $W$. This motivates consideration of the following $\mathcal{M}^2$ stochastic evolution equation

$$dz(t) = \tilde{A}z(t)\,dt + \tilde{D}\,dw(t)$$
$$z(0) = \tilde{h}, \tag{3.8}$$

where $\tilde{A}$ is given by (2.11) and (2.12), $\tilde{h} = (h(0), h(\cdot))$ is a random variable in $\mathcal{M}^2$, and $\tilde{D}$ satisfies

$$(\tilde{D}y)(\theta) = \begin{cases} Dy, & \theta = 0 \\ 0, & \theta \neq 0 \end{cases} \tag{3.9}$$

It is clear that to consider stochastic evolution equations we need some basic abstract probability theory (see [3, 27, 42, 21]).

We suppose throughout that $(\Omega, \mathcal{P}, p)$ is a complete probability space, $X$ is a Banach space and $H, K$ are real separable Hilbert spaces.

**Definition 3.1** An $X$-valued random variable is a map $x: \Omega \to X$, which is strongly measurable with respect to $p$ measure. If $x$ is integrable on $X$ we define its expectation by

$$E\{x\} = \int_\Omega x\,dp$$

(where the integral is a Bochner integral [27]).

Most of our abstract spaces are Hilbert spaces and in this case we can define the covariance operator of a random variable, which generalizes the covariance matrix $\text{Cov}\{u\} = E\{(u - E\{u\})(u - E\{u\})'\}$ for vector-valued random variables.

**Definition 3.2** The covariance operator of an $H$-valued random variable $h \in L_2(\Omega, p; H)$ is defined by

$$\text{Cov}\{h\} = E\{(h - E\{h\})_0 (h - E\{h\})\},$$

where the tensor product $h \circ k \in \mathcal{L}(H)$ is defined for all $h, k \in H$ by

$$(h \circ k)u = h\langle k, u\rangle, \quad u \in H.$$

Consequently, $\text{Cov}\{h\}$ is symmetric, positive, and nuclear.

If we consider the covariance operator $P_0$ of Example 3.1, given by (3.5), we see that for $P_0$ to be positive and nuclear, we require that $\alpha_i \geq 0$ and $\sum_{i=1}^\infty \alpha_i = \text{tr}\, P_0 < \infty$.

As for real random variables, the random variable $x: \Omega \to X$ induces a measure $p_x$ on $\mathcal{B}(X)$, the Borel sets of $X$, by virtue of

$$p_x(A) = p\{\omega: x(\omega) \in A\} \quad \text{for} \quad A \in \mathcal{B}(X),$$

and $(X, \mathscr{B}(X), p_x)$ is a complete probability space. Of special interest are Gaussian measures or random variables.

**Definition 3.3** An $X$-valued random variable $x$ is Gaussian if $f(x)$ is a real Gaussian random variable for all $f \in X^*$, the dual of $X$.

There are many types of convergence one can introduce for random variables, but we shall only need the following.

**Definition 3.4** (a) A sequence $\{x_n\}$ of $X$-valued random variables converges to $x$ in mean square if

$$E\{\|x_n - x\|^2\} \to 0 \quad \text{as} \quad n \to \infty.$$

(b) Let $p_n$ and $p_x$ be the measures induced on $\mathscr{B}(X)$ by $x_n$ and $x$, respectively. Then $p_n$ converges weakly to $p_x$ [in the space of all measures on $\mathscr{B}(X)$] if

$$\int_X f \, dp_n \to \int_X f \, dp_x \quad \text{as} \quad n \to \infty$$

for all continuous bounded functions $f: X \to R$.

**Definition 3.5** $X$-valued random variables $x$ and $y$ are independent if $\{\omega: x(\omega) \in A\}$ and $\{\omega: y(\omega) \in B\}$ are independent sets in $\mathscr{P}$ for any Borel sets $A, B$ in $X$.

If $h$ and $K$ are in $L_1(\Omega, p; H)$ and are independent, then

$$E\{\langle h, k \rangle\} = \langle E\{h\}, E\{k\} \rangle.$$

We find it convenient to use the following less general definition for abstract stochastic processes, requiring measurability on the parameter $t$.

**Definition 3.6** An $X$-valued stochastic process is a map $u: (0, t_1) \times \Omega \to X$, which is measurable in the product measure on $(0, t_1) \times \Omega$ using Lebesgue measure on $(t_0, t_1)$.

There are several types of continuity one can define (see [3, 27]), but all we require are the following.

**Definition 3.7** Let $u(t)$ be an $X$-valued stochastic process on $(0, t_1)$, then

(a) $u(t)$ is continuous in mean square if

$$E\{\|u(t+\delta) - u(t)\|^2\} \to 0 \quad \text{as} \quad \delta \to 0.$$

(b) $u(t)$ has continuous sample paths if

$$p\{\sup_{0 \leq t \leq t_1} \|u(t+\delta) - u(t)\| > 0\} \to 0 \quad \text{as} \quad \delta \to 0.$$

Of course (b) implies (a).

ESTIMATION AND STOCHASTIC CONTROL FOR LINEAR SYSTEMS

In applications we shall be concerned with finding estimates of certain stochastic processes based on related ones, which involves us with the concept of the conditional expectation for an $X$-valued random variable. It is always well-defined and has analogous properties to the scalar conditional expectation.

**Definition 3.8** The conditional expectation $E\{x|\mathscr{F}\}$ of an $X$-valued random variable $x \in L_1(\Omega, p; X)$ relative to a subsigma field $\mathscr{F} \subset \mathscr{P}$ is such that

$$\int_C x(\omega)\, dp = \int_C E\{x|\mathscr{F}\}(\omega)\, dp \qquad \forall C \in \mathscr{F}$$

$E\{x|\mathscr{F}\}$ is uniquely defined by this relationship and is measurable relative to $\mathscr{F}$.

Usually $\mathscr{F}$ is the sigma field generated by a Banach-space-valued random variable $y$ on $(\Omega, \mathscr{P}, p)$ and in this case we write $E_y\{h\}$ for the conditional expectation of $h$, given $y$, which has the physical interpretation as the best global estimate of $h$, given $y$.

If $h$ and $y$ are second-order Hilbert-space-valued random variables, that is $h \in L_2(\Omega, p; H) = \mathscr{H}$ and $y \in L_2(\Omega, p; K)$, then $E_y\{h\}$ also has the geometric interpretation as the projection of $h$ on a subspace $\mathscr{H}_y$ of $\mathscr{H}$. To see this we denote by $p_y$ the probability measure induced by $y$ on $(K, \mathscr{B}(K))$ and by $L_2(K, p_y; H)$ we mean the space of $p_y$-measurable maps from $K$ to $H$ with norm

$$\|\phi\| = \left(\int_K \|\phi(K)\|_H^2\, dp_y(K)\right)^{\frac{1}{2}}$$

Then $L_2(K, py; H)$ is isomorphic to a subspace of $\mathscr{H} = L_2(\Omega, p; H)$ which we denote by $\mathscr{H}_y$. More precisely,

$$\mathscr{H}_y = \begin{cases} u(\omega) \in \mathscr{H}: u(\omega) = \lambda(y(\omega)) & \text{and} \quad \lambda: K \to H \\ \text{is measurable relative to } p_y \end{cases}. \quad (3.10)$$

We shall also be concerned with linear estimates.

**Definition 3.9** The best linear estimate $\hat{h}$ of $h \in L_2(\Omega, p; H)$ based on $y \in L_2(\Omega, p; K)$ is $\hat{h} = \Lambda y$, where $\Lambda \in \mathscr{L}(K, H)$ is such that $E\{\|h - Gy\|^2\}$ is minimized over all $G \in \mathscr{L}(K, H)$.

As for scalar random variables, if $h$ and $y$ are Gaussian and the best linear estimate of $h$, given $y$, exists, then it equals the best global estimate $E_y\{h\}$.

In applications we need to consider estimates of stochastic processes and so we extend the concept of conditional expectations to stochasic processes.

If the stochastic process $y: (0, t_1) \times \Omega \to K$ has continuous sample paths, we define $y_t$ to be the restriction of $y$ on $(0, t)$. $y_t$ is then a random variable with values in the Banach space $C(0, t; K)$. For each $t$, $\mathcal{H}y_t$ is defined by (3.10) and we consider the Hilbertian sum $\int^{\oplus} \mathcal{H}_{y_t} dt$ defined by

$$\int^{\oplus} \mathcal{H} y_t \, dt = \{h(t, \omega) \in L_2(0, t_1; \mathcal{H}): h(t) \in \mathcal{H}_{y_t} \text{ for almost all } t\}. \tag{3.11}$$

Then the conditional expectation of $h(t)$ with respect to $y(s)$; $0 \leqslant s \leqslant t$ is $E_{y_t}\{h(t)\}$ and it can be shown that $E_{y_t}\{h(t)\} \in \int^{\oplus} \mathcal{H} y_t \, dt$ except on a null set. A special class of stochastic processes, which occur frequently in various applications, are martingales.

**Definition 3.10** Let $m(\cdot, \cdot)$ be an $X$-valued stochastic process on $(0, t_1)$ such that $m(t, \cdot) \in L_1(\Omega, p; X)$ for almost all $t \in (0, t_1)$. Suppose that $\{\mathcal{F}_t\}$ is an increasing family of sigma fields, $\mathcal{F}_s \subset \mathcal{F}_t \subset \mathcal{P}$ for $s < t$, and $m(t, \cdot)$ is measurable relative to $\mathcal{F}_t$ for almost all $t \in (0, t_1)$. Then $\{m(t, \cdot), \mathcal{F}_t, (0, t_1)\}$ is a martingale if

$$E\{m(t, \cdot) | \mathcal{F}_s\} = m(s, \cdot) \quad \text{w.p. 1 for} \quad s \in (0, t_1).$$

A particular example of a martingale is a Wiener process, which is used for modeling white noise disturbances in engineering systems. The following is one of several equivalent definitions [3].

**Definition 3.11** *Wiener Process* $w(t)$ is an $H$-valued Wiener process on $(0, t_1)$ if it is an $H$-valued process on $(0, t_1)$, such that

$$w(t) \in L_2(\Omega, p; H) \quad \text{for all} \quad t \in (0, t_1)$$

and

(i) $E\{w(t) - w(s)\} = 0$,
(ii) $\text{Cov}\{w(t) - w(s)\} = (t-s)W$,

where $W \in \mathcal{L}(H)$ and is positive and nuclear. ($W$ is called the incremental covariance operator).

(iii) $w(s_4) - w(s_3)$ and $w(s_2) - w(s_1)$ are independent whenever $0 \leqslant s_1 \leqslant s_2 \leqslant s_3 \leqslant s_4 \leqslant t_1$.
(iv) $w(t)$ has continuous sample paths on $[0, t_1]$.

This clearly generalizes the $n$-dimensional Wiener process

$$w(t) = \begin{pmatrix} \beta_1(t) \\ \vdots \\ \beta_n(t) \end{pmatrix},$$

where $\beta_i(t)$ are mutually independent real Wiener processes with incremental variance parameter $\lambda_i$ and $W = \text{diag}(\lambda_1, \ldots, \lambda_n)$. It is easily verified that in Example 3.1, (3.4) defines a stochastic process satisfying properties (i)–(iv) provided that the incremental covariance operator $W$ is positive and nuclear. Now $W$ is given by

$$We_i = \lambda_i e_i, \quad \lambda_i \geq 0,$$

and so $W$ is positive and is nuclear provided that $\sum_{i=1}^{\infty} \lambda_i < \infty$.

In fact, one can show that as a consequence of Definition 3.11, there exists a complete orthonormal basis $\{e_i\}_{i=1}^{\infty}$ for $H$ such that

$$w(t) = \sum_{i=0}^{\infty} \beta_i(t) e_i \quad \text{w.p. 1},$$

where $\beta_i(t)$ are mutually independent real Wiener processes with incremental variance parameter $\lambda_i$ and $\sum_{i=1}^{\infty} \lambda_i < \infty$. Furthermore, $w(t)$ is Gaussian and

$$E\{\|w(t)-w(s)\|^2\} = \sum_{i=0}^{\infty} \lambda_i(t-s) = \operatorname{tr} W(t-s). \tag{3.12}$$

We now develop a theory of stochastic integration for the Hilbert space-valued Wiener process, restricting ourselves to the case of nonrandom integrals* in the space

$$\mathscr{B}_2(0, t_1; \mathscr{L}(K, H)) = \left\{ \begin{array}{l} D: [0, t_1] \to \mathscr{L}(K, H) \\ \text{such that } D(\cdot) \text{ is strongly measurable and} \\ \int_0^{t_1} \|D(t)\|_{\mathscr{L}(K,H)}^2 \, dt < \infty \end{array} \right\}.$$

We define the integral with respect to a Hilbert-space-valued Wiener process in terms of that with respect to a real Wiener process.

**Definition 3.12**

$$\int_{t_0}^{t} D(s) \, dw(s) = \sum_{i=0}^{\infty} \int_{t_0}^{t} D(s) e_i \, d\beta_i(s), \quad 0 \leq t_0 \leq t_1 \leq t,$$

where $w(s) = \sum_{i=0}^{\infty} \beta_i(s) e_i$, $D \in \mathscr{B}_2(0, t_1; \mathscr{L}(K, H))$, and the limit is in $L_2(\Omega, p; H)$.

In order to justify this definition we must first define integrals of the form $\int_{t_0}^{t} f(s) \, d\beta(s)$, where $\beta$ is a real Wiener process of incremental covariance $\lambda$ and $f \in L_2(0, t_1; H)$. This is easily done along the lines of Doob [26] for real integrals by first defining the integral when $f$ is a step function of time, that is

$$f(s) = f_i \quad \text{on} \quad (s_i, s_{i+1}), \quad t_0 = s_0 < s_1 < \cdots < s_k = t.$$

---

* For stochastic integrals with random integrals see [6, 38].

If we define

$$\int_{t_0}^t f(s)\,d\beta(s) = \sum_{i=0}^{k-1} f_i(\beta(s_{i+1}) - \beta(s_i)),$$

the following holds:

$$E\left\{\int_{t_0}^t f(s)\,d\beta(s)\right\} = 0, \tag{3.13}$$

$$E\left\{\left\langle \int_{t_0}^t f(s)\,d\beta(s), \int_{t_0}^t f(s)\,d\beta(s) \right\rangle\right\} = \lambda \int_{t_0}^t \langle f(s), f(s)\rangle\,ds. \tag{3.14}$$

Since the step functions are dense in $L_2(t_0, t; H)$, we can extend this integral to arbitrary $f \in L_2(t_0, t; H)$ by

$$\int_{t_0}^t f(s)\,d\beta(s) = \lim_{n\to\infty} \int_{t_0}^t f_n(s)\,d\beta(s),$$

where the limit is in $L_2(\Omega, p; H)$ and $f_2$ is a sequence of step functions converging to $f$ in $L_2(t_0, t; H)$. Consequently, $\int_{t_0}^t f(s)\,d\beta(s)$ is a well-defined $H$-valued random variable satisfying (3.13) and (3.14). It is then straightforward to show that $\int_{t_0}^t D(s)\,dw(s)$ of Definition 3.12 is in $L_2((t_0, t_1) \times \Omega; H)$ and satisfies

$$E\left\{\int_{t_0}^t D(s)\,dw(s)\right\} = 0, \tag{3.15}$$

$$E\left\{\left\|\int_{t_0}^t D(s)\,dw(s)\right\|^2\right\} = \operatorname{tr}\left\{\int_{t_0}^t D(s)WD^*(s)\,ds\right\} \tag{3.16}$$

$$\leq \operatorname{tr} W \int_{t_0}^t \|D(s)\|^2\,ds.$$

Furthermore, the indefinite integral $y(t) = \int_0^t D(s)\,dw(s)$ is a well-defined $H$-valued stochastic process, which has a version with continuous sample paths. It is also a martingale relative to the sigma field generated by

$$\{w(s)e_i;\ 0 \leq s \leq t;\ i = 0, 1, \ldots\}.$$

Finally, we consider the stochastic evolution equation

$$dz(t) = Az(t)\,dt + D(t)\,dw(t) + g(t)\,dt;\quad z(t_0) = z_0, \tag{3.17}$$

where, by (3.17), we mean the integral equation

$$z(t) = z_0 + \int_0^t [Az(s) + g(s)]\,ds + \int_{t_0}^t D(s)\,dw(s), \tag{3.18}$$

where $A$ is the infinitesimal generator of a strongly continuous semigroup

$T_t$, $D \in \mathcal{B}_2(t_0, t; \mathcal{L}(K, H))$, $g \in L_1(t_0, t_1; H)$ w.p. 1, $z_0$ is an $H$-valued random variable and $w$ is a $K$-valued Wiener process. We are interested in under what conditions (3.17) has the solution

$$z(t) = T_{t-t_0} z_0 + \int_{t_0}^{t} T_{t-s} D(s)\, dw(s) + \int_{t_0}^{t} T_{t-s} g(s)\, ds, \qquad (3.19)$$

where by solution we mean the following.

**Definition 3.13** $z(t)$ is a strong solution of (3.17) if $z(t) \in \mathcal{D}(A)$ w.p. 1. $z(t)$ satisfies (3.18) almost everywhere on $[t_0, t_1] \times \Omega$ and $z(t)$ has continuous sample paths.

$z(t)$ is unique if whenever $z_1(t)$ is another solution

$$p\{\sup_{t_0 \leqslant t \leqslant t_1} \|z(t) - z_1(t)\| \neq 0\} = 0.$$

In general (3.19) is not a strong solution and does not have continuous sample paths. However, it is a well-defined stochastic process and is continuous in mean square. We call it the *mild solution* of (3.17) and it is the strong solution under the following extra assumptions.

**Theorem 3.1** *If* $T_{t-s} D(s) e_i$, $T_{t-s} g(s)$, *and* $T_{t-t_0} z_0 \in \mathcal{D}(A)$ *w.p. 1 for all i,* *then* $t_0 \leqslant s \leqslant t \leqslant t_1$,

$$\int_{t_0}^{t} \|A T_{t-s} g(s)\|\, ds < \infty \qquad \text{w.p. 1}, \qquad (3.20)$$

*and*

$$\sum_{i=0}^{\infty} \lambda_i \int_{t_0}^{t} \|A T_{t-s} D(s) e_i\|^2\, ds < \infty, \qquad (3.21)$$

*then* $z(t)$ *given by* (3.19) *is the unique strong solution of* (3.17).

We remark that assumptions (3.20) and (3.21) always hold in the finite-dimensional case. To see their significance in infinite dimensions we consider Example 3.1.

Now $T_t: L_2(0, 1) \to \mathcal{D}(A)$ and so we must verify that

$$\sum_{n=1}^{\infty} \lambda_n \int_0^t \|A T_{t-s} \sin n\pi(\cdot)\|^2\, ds < \infty.$$

But

$$A T_{t-s} \sin n\pi(\cdot) = -\pi^2 n^2 e^{-n^2 \pi^2 (t-s)} \sin n\pi(\cdot)$$

and so we require $\sum_{n=1}^{\infty} n^2 \lambda_n < \infty$, an extra restriction on the variance of the noise.

Let us now consider the $\mathcal{M}^2$ equation (3.8) of Example 3.2. We need the following representation of $\tilde{T}_t$ from (2.4):

$$(\tilde{T}_t h)(\theta) = \begin{cases} \Phi^0(t+\theta)h(0); & t+\theta \geq 0 \\ 0 & ; t+\theta < 0 \end{cases} + \begin{cases} \Phi^1(t+\theta)h(\theta); & t+\theta \geq 0 \\ h(t+\theta) & ; t+\theta < 0 \end{cases}. \tag{3.22}$$

where $\Phi^0(t) \in \mathscr{L}(R^n)$ is the unique solution of

$$\frac{d}{dt}\Phi^0(t) = A_0 \Phi^0(t) + \sum_{i=1}^{N} A_i \begin{cases} \Phi^0(t+\theta_i); & t+\theta_i \geq 0 \\ 0 & ; t+\theta_i < 0 \end{cases}$$

$$+ \int_{-b}^{0} A_{01}(\theta) \begin{cases} \Phi^0(t+\theta); & t+\theta \geq 0 \\ 0 & ; t+\theta < 0 \end{cases} d\theta,$$

$$\Phi^0(0) = I, \tag{3.23}$$

and $\Phi^1(t) \in \mathscr{L}[L_2(-b, 0; R^n)]$, but we do not need its representation. The mild solution of (3.8) is

$$z(t) = \tilde{T}_t h + \int_0^t \tilde{T}_{t-s} \tilde{D} \, dw(s)$$

and from Eqs. (3.20) and (3.21) we deduce that

$$z(t)(\theta) = \Phi^0(t)h(\theta) + \int_0^{t+\theta} \Phi^0(t+s) D \, dw(s). \tag{3.24}$$

Now for $z(t) \in \mathscr{D}(\tilde{A})$, we need $z(t)(\theta)$ to be differentiable in $\theta$, but since the Wiener process is not differentiable, this will never happen. So it is futile to seek strong solutions of (3.8) in the sense of Definition 3.13. However, it turns out that it is still useful to consider an $\mathscr{M}^2$ representation for the delay equation (3.7), namely (3.24). It is possible to show that it has continuous sample paths in $\mathscr{M}^2$ and that $z(t)(0)$ satisfies the integrated form of (3.7) w.p. 1, and so $z(t)(0)$ agrees with a version of the unique solution of (3.7). (For the proof of this and the details of this theory of stochastic evolution equations see [13].)

## IV. Deterministic Quadratic Cost-Control Problem

Before considering the stochastic control problem, we summarize the main results for the deterministic regulator problem from [14]. To avoid the difficulties of the existence of solutions of inhomogeneous abstract evolution equations, we define our system by the integral input–output relationship:

$$z(t) = T_t z_0 + \int_0^t T_{t-s} Bu(s) \, ds, \quad 0 \leq t \leq t_1 < \infty, \tag{4.1}$$

where $H$ and $U$ are real Hilbert spaces, $B \in \mathscr{L}(U, H)$, $z_0 \in H$, and $\{T_t; t \geq 0\}$

is a semigroup on $H$ with generator $A$. From Section II we know that $z(t)$ is strongly continuous on $[0, t_1]$ for all inputs $u \in L_2(0, t, \cdot, U)$ and is associated with the evolution equation,

$$\dot{z}(t) = Az(t) + Bu(t), \qquad z(0) = z_0. \tag{4.2}$$

For the cost functional, we take

$$J(u; z_0) = \langle z(t_1), Gz(t_1) \rangle + \int_0^{t_1} (\langle z(s), Mz(s) \rangle + \langle u(s), Ru(s) \rangle) \, ds, \tag{4.3}$$

where $M, G \in \mathcal{L}(H)$ are self-adjoint and nonnegative, and $R, R^{-1} \in \mathcal{L}(H)$ are self-adjoint and strictly positive.

Our control problem is to find an optimal control that minimizes $J(u; z_0)$.

In finite dimensions this is called the "regulator problem," the motivation being to bring the state $z(t_1)$ close to the zero state, and as in finite dimensions the optimal control is feedback

$$u^*(t) = -R^{-1}B^*Q(t)z(t), \tag{4.4}$$

where $Q(t) \in \mathcal{L}(H)$ and the minimum cost is $\langle z_0, Q(0)z_0 \rangle$. $Q(t)$ is the unique solution of the following inner product Riccati equation in the class of weakly absolutely continuous self-adjoint operators in $\mathcal{L}(H)$.

$$\frac{d}{dt} \langle Q(t)h, k \rangle + \langle Q(t)h, Ak \rangle + \langle Ah, Q(t)k \rangle + \langle h, Mk \rangle$$

$$= \langle Q(t)BR^{-1}B^*Q(t)h, k \rangle \qquad \text{on} \quad [t_0, t_1]$$

$$Q(t_1) = G, \tag{4.5}$$

where $h, k \in \mathcal{D}(A)$.

In finite dimensions the inner product is unnecessary and (4.5) simplifies to

$$\dot{Q}(t) + A^*Q(t) + Q(t)A = Q(t)BR^{-1}B^*Q(t), \tag{4.6}$$

but the inner products are necessary in the general case, since $A^*Q(t)$ need not be defined when $A$ is unbounded. It is constructive to consider some typical examples.

**Example 4.1** *Heat Equation* (cf. Example 2.3)

$$z_t = z_{xx} + u(t, x),$$

$$z(0, t) = 0 = z(1, t), \qquad z(x, 0) = z_0(x), \tag{4.7}$$

where we seek to minimize

$$J(u) = \int_0^1 z^2(t_1, x) \, dx + \int_0^{t_1} \int_0^1 (\lambda_1 z^2(t, x) + \lambda_2^2 u^2(t, x)) \, dx \, dt,$$

where $\lambda_1, \lambda_2 > 0$. Then defining $Q(t)$ by

$$Q(t)h = \sum_{i,j=1}^{\infty} q_{ij}(t) \langle h, \phi_i \rangle \phi_j, \qquad \phi_i = r_2 \sin \pi i x,$$

(4.25) yields the equations

$$\dot{q}_{ij} - \pi^2(i^2+j^2)q_{ij} + \delta_{ij}\lambda_1 = \frac{1}{\lambda_2} \sum_{k=1}^{\infty} q_{ik} q_{jk},$$

$$q_{ij}(t_1) = \delta_{ij}. \tag{4.8}$$

Now $q_{ij}(t) = 0$ is a solution to (4.8) and since the solution is unique, we do have $q_{ij}(t) \equiv 0$ for $i \neq j$ and

$$\dot{q}_{ii} - 2\pi^2 i^2 q_{ii} + \lambda_1 = \frac{1}{\lambda_2} q_{ii}^2, \qquad q_{ii}(t_1) = 1.$$

In this case we can solve explicitly for $q_{ii}(t)$, obtaining

$$q_{ii}(t) = \frac{a(1-b) - b(1-a)e^{\alpha(t-t_1)}}{(1-b) - (1-a)e^{\alpha(t-t_1)}},$$

where

$$\alpha = -2/\lambda_2 (\pi^4 i^4 \lambda_2 + \lambda_1 \lambda_2)^{\frac{1}{2}}; \qquad a = -\pi^2 i^2 \lambda_2 - (\lambda_2 \alpha/2),$$

and

$$b = -\pi^2 i^2 \lambda_2 + (\lambda_2 \alpha/2).$$

Then if we expand $u(t) = \sum_{i=1}^{\infty} u_i(t) \phi_i$, $z(t) = \sum_{i=1}^{\infty} z_i(t) \phi_i$, we have $u_i(t) = -(1/\lambda_2) q_{ii}(t) z_i(t)$.

**Example 4.2** *Delay Equations* (cf. Example 2.2)

$$\dot{x}(t) = A_0 x(t) + \sum_{i=1}^{N} A_i \begin{cases} x(t+\theta_i); & t+\theta_i \geq 0 \\ h(t+\theta_i); & t+\theta_i < 0 \end{cases}$$

$$+ \int_{-b}^{0} A_{01}(\theta) \begin{cases} x(t+\theta) \, d\theta; & t+\theta \geq 0 \\ h(t+\theta) & ; \ t+\theta < 0 \end{cases} + Bu(t),$$

$$x(0) = h(0). \tag{4.9}$$

Where we make the same assumptions as in Example 2.2 and $B \in \mathscr{L}(R^m, R^n)$. We suppose that we wish to choose a control $u \in L_2(0, t_1; R^m)$, so that the following cost functional is minimized:

$$J(u) = \langle x(t_1), Gx(t_1) \rangle_{R^n} + \int_0^{t_1} (\langle x(s), Mx(s) \rangle_{R^n} + \langle u(s), Ru(s) \rangle_{R^m}) \, ds,$$

$$\tag{4.10}$$

where $G, M \in \mathscr{L}(R^n)$ are self-adjoint, and positive, and $R, R^{-1} \in \mathscr{L}(R^m)$ are

ESTIMATION AND STOCHASTIC CONTROL FOR LINEAR SYSTEMS

strictly positive and self-adjoint. As in Example 2.2 we reformulate Eqs. (4.9) and (4.10) on $\mathcal{M}^2(-b, 0; R^n)$, obtaining

$$\dot{z} = \tilde{A}z + \tilde{B}u, \quad z(0) = b, \tag{4.11}$$

$$J(u) = \langle z(t_1), \tilde{G}z(t_1)\rangle_{\mathcal{M}^2} + \int_0^{t_1} \langle z(s), \tilde{M}z(s)\rangle_{\mathcal{M}^2} \, ds$$

$$+ \int_0^{t_1} \langle u(s), Ru(s)\rangle_{R^m} \, ds, \tag{4.12}$$

where $\tilde{B}$, $\tilde{G}$, and $\tilde{M}$ are degenerate maps onto $\mathcal{M}^2$ given by

$$(\tilde{B}u)(\theta) = \begin{cases} Bu; & \theta = 0 \\ 0; & \theta \neq 0 \end{cases} \quad \text{for } u \in R^m,$$

$$(\tilde{M}h)(\theta) = \begin{cases} Mh(0); & \theta = 0 \\ 0; & \theta \neq 0 \end{cases} \quad \text{for } h \in \mathcal{M}^2(-b, 0; R^n),$$

$$(\tilde{G}h)(\theta) = \begin{cases} Gh(0); & \theta = 0 \\ 0; & \theta \neq 0 \end{cases} \quad \text{for } h \in \mathcal{M}^2(-b, 0; R^n),$$

and $\tilde{A}$ is as before. In this case the Riccati equation (4.5) is very complicated; however, it is possible to decompose it into simpler ones by introducing the decomposition

$$Q(t) = \begin{pmatrix} Q_{11}(t) & Q_{12}(t) \\ Q_{21}(t) & Q_{22}(t) \end{pmatrix}, \tag{4.13}$$

where

$$Q_{11}(t) \in \mathcal{L}(R^n), \quad Q_{22}(t) \in \mathcal{L}(L_2(-b, 0; R^n)),$$

$$Q_{21}(t) \in \mathcal{L}(R^n, L_2(-b, 0; R^n)),$$

and

$$Q_{12}(t) = Q_{21}^*(t).$$

We also define $Q_{21}(t, \theta)$ and $Q_{22}(t, \theta)$ by

$$Q_{21}(t, \theta) x = [Q_{21}(t) x](\theta) \quad \text{for } x \in R^n, \tag{4.14}$$

$$[Q_{22}(t) h](\theta) = \int_{-b}^{0} Q_{22}(t, \theta, \alpha) h(\alpha) \, d\alpha. \tag{4.15}$$

Substituting (4.13)–(4.15) into (4.5) yields a first-order ordinary differential equation for $Q_{11}(t)$ and first-order partial differential equations for $Q_{21}(t, \theta)$ and $Q_{22}(t, \theta, \alpha)$. For details of this analysis see [25].

Finally, we consider the infinite time quadratic cost control problem, that is (4.2) with the cost functional

$$J_\infty(u; z_0) = \int_0^\infty (\langle z(s), Mz(s)\rangle + \langle u(s), Ru(s)\rangle) \, ds, \tag{4.16}$$

where we impose the same assumptions on $T_t$, $B$, $u$, $M$, $R$, and $z_0$ as before, and in addition we assume that $(A, B)$ and $(A^*, M^{1/2})$ are stabilizable.

Again the problem is to find an optimal control $u^* \in L_2(0, \infty; U)$, which minimizes $J_\infty(u; z_0)$, and again the optimal control is the feedback control

$$u^*(t) = -R^{-1}B^*Q_\infty z(t), \qquad (4.17)$$

where $Q_\infty$ is the unique solution of the algebraic Riccati equation

$$\langle Q_\infty h, Ak \rangle + \langle Ah, Q_\infty k \rangle = \langle Q_\infty BR^{-1}B^*Q_\infty h, k \rangle \qquad \text{for} \quad h, k \in \mathscr{D}(A) \qquad (4.18)$$

in the class of self-adjoint positive operators on $H$.

## V. State Estimation

As in Section IV, to avoid problems of existence and uniqueness of abstract evolution operations, we specify our system in integral form,

$$z(t) = T_t z_0 + \int_0^t T_{t-s} D \, dw(s), \qquad 0 \leq t \leq t_1 \qquad (5.1)$$

$$y(t) = \int_0^t Cz(s) \, ds + Fv(t), \qquad (5.2)$$

where $T_t$ is a strongly continuous semigroup on a separable Hilbert space $H$, $w(t)$ is a Wiener process on a separable Hilbert space $K$ with incremental covariance operator $W$, $D \in \mathscr{L}(K, H)$, $z_0 \in L_2(\Omega, p; H)$ and is Gaussian with zero mean, and covariance operator $P_0 \cdot v(t)$ is a vector valued Wiener process on $R^k$ and has incremental covariance matrix $V$. $V, V^{-1}, F, F^{-1} \in \mathscr{L}(R^k)$, $C \in \mathscr{L}(H, R^k)$ and $z_0$, $w$, and $v$ are mutually independent

The state estimation problem is to find the best global estimate of the state $z(t)$ at time $t$, based on the observation process $y_{t_0} = \{y(s); 0 \leq s \leq t_0\}$. If $t = t_0$, this is called the filtering problem, if $t < t_0$ it is the smoothing problem, and for $t > t_0$ the prediction problem.

We remark that we must assume a finite-dimensional observation space because $V$ must be simultaneously nuclear and invertible. As in practice, observations are necessarily finite dimensional; this is no real restriction.

We follow the approach in [11] and first find the linear estimate of the form

$$\hat{z}(t/t_0) = \int_0^{t_0} K(t, s) \, dy(s), \qquad (5.3)$$

where $K(t, \cdot) \in \mathscr{B}_2(0, t_1 \cdot, \mathscr{L}(R^k, H))$ minimizes $E\{\langle h, z(t) - \hat{z}(t/t_0) \rangle^2\}$ for all $h \in H$.

Then since $z_0$, $v$, and $w$ are Gaussian, it can be shown that this estimate is the best global estimate, i.e.,

$$\hat{z}(t/t_0) = E_{y_{t_0}}\{z(t)\}. \tag{5.4}$$

There are three key results used in the solution of the filtering problem, all of which are well known in finite-dimensional theory.

**Lemma 5.1** *Orthogonal projections lemma:* $\hat{z}(t/t_0)$ *is the best linear estimate of type* (5.3) *if and only if*

$$E\{\tilde{z}(t/t_0) \circ y(\sigma) - y(\tau)\} = 0 \quad \text{for} \quad 0 \leqslant \tau \leqslant \sigma \leqslant t_0 \leqslant t_1,$$

*where* $\tilde{z}(t/t_0) = z(t) - \hat{z}(t/t_0)$ *is the error process.*

**Lemma 5.2** *Generalized Wiener–Hopf equation:* Under the assumptions of our problem the following integral equation has a unique solution $K(t, \cdot) \in \mathcal{B}_2(0, t; \mathcal{L}(R^k, H))$ for $t \in (0, t_1)$:

$$\int_0^t K(t,s) C\Lambda(s,\sigma) C^* x \, ds + K(t,\sigma) FVF^* x = \Lambda(t,\sigma) C^* x, \quad x \in R^k. \tag{5.5}$$

*Furthermore, there is a solution* $\hat{z}(t) = \int_0^t K(t,s) \, dy(s)$ *to the linear filtering problem if and only if* (5.5) *has a solution* $K_0(t, s)$.

**Lemma 5.3** $K(t,s) = Y(t,s) P(s) C^*(FVF^*)^{-1}$ *is the unique solution to* (5.5), *where* $P(t)$ *is the unique solution of the Riccati equation,*

$$\frac{d}{dt}\langle P(t)h, k\rangle - \langle P(t)h, A^*k\rangle - \langle A^*h, P(t)k\rangle$$
$$- \langle DWD^*h, k\rangle + \langle P(t) C^*(FVF^*)^{-1} CP(t)h, k\rangle = 0,$$
$$P(0) = P_0, \quad h, k \in \mathcal{D}(A^*), \tag{5.6}$$

*and* $Y(t, s)$ *is the perturbation evolution operator of* $T_t$ *by* $-P(t) C^*(FVF^*)^{-1}C$.

These three lemmas and our Gaussian assumption yield that there exists an optimal filter given by

$$E_{y_t}\{z(t)\} = \hat{z}(t) = \int_0^t Y(t,s) P(s) C^*(FVF^*)^{-1} \, dy(s). \tag{5.7}$$

We remark that the existence and uniqueness of the solution to (5.6) follows from the results of Section IV for the dual Riccati equation (4.5). To see this replace $t$ by $t_1 - s$, $A$ by $A^*$, $G$ by $P_0$, $B$ by $C^*$, $R$ by $FVF^*$, and $M$ by $DWD^*$. $P(t)$ also has an interpretation as the covariance of the error process

$$P(t) = E\{\tilde{z}(t) \circ \tilde{z}(t)\}. \tag{5.8}$$

To examine filter stability, we consider the observation process when our initial state is an arbitrary $H$-valued random variable $h$,

$$y(h,t) = \int_0^t C\left(T_s h + \int_0^s T_{s-\alpha} D\, dw(\alpha)\right) ds + Fv(t) \tag{5.9}$$

and our estimate

$$\tilde{K}y(h,t) = \int_0^t Y(t,s) P(s) C^*(FVF^*)^{-1} dy(s). \tag{5.10}$$

Of course if $h$ is second order and Gaussian, (5.10) will be the best global estimate. However, we allow $h$ to be an arbitrary $H$-valued random variable and let $\mu_t(h)$ be the measure induced on $\mathcal{B}(H)$ by the error process $e(h,t)$,

$$e(h,t) = T_t h + \int_0^t T_{t-\alpha} D\, dw(\alpha) - \tilde{K}y(h,t). \tag{5.11}$$

Vinter [42] shows that if $(A, DW^{\frac{1}{2}})$ and $(A^*, C^*)$ are stabilizable, then as $t \to \infty$, $\mu_t(h)$ converges weakly to a measure $\mu$ on $\mathcal{B}(H)$. This reduces to the sharpest conditions for filtering stability in finite dimensions [23] and so is an appropriate definition for filter stability in infinite dimensionals.

To consider the prediction and smoothing problems, as in finite dimensionals, it is useful to define the innovations process for (5.1) and (5.2) by

$$\rho(t,\omega) = y(t,\omega) - \int_0^t C\hat{z}(s,\omega)\, ds, \tag{5.12}$$

as it can be shown that the innovations process contains the same information as the observation process. Furthermore, $\rho(t)$ is a $k$-dimensional Wiener process with incremental covariance $FVF^*$. Using these results and Lemma 5.1, the following equations are obtained for the optimal predictor,

$$\begin{aligned}\hat{z}(t/t_0) &= E_{y_{t_0}}\{z(t)\} \quad \text{for } t > t_0 \\ &= T_{t-t_0}\hat{z}(t_0),\end{aligned} \tag{5.13}$$

and the optimal smoother,

$$\begin{aligned}\hat{z}(t/t_0) &= E_{y_{t_0}}\{z(t)\} \\ &= \hat{z}(t) + P(t)\lambda(t) \quad \text{for } t < t_0,\end{aligned} \tag{5.14}$$

where $P(t)$ is the unique solution of (5.6) and $\lambda(t)$ is given by

$$\lambda(t) = \int_t^{t_0} Y^*(s,t) C^*(FVF^*)^{-1} d\rho(s,\omega),$$

where $Y(t,s)$ is the perturbation of $T_t$ by $-P(t)C^*(FVF^*)^{-1}C$. The optimal filter is also given by

$$\hat{z}(t) = E_{y_t}\{z(t)\} = \int_0^t T_{t-s} P(s) C^*(FVF^*)^{-1} d\rho(s,\omega). \tag{5.15}$$

The above estimation results represent a complete generalization of finite-dimensional theory [23], except that we have expressed our estimators in integral form (5.13)–(5.15), rather than differential form, which is more common in finite-dimensional theory. The optimal predictor is the strong solution of the stochastic evolution equation

$$d\hat{z}(t/t_0) = A\hat{z}(t/t_0)\,dt, \qquad \hat{z}(t_0/t_0) = \hat{z}(t_0). \tag{5.16}$$

However, to express the optimal filter and the optimal smoother in differential form we need the additional assumptions that

$$T_t P_0 \quad \text{and} \quad T_t DW: H \to \mathscr{D}(A) \quad \text{for } t > 0 \text{ and all } i, \text{ and}$$

$$\sum_{i=0}^{\infty} \lambda_i^2 \int_0^{t_1} \|AT_t De_i\|^2\, dt < \infty,$$

$$\sum_{i=0}^{\infty} \mu_i^2 \int_0^{t_1} \|AT_t f_i\|^2\, dt < \infty, \tag{5.17}$$

where $(\lambda_i, e_i)$ and $(\mu_i, f_i)$ are the eigenvalues and eigenvectors of the operators $W$ and $P_0$, respectively. Then the optimal filter $\hat{z}(t)$ is the unique solution of the stochastic evolution equation

$$d\hat{z}(t) = A\hat{z}(t)\,dt + P(t)C^*(FVF^*)^{-1}\,d\rho(t)$$
$$\hat{z}(0) = 0, \tag{5.18}$$

and the optimal smoother is the unique solution of

$$d\hat{z}(t/t_0) = A\hat{z}(t/t_0)\,dt + DWD^*\lambda(t)\,dt$$
$$\hat{z}(t/t_0) = \hat{z}(t_0). \tag{5.19}$$

Of course these assumptions are automatically satisfied in the finite-dimensional situation, but they do represent restrictions on the noise disturbances for distributed systems as we see in the following example.

**Example 5.1** *Filtering for the Heat Equation* We consider the filtering problem for the heat equation of Example 3.1 and suppose that we can observe the process by

$$dy(t) = Cz(t)\,dt + dv(t), \tag{5.20}$$

where $v$ is a scalar Wiener process of unit incremental covariance and the observation map $C$ is defined by

$$Ch = \int_0^1 b(x)h(x)\,dx \qquad \text{for some fixed } b \in L_2(0,1). \tag{5.21}$$

The solution of the Riccati equation (5.6) can be expressed as

$$P(t)h = \sum_{i,j=0}^{\infty} p_{ij}(t) \phi_i \langle h, \phi_j \rangle \qquad \forall h \in H, \qquad (5.22)$$

where $\phi_i(x) = \sqrt{2} \sin \pi i x$, $i = 1, 2, \ldots$. Then (5.22) can be reduced to the infinite system

$$\dot{p}_{ij}(t) + \pi^2(i^2 + j^2) p_{ij}(t) - \lambda_i \delta_{ij} + 2 \sum_{m,n=0}^{\infty} p_{im}(t) p_{jn}(t) \alpha_{mn} = 0,$$

$$p_{ij}(0) = \delta_{ij} \alpha_i, \qquad i, j = 1, 2, \ldots. \qquad (5.23)$$

where

$$\alpha_{mn} = \left[ \int_0^1 b(x) \sin \pi m x \, dx \right] \left[ \int_0^1 b(x) \sin n \pi x \, dx \right].$$

The optimal estimators are then given by (5.13)–(5.15). For the differential forms of the estimators to be valid we need that $T_t: L_2(0, 1) \to \mathcal{D}(A)$ and $AT_{t-s}\phi_i = -\pi^2 i^2 \exp[-i^2 \pi^2 (t-s)] \phi_i$. Hence if the noise parameters satisfy $\sum_{i=1}^{\infty} \lambda_i i^2 < \infty$ and $\sum_{i=1}^{\infty} \alpha_i i^2 < \infty$, $\hat{z}(t)$ is the unique solution of

$$d\hat{z}_i(t) = -\pi^2 i^2 \hat{z}_i(t) \, dt - 2 \sum_{r=0}^{\infty} \hat{z}_r(t) \alpha_{jr} \sum_{j=0}^{\infty} p_{ji}(t) \, dt$$

$$+ 2 \sum_{j=1}^{\infty} p_{ij}(t) \alpha_{ij} \, dy(t),$$

$$\hat{z}_i(0) = 0, \qquad (5.24)$$

where $\hat{z}(t) = \sum_{i=1}^{\infty} \hat{z}_i(t) \phi_i$. The filter is stable as $A = A^*$ generates a stable semigroup.

**Example 5.2** *Filtering for the Delay Equation* We consider the $\mathcal{M}^2$ filtering problem,

$$z(t) = \tilde{T}_t h + \int_0^t \tilde{T}_{t-s} \tilde{D} \, dw(s), \qquad (5.25)$$

$$y(t) = \int_0^t Cz(s) \, ds + v(t), \qquad (5.26)$$

where $\tilde{T}_t$, $\tilde{D}$ are as in Example 3.2, $h \in L_2(\Omega, p; \mathcal{M}^2)$, $v(t)$ is a $k$-dimensional Wiener process with unit incremental covariance, and $C$ is given by

$$Ch = \int_{-b}^{0} k(\theta) h(\theta) \, d\theta \qquad \text{for} \quad h \in \mathcal{M}^2, \qquad (5.27)$$

where $k \in L_2[-b, 0; \mathcal{L}(R^n, R^k)]$. We remark that as we require $C$ to be bounded from $\mathcal{M}^2$ to $R^k$ we cannot allow for delayed observations, although we can approximate them by suitably choosing $k$.

As for the control problem of Example 4.2, by introducing the decomposition,

$$P(t) = \begin{pmatrix} P_{11}(t) & P_{21}^*(t) \\ P_{21}(t) & P_{22}(t) \end{pmatrix},$$

and using the results on the adjoint of $\tilde{A}$ in $\mathcal{M}^2$ from [40], it is possible to decompose (5.6) into three coupled integrodifferential equations for the matrix-valued functions $P_{11}(t)$, $P_{21}(t,\theta)$, and $P_{22}(t,\theta,\alpha)$, where

$$P_{21}(t,\theta)x = [P_{21}(t)x](\theta) \quad \text{for} \quad x \in R^n \tag{5.28}$$

$$[P_{22}(t)h](\theta) = \int_{-b}^{0} P_{22}(t,\theta,\alpha)h(\alpha)\,d\alpha \tag{5.29}$$

(see [34] for details). So the $\mathcal{M}^2$ filtering problem has a unique solution and we now examine the implications for filtering for the delay equation (3.7) of Example 3.2, whose solution is $x(t) = z(t)(0)$. The observation process is then

$$y(t) = \int_0^t \int_{-b}^{0} k(\theta) \begin{cases} x(t+\theta); & t+\theta \geq 0 \\ h(t+\theta); & t+\theta < 0 \end{cases} d\theta\, dt,$$

an "averaged" delay observation, which can approximate observations at $x(t+\theta_i)$; $-b \leq \theta_i \leq 0$ by suitably shaping $k$. In Section III we have already seen that $z(t)$ given by (5.25) has continuous sample paths in $\mathcal{M}^2$ and if $\hat{z}(t)$ also has continuous sample paths in $\mathcal{M}^2$, we can identify $\hat{z}(t)(0) = \hat{x}(t)$ as the best estimate of $z(t)(0) = x(t)$. In [21], it is shown that $\hat{z}(t)$ has continuous sample paths provided $h \in L_2[\Omega, p; \mathcal{D}(A)]$, which means the initial state, though random, is regular in $t$.

## VI. The Separation Principle for Stochastic Optimal Control

Consider the abstract stochastic optimal control problem

$$z(t) = T_t z_0 + \int_0^t T_{t-s} Bu(s)\,ds + \int_0^t T_{t-s} D\,dw(s), \tag{6.1}$$

$$y(t) = \int_0^t Cz(s)\,ds + Fv(t), \tag{6.2}$$

where the cost functional to be minimized is

$$J(u) = E\left\{ \langle z(t_1), Gz(t_1) \rangle + \int_0^{t_1} \langle z(s), Mz(s) \rangle\,ds + \int_0^{t_1} \langle u(s), Ru(s) \rangle\,ds \right\}. \tag{6.3}$$

$(\Omega, \mathscr{P}, p)$, $H$, $K$, $T_t$, $D$, $C$, $F$, $v$, $w$, and $z_0$ are defined as in Section V and $t_1$, $U$, $B$, $G$, $M$, and $R$, as in Section IV. Our control problem is to minimize (6.3) over a certain admissible subset $\mathscr{U}_{ad}$ of controls from $L_2(0, t_1; \mathscr{U})$, where we write $\mathscr{U} = L_2(\Omega, p; U)$. From physical considerations we would seek controls that depend only on the observation process $y_t$, say $u \in \int^{\oplus} \mathscr{U}_{y_t} dt$, where by $y_t$ we mean the restriction of $y(\cdot, w)$ on $(0, t)$. However, for arbitrary $u \in \int^{\oplus} \mathscr{U}_{y_t} dt$, $\mathscr{U}_{y_t}$ depends on the choice of $u$. To avoid the dependence of $\mathscr{U}_{y_t}$ on the control strategy Bensoussan and Viot [4] introduced the admissible control set

$$\mathscr{U}_{ad} = \int^{\oplus} \mathscr{U}_{\eta_t} dt \cap \int^{\oplus} \mathscr{U}_{y_t} dt, \tag{6.4}$$

where $\eta_t$ is the solution of (6.1) with zero control input. $\mathscr{U}_{y_t}$ is then the same for all $u \in \mathscr{U}_{ad}$ and $\mathscr{U}_{ad}$ contains all the controls depending on the observation process, up to time $t - \varepsilon$. In fact $\int^{\oplus} \mathscr{U}_{y_t} dt$ is dense in $\mathscr{U}_{ad}$ and $\mathscr{U}_{ad}$ is dense in $\int^{\oplus} \mathscr{U}_{\eta_t} dt$. This is an open loop class of controls, however, one can show that it contains the feedback controls $u(t, w) = \psi(t, y_t)$, if $\psi$ is nonanticipative and Lipschitzian.

In particular, the linear controls,

$$u(t) = \bar{u}(t) + \int_0^t K(t, s) \, dy(t), \tag{6.5}$$

are admissible for all $K(t, \cdot) \in \mathscr{B}_{\infty}(0, t; \mathscr{L}(R^k, U))$ and nonrandom $\bar{u} \in L_2(0, t; U)$.

So we consider the control problem of minimizing (6.3) over all controls in $\mathscr{U}_{ad}$ for the system defined by (6.1) and (6.2). Following the theory in [16] we decompose (6.1) and (6.2) according to

$$z(t) = \zeta(t) + z_u(t), \tag{6.6a}$$

$$y(t) = \eta(t) + y_u(t), \tag{6.6b}$$

where

$$\zeta(t) = T_t z_0 + \int_0^t T_{t-s} D \, dw(s), \tag{6.7}$$

$$\eta(t) = \int_0^t C\zeta(s) \, ds + Fv(t), \tag{6.8}$$

and

$$z_u(t) = \int_0^t T_{t-s} Bu(s) \, ds, \tag{6.9}$$

$$y_u(t) = \int_0^t Cz_u(s) \, ds. \tag{6.10}$$

Let $\rho(t, w)$ be the innovations process for (6.7) and (6.8), that is

$$\rho(t) = \eta(t) - \int_0^t C\hat{\zeta}(s)\, ds. \tag{6.11}$$

Then, from Section V,

$$\hat{\zeta}(t) = E_{\eta_t}\{\zeta(t)\} = E_{\rho_t}\{\zeta(t)\} \tag{6.12}$$

and so

$$\hat{z}(t) = E_{\rho_t}\{z(t)\} = E_{\eta_t}\{z(t)\} = \hat{\zeta}(t) + z_u(t) \tag{6.13}$$

since $u \in \mathcal{U}_{\text{ad}}$.

Introducing the error process

$$e(t) = z(t) - \hat{z}(t) = \zeta(t) - \hat{\zeta}(t), \tag{6.14}$$

we can show that

$$E\{\langle Mz(t), z(t)\rangle\} = E\{\langle M\hat{z}(t), \hat{z}(t)\rangle + \langle Me(t), e(t)\rangle\},$$

and so the problem of minimizing (6.3) reduces to minimizing

$$J_0(u) = E\left\{\int_0^{t_1} (\langle M\hat{z}(t), \hat{z}(t)\rangle + \langle Ru, u\rangle)\, dt \right.$$
$$\left. + \langle G\hat{z}(t_1), \hat{z}(t_1)\rangle\right\}, \tag{6.15}$$

where $\hat{z}(t)$ is given by

$$\hat{z}(t) = \int_0^t T_{t-s} Bu(s)\, ds + \int_0^t T_{t-s} P(s) C^*(FVF^*)^{-1}\, d\rho(s, \omega) \tag{6.16}$$

from (6.13) and (5.15).

This is a stochastic optimal control problem with complete observations, which, by an adaptation of the results of Section IV, can be shown to have the solution

$$u_*(t) = -R^{-1}B^*(t)Q(t)\hat{z}_*(t), \tag{6.17}$$

$$\hat{z}_*(t) = \int_0^t U_Q(t,s) P(s) C^*(FVF^*)^{-1}\, d\rho(s), \tag{6.18}$$

where $Q(t)$ is the unique solution of (4.5), $U_Q(t,s)$ is the perturbation of $T_t$ by $-BR^{-1}B^*Q(t)$, and $P(t)$ is the unique solution of (5.6). $u_*(t)$ is admissible and so (6.51) and (6.16) present the solution to our stochastic control problem, which separates into two distinct parts, namely a filtering problem and a deterministic quadratic cost-control problem. This is the separation principle generalized to infinite dimensions (cf. [43]). Equation (6.18) is more usually written in terms of the observation

$$\hat{z}_*(t) = \int_0^t \bar{U}(t,s) P(s) C^*(FVF^*)^{-1}\, dy(s), \tag{6.19}$$

where $\bar{U}$ is the perturbation of $T_t$ by $-BR^{-1}B^*Q(t)-P(t)C^*(FVF^*)^{-1}C$. Finally, the optimal cost is given by

$$J(u_*) = \text{tr}\{GP(t_1)\} + \int_0^{t_1} \text{tr}\{MP(s)\}\,ds$$
$$+ \int_0^{t_1} \text{tr}\{Q(s)P(s)C^*(FVF^*)^{-1}CP(s)\}\,ds. \quad (6.20)$$

**Example 6.1** *Heat Equation* Consider the controlled noisy heat equation

$$dz(t) = Az(t)\,dt + dw(t) + u(t)\,dt, \quad z(0) = z_0, \quad (6.21)$$

with observation process

$$dy(t) = Cz(t)\,dt + dv(t), \quad (6.22)$$

where $H = K = U = L_2(0,1)$ and $A$, $w$, $C$, $P_0$, $v$, and $z_0$ are defined as in Example 5.1. For our cost functional we take

$$J(u) = E\left\{\int_0^1 z^2(t,x)\,dx + \int_0^{t_1}\int_0^1 (\lambda_1 z^2(t,x) + \lambda_2 u^2(t,x))\,dx\,dt\right\}, \quad (6.23)$$

where $\lambda_1, \lambda_2 > 0$. Then the optimal control is given by

$$u_*(t) = -(1/\lambda_2)Q(t)\hat{z}_*(t)$$

and

$$\hat{z}_*(t) = \int_0^t \bar{U}(t,s)P(s)C^*\,dy(s), \quad (6.24)$$

where $P$ and $Q$ are as in Examples 5.1 and 4.1, respectively.

**Example 6.2** *Delay Equation* Consider the controlled stochastic delay equation

$$dx(t) = A_0 x(t)\,dt + \sum_{i=1}^N A_i \begin{Bmatrix} x(t+\theta_i); & t+\theta_i \geq 0 \\ h(t+\theta_i); & t+\theta_i < 0 \end{Bmatrix} dt$$
$$+ \int_{-b}^0 A_{01}(t,\theta) \begin{Bmatrix} x(t+\theta); & t+\theta \geq 0 \\ h(t+\theta); & t+\theta < 0 \end{Bmatrix} d\theta\,dt + Bu(t)\,dt + D\,dw(t),$$
$$x(0) = h(0), \quad (6.25)$$

where the assumptions are as in Examples 4.2 and 5.2. We take the observation process of Example 5.2 and the cost functional

$$J(u) = E\left\{\langle x(t_1), Gx(t_1)\rangle_{R^n} + \int_0^{t_1} \langle x(s), Mx(s)\rangle_{R^n}\,ds\right.$$
$$\left. + \int_0^{t_1} \langle u(s), Ru(s)\rangle_{R^n}\,ds\right\}. \quad (6.26)$$

ESTIMATION AND STOCHASTIC CONTROL FOR LINEAR SYSTEMS

We abstract this problem to obtain the $\mathcal{M}^2$ version,

$$z(t) = T_t h + \int_0^t T_{t-s} \tilde{D} \, dw(s) + \int_0^t T_{t-s} \tilde{B} u(s) \, ds, \tag{6.27}$$

$$y(t) = \int_0^t Cz(s) \, ds + v(t), \tag{6.28}$$

$$J(u) = E\left\{ \langle z(t_1), \tilde{G} z(t_1) \rangle_{\mathcal{M}^2} + \int_0^{t_1} \langle z(s), \tilde{M} z(s) \rangle_{\mathcal{M}^2} + \langle u(s), Ru(s) \rangle_{R^n} \, ds \right\}, \tag{6.29}$$

where $\tilde{B}$, $\tilde{G}$, and $\tilde{M}$ are defined as in Example 4.2 and $\tilde{D}$ as in Example 5.2. This $\mathcal{M}^2$ stochastic control problem has the unique solution

$$u_*(t) = -R^{-1} \tilde{B}^* Q(t) \hat{z}_*(t)$$

$$\hat{z}_*(t) = \int_0^t U_Q(t, s) P(s) C^* \, dp(s). \tag{6.30}$$

Under the assumption that $h \in L_2(\Omega, p; \mathcal{D}(A))$, it can be shown that $\hat{z}_*(t)$ has continuous sample paths and $\hat{z}_*(t)(0)$ and $u_*(t)$ solve the stochastic control problem for the original system (6.25)–(6.26) (see [21] for details).

## VII. Extensions

### A. Related Control and Estimation Problems

In Section IV we have given the results for the so-called regulator problem, but it is also possible to solve the tracking problem, where the cost (4.3) is replaced by

$$J_0(u) = \langle z(t_1) - r(t_1), G[z(t_1) - r(t_1)] \rangle + \int_0^{t_1} \langle [z(s) - r(s)], M[z(s) - r(s)] \rangle \, ds$$

$$+ \int_0^{t_1} \langle u(s), Ru(s) \rangle \, ds, \tag{7.1}$$

where $r(t)$ is some desired trajectory. The estimation problem of Section V can be generalized to the signal process

$$z(t) = T_t z_0 + \int_0^t T_{t-s} D \, dw(s) + \int_0^t T_{t-s} g(s) \, ds, \tag{7.2}$$

where $g \in L_2(0, t_1; H)$. Combining these two extensions it is then possible to consider the following general stochastic control problem (see [16]).

$$z(t) = T_t z_0 + \int_0^t T_{t-s} Bu(s) \, ds + \int_0^t T_{t-s} D \, dw(s) + \int_0^t T_{t-s} g(s) \, ds, \tag{7.3}$$

$$y(t) = \int_0^t Cz(s)\,ds + Fv(t), \tag{7.4}$$

$$J(u) = E\{J_0(u)\}. \tag{7.5}$$

For this problem there exists an optimal control in $\mathcal{U}_{\text{ad}}$ given by

$$u_*(t) = -R^{-1}B^*(Q(t)\hat{z}_*(t)+s(t)), \tag{7.6}$$

$$\hat{z}_*(t) = \int_0^t \bar{U}(t,s)g(s)\,ds + \int_0^t \bar{U}(t,s)P(s)C^*(FVF^*)^{-1}\,dy(s), \tag{7.7}$$

$$s(t) = -U_Q^*(t_1,t)Gr(t_1) + \int_t^{t_1} U_Q^*(s,t)(Q(s)g(s)-Mr(s))\,ds, \tag{7.8}$$

where $P$, $Q$, $U_Q$, and $\bar{U}$ are as defined in Section VI.

### B. Time-Dependent System Operators

Although we have limited ourselves to time invariant systems in the exposition of Sections II–VI, all of the results generalize to time-dependent systems defined in terms of the mild evolution operator $U(t,s)$ of Definition 2.4; that is

$$z(t) = U(t,0)z_0 + \int_0^t U(t,s)B(s)u(s)\,ds + \int_0^t U(t,s)D(s)\,dw(s), \tag{7.9}$$

where $B \in \mathcal{B}_\infty(0,t_1;\mathcal{L}(U,H))$, $D \in \mathcal{B}_\infty(0,t;\mathcal{L}(K,H))$, and $u$, $w$, and $z_0$ are specified as before.

Mild evolution operators generalize most of the properties of $T_{t-s}$, including the fact that they are stable under bounded perturbations in the sense that the unique solution of

$$U_v(t,s)h = U(t,s)h + \int_s^t U(t,r)D(r)U_D(r,s)h\,dr \tag{7.10}$$

is also a mild evolution operator. So the theory for the time-dependent case follows along lines similar to Sections IV–VI, the main modification being that $Q(t)$ and $P(t)$ are defined by integral rather than differential equations. For the treatment of these time-dependent systems see [11] and [16], and in particular [14] for several examples of systems defined by mild evolution operators. These include time-dependent delay equations of the type considered in [24], parabolic partial differential equations of Lions and Kato types [35, 36], hyperbolic equations including the nonsymmetric class in [46], and of course linear integro-partial differential equations.

Usually the motivation for considering (7.9) is the abstract stochastic evolution equation

$$dz(t) = A(t)z(t)\,dt + B(t)u(t)\,dt + D(t)\,dw(t), \qquad z(0) = z_0,$$
(7.11)

where $A(t)$ is associated with $U(t,s)$. Most systems described by delay equations and partial differential equations are defined in terms of an operator $A(t)$, which generates a quasi-evolution operator.

**Definition 7.1** *Quasi-Evolution Operators* A quasi-evolution operator $U(t,s)$ is a mild evolution operator such that there exists a nonzero $z_0 \in H$ and a closed linear operator $A(s)$ on $H$ for almost all $s \in [0, t_1]$ satisfying

$$\langle h, U(t,s)z_0 - z_0 \rangle = \int_s^t \langle h, A(\alpha) z_0 \rangle \, d\alpha \qquad \text{for all} \quad h \in H. \quad (7.12)$$

We denote the set of $z_0 \in H$ for which (7.12) is valid as $\mathscr{D}_A$, and we call $A(t)$ the quasi generator of $U(t,s)$.

In Section II we remarked that perturbations of semigroups produce quasi-evolution operators and this is a special case of the fact that bounded perturbations of quasi-evolution operators defined by (7.10) are also quasi. If $U(t,s)$ is quasi it can be shown that $Q(t)$ satisfies a differential Riccati equation and if the dual evolution operator $U^*(T-t, T-s)$ is quasi, then $P(t)$ satisfies a differential Riccati equation. However, we only get uniqueness if these operators are strong, but first we define

**Definition 7.2** *Almost Strong Evolution Operators* An almost strong evolution operator is a mild evolution operator on $H$ for which there exists an associated closed linear operator $A(t)$ on $H$ for almost all $t \in [0, t_1]$ such that

$$U(t,s) \colon \mathscr{D}(A(s)) \to \mathscr{D}(A(t)) \qquad \text{for almost all} \quad t > s \in [0, t_1],$$

and

$$\int_s^t A(r) U(r,s) z_0 \, dr = (U(t,s) - I) z_0 \qquad \text{for} \quad z_0 \in \mathscr{D}[A(s)]. \quad (7.13)$$

Equation (7.13) implies that for almost all $t > s \in [0, t_1]$,

$$\frac{\partial}{\partial t} U(t,s) z_0 = A(t) U(t,s) z_0 \qquad \text{for} \quad z_0 \in \mathscr{D}(A(s)). \quad (7.14)$$

This means that the time-dependent abstract evolution equation,

$$\dot{z}(t) = A(t) z(t), \qquad z(0) = z_0, \quad (7.15)$$

has the unique solution $U(t,0) z_0$, which satisfies (7.15) almost everywhere. If (7.14) holds everywhere, $U(t,s)$ is called a *strong evolution operator*

[14], but for stochastic evolution equations the almost strong concept is more appropriate [13]. If $U(t,s)$ is almost strong, it is then possible to give conditions for expressing the estimations in differential form [11]. Examples of systems described by almost strong evolution operators are delay equations and some parabolic partial differential equations of the Kato type [35].

## C. Non-Gaussian Noise Processes

It is possible to consider stochastic evolution equations where $w(t)$ is a non-Gaussian stochastic process of orthogonal increments type:

$$w(t) = \sum_{i=0}^{\infty} \alpha_i(t) e_i, \tag{7.16}$$

where $\{e_i\}$ is an orthonormal basis for $K$ and $\alpha_i(t)$ are real independent increments processes, such that

$$E\{\alpha_i(t)\} = g(t)\mu_i,$$

$$E\{[\alpha_i(t)-\alpha_i(s)][\alpha_j(t)-\alpha_j(s)]\} = f(t-s)\lambda_{ij}, \tag{7.17}$$

where $\sum_{i=0}^{\infty} \mu_i < \infty$, $\sum_{i=0}^{\infty} \lambda_{ii} < \infty$, and $f, g$ are nonincreasing and monotonic.

Then $w(t) \in L_2(\Omega, p; K)$, with

$$E\{w(t)\} = g(t) \sum_{i=0}^{\infty} \mu_i e_i,$$

$$\mathrm{Cov}\{w(t)-w(s)\} = \Lambda f(t-s), \tag{7.18}$$

where $\Lambda e_i = \sum_{j=0}^{\infty} \lambda_{ij} e_j$ and $\Lambda$ is nuclear, positive, and self adjoint. Then $y(t) = \int_0^t D(s) \, dw(s)$ can be defined for $D \in \mathcal{B}_2(0, t_1; \mathcal{L}(K, H))$ just as in Section V, the only difference being that $y(t) - \int_0^t D(s) \, d(E\{w(s)\})$ is a martingale and, in general, $y(t)$ does not have continuous sample paths. If we only require solutions of stochastic evolution equations to be continuous in mean square, then Theorem 3.1 remains valid for zero-mean orthogonal-increments processes $w(t)$. (See [13].)

**Example 7.1** In [15] the following model described the pollution of a river due to dumping of waste along its length $0 \leqslant x \leqslant l$:

$$dy(t) = \left(D \frac{\partial^2 y}{\partial x^2} - V \frac{\partial y}{\partial x}\right) dt + dq(t, x), \tag{7.19}$$

$$y_x(0, t) = 0 = y_x(l, t),$$

where $y(t, x)$ is the amount of pollution at time $t$ at a point distance $x$ along the river, $D$ is the dispersion coefficient of the river, $l$ is the length of the river, $V$ is the river velocity, and $q(t, x)$ the increase of pollution concentration at $(t, x)$ is modeled by

$$q(t, x) = \sum_{k=1}^{\infty} \alpha_k(t) \left(\frac{2}{l}\right)^{\frac{1}{2}} \sin\left(\frac{k\pi}{l} + \varepsilon_k\right), \tag{7.20}$$

where $\tan \varepsilon_k = -(2\pi k D/lV)$ and $\alpha_k(t)$ is a compound Poisson process with $E\{\alpha_k(t)\} = \mu_k t$, $E\{[\alpha_k(t) - \mu_k t]^2\} = \lambda_k t$, and $\sum_{k=0}^{\infty} \mu_k < \infty$, $\sum_{k=0}^{\infty} \lambda_k < \infty$.

The estimation theory of Section V is valid for non-Gaussian $z_0$ and $w(t)$ of orthogonal-increments type, except that the estimates are now the best *linear* estimates (see [12]). The stochastic optimal control problem of Section VI can only be solved for complete observations, although the control scheme (6.17)–(6.18) does suggest a useful suboptimal control law. For applications of these results to pollution monitoring and control see [20].

### D. Limited Sensing and Control

For most systems described by partial differential equations there are severe limitations on the observation and control processes. For example, it is usually impossible to sense or control the state of the system at all points of the system region $\Omega$. Instead, sensing and control is restricted to submanifolds within $\Omega$, the boundary $\Gamma$, or even points, and this gives rise to unbounded operators $B$ and $C$ in (6.1) and (6.2), respectively. If we are observing at points and controlling on certain submanifolds we may wish to penalize for this in the cost, which means that $M$ in (6.3) will also be unbounded. Finally, we may wish to allow for boundary noise, which gives rise to unbounded $D$ in (6.1). Recently [17, 18] a semigroup theory has been formulated that allows for these "unbounded" operators (see [2] for a different approach). Although the theory covers time-dependent systems, for simplicity we describe the theory for the time-invariant case.

Suppose now we wish to consider the estimation problem for a distributed system where the noise is restricted to a submanifold of the system manifold $\Omega \subset R^n$. Then the formal stochastic differential equation is

$$\dot{z}(t) = Az(t),$$
$$z(t) = w(t) \quad \text{on} \quad \Gamma, \tag{7.21}$$

where $w(t)$ is a Wiener process on $K$. To relate (7.21) to an abstract model,

$$z(t) = T_t z_0 + \int_0^t T_{t-s} D \, dw(s), \tag{7.22}$$

we need to introduce a Banach space $V \supset H$ such that

(a) $D \in \mathscr{L}(K, V)$,
(b) $T_t \in \mathscr{L}(V, H)$,
(c) $\|T_t h\|_H \leq g_1(t) \|h\|_V$ for $h \in V$ and $g_1 \in L_2(0, t_1)$. (7.23)

We also suppose that associated with (7.21) is Green's formula

$$\langle Ah, k \rangle_H - \langle h, A^*K \rangle_H = \langle (\gamma h)_\Gamma, \alpha k \rangle_{KK^*} - \langle (\alpha_* h)_\Gamma, \gamma_* k \rangle_{KK^*}, \quad (7.24)$$

for $k \in \mathcal{D}(A^*)$ on $\bar{\Omega}$, $h \in \mathcal{D}(A)$ on $\bar{\Omega} - \Gamma$, where $\langle \cdot, \cdot \rangle_{KK^*}$ denotes the duality pairing between $K$ and $K^*$, and $\alpha$, $\gamma$, $\alpha_*$ and $\gamma_*$ are certain bounded operators. Under these assumptions, it is shown in [18] that (7.22) is the weak solution of (7.21) if we define $D$ by

$$\langle Dx, y \rangle_{VV^*} = \langle x, \delta y \rangle_{KK^*}, \quad (7.25)$$

where by weak solution we mean that (7.22) satisfies

$$\int_0^{t_1} \langle z(s), \dot{\zeta}(s) + A^*\zeta(s) \rangle_H \, ds + \int_0^{t_1} \langle dz(s), \delta\zeta(s) \rangle_{KK^*} + \langle z_0, \zeta(0) \rangle_H = 0 \quad \text{w.p. 1} \quad (7.26)$$

where $\zeta \in C(0, t_1; V) \cap \mathcal{D}(A^*)$, $\dot{\zeta} \in H$ and $\dot{\zeta}, A^*\zeta$ are integrable with $\zeta(t_1) = 0$.

**Example 7.2** Consider the formal stochastic differential equation

$$z_t = z_{xxxx}, \quad z_x(0) = 0 = z_x(1), \quad z_{xxx}(0) = 0$$

$$z_{xxx}(1) = \beta(t), \quad z(0, x) = z_0, \quad (7.27)$$

where $\beta(t)$ is a real Wiener process.

Introducing $A = (\partial^4/\partial x^4)$ on $H = L_2(0, 1)$ with domain

$$\mathcal{D}(A) = \left\{ z \in H: z_{xxxx} \in H, \quad \begin{array}{l} z_x(0) = 0 = z_x(1) \\ z_{xxx}(0) = 0 = z_{xxx}(1) \end{array} \right\},$$

we see that $A$ generates a semigroup $T_t$ on $H$ given by

$$T_t h = \sum_{n=1}^{\infty} 2 \exp(-n^4 \pi^4 t) \cos n\pi x \int_0^1 \cos n\pi r h(r) \, dr \quad (7.28)$$

and we have Green's formula

$$\int_0^1 (h_{xxxx} k - k_{xxxx} h) \, dx = -h_{xxx}(0) k(0) + k_{xxx}(0) h(0), \quad (7.29)$$

where $\alpha k = -k(0)$. If we let $V^* = H^{1/2}(0, 1)$, then $\alpha \in \mathcal{L}(H^{1/2}(0, 1), R)$ and $D$ is the Dirac delta function $\delta$. It is easily verified that

$$\|T_t h\|_H \leq (M/t^{1/8}) \|h\|_V$$

and so (7.27) has the weak solution

$$z(t) = T_t z_0 + \int_0^t T_{t-s} \delta \, d\beta(s). \quad (7.30)$$

Returning to our general estimation problem, we wish to allow for limited sensing as well as restricted noise; that is $D$ is unbounded in (7.22) and $C$

ESTIMATION AND STOCHASTIC CONTROL FOR LINEAR SYSTEMS

is unbounded in
$$y(t) = \int_0^t Cz(s)\,ds + v(t). \tag{7.31}$$

In [18] it is shown how the results of Section V can be extended to allow for unbounded $D$ and $C$, provided that $D$ satisfies Eq. (7.23) and there exists a Banach space $W$ such that

(a) $H \supset \mathscr{D}(C) \supset W$ and $\overline{W} = H$, (7.32a)

(b) $C \in \mathscr{L}(W, R^k)$, (7.32b)

(c) $T_t \in \mathscr{L}(H, W)$ for $t > 0$, (7.32c)

(d) $\|T_t h\|_W \leq g(t) \|h\|_H$ for $h \in H$ and $g_1 g_2 \in L_2(0, t_1)$. (7.32d)

**Example 7.3** Consider the estimation problem for the system (7.27) in Example 7.2 with the observation map $C$ and the evaluation at $x_1$; $0 < x_1 < 1$. Then if $H = L_2(0, 1)$, $W = H^{\frac{1}{2} + \varepsilon}(0, 1)$, $C \in \mathscr{L}(W, R)$, and $\varepsilon > 0$, we have the estimate
$$\|T_t h\|_W \leq [M/t^{1/8 + \varepsilon/4}] \|h\|_H.$$

So assumptions (7.32) are satisfied and the estimation problem for this system has a unique solution.

In fact the semigroup theory for unbounded sensing and control was first motivated by boundary value control problems [17] and we can consider stochastic control problems [17] and we can consider stochastic control problems with control action and noise on the boundary.

**Example 7.4** Suppose we wish to implement control action $u$ on (7.27) at $x = 1$, then the formal stochastic differential equation to consider is
$$z_t = z_{xxxx}, \quad dz_{xxx}(1) = u(t)\,dt + d\beta(t), \tag{7.33}$$

and it has the weak solution
$$z(t) = T_t z_0 + \int_0^t T_{t-s}\delta\,d\beta(s) + \int_0^t T_{t-s}\delta u(s)\,ds. \tag{7.34}$$

For Gaussian $z_0$, the separation principle holds for the cost,
$$J(u) = \int_0^1 z^2(t_1, x)\,dx + \int_0^{t_1} u^2(t)\,dt + \int_0^{t_1}\int_0^1 z^2(t, x)\,dx\,dt. \tag{7.35}$$

This illustrates the general result proved in [18], namely that the separation principle can be extended to stochastic control systems with limited sensing and control and noise restricted to a submanifold of the spatial domain, provided $D$ satisfies (7.23) and (7.25), $C$ satisfies (7.32), and $B$ satisfies analogs of (7.23) and (7.25).

This abstract theory has been used to find the optimal location of point sensors for filtering for distributed systems in [19] and has been used in [20] to obtain a more realistic model for the river pollution problem of Example 7.1 by allowing for dumping at point sites.

### E. Delayed Observation and Delayed Control Action

At the end of Example 5.2 on the filtering problem for linear delay systems we remarked that the theory in Section V does not allow for delayed observations of the type

$$y(t) = x(t-a), \quad -b \leq -a \leq 0,$$

as this corresponded to an unbounded $C$ map. Recently, a semigroup theory that solves these problems has been developed by Ichikawa [34]. He considers the following abstract signal and observation process

$$x(t) = T_t h(0) + \int_0^t T_{t-s} D \, dw(s), \tag{7.36}$$

$$x(\theta) = h(\theta), \quad -b \leq \theta \leq 0, \tag{7.37}$$

$$y(t) = \int_0^t Cx(s) \, ds + Fv(t), \tag{7.38}$$

$$Ch = \sum_{i=0}^{k} C_i h(-b_i) + \int_{-b}^{0} C_{01}(\theta) h(\theta) \, d\theta \quad \text{for} \quad h \in W^{1,2}(-b, 0; H), \tag{7.39}$$

where $H$ and $K$ are real separable Hilbert spaces, $T_t$ is a strongly continuous semigroup on $H$ with generator $A$, $D \in \mathscr{L}(K, H)$, $w$ is a $K$-valued Wiener process with incremental covariance operator $W$, $v$ is a $k$-dimensional Wiener process with incremental covariance matrix $V$, $C_i \in \mathscr{L}(H, R^k)$, $F, F^{-1} \in \mathscr{L}(R^k)$, $C_{01} \in \mathscr{B}_\infty(0, t_1; \mathscr{L}(H, R^k))$, $h \in L_2(\Omega x[-b, 0]; H)$, and $h$, $w$, and $v$ are mutually independent.

The filtering problem for (7.36)–(7.39) is to find the best estimate of the state $x(t)$ based on the observation, $y(t)$; $0 \leq s \leq t$, which has the form $\hat{x}(t) = \int_0^t K(t, s) \, dy(s)$, where $K(t, \cdot) \in \mathscr{B}_2(0, t; \mathscr{L}(H, R^k))$. This filtering problem can be reformulated so that it is mathematically equivalent to one of filtering with point observations described in Section VII.F. To do this we introduce a first-order hyperbolic system on $H$,

$$\frac{\partial u}{\partial t}(t, \theta) = \frac{\partial u}{\partial \theta}(t, \theta), \quad u(0, \theta) = h(\theta).$$

$$u(t, 0) = x(t). \tag{7.40}$$

The solution of (7.39) can be expressed in terms of the left-shift semigroup

ESTIMATION AND STOCHASTIC CONTROL FOR LINEAR SYSTEMS

$S_t$ on $V = L_2(-b, 0; H)$ with generator $B$ and domain,
$$\mathscr{D}(B) = \{h \colon h \in W^{1,2}(-b, 0; H); h(0) = 0\}.$$
Then
$$\tilde{A} = \begin{pmatrix} A & 0 \\ 0 & B \end{pmatrix}$$
generates a strongly continuous semigroup $\tilde{T}_t$ on $H \times V$ and
$$\tilde{T}_t \begin{pmatrix} h(0) \\ h \end{pmatrix}$$
is the mild solution of
$$\dot{x}(t) = Ax(t), \quad x(0) = h(0),$$
$$\frac{\partial u}{\partial t} = \frac{\partial u}{\partial \theta}, \quad u(0) = h,$$
$$u(t, \theta) = x(t). \tag{7.41}$$

This means we can transform the filtering problem (7.36)–(7.38) to
$$z(t) = \tilde{T}_t z_0 + \int_0^t \tilde{T}_{t-s} \tilde{D} \, dw(s) \tag{7.42}$$
$$y(t) = \int_0^t \tilde{C}z(s) \, ds + Fv(t), \tag{7.43}$$
where
$$z(t) = \begin{pmatrix} x(t) \\ u(t) \end{pmatrix}, \quad z_0 = \begin{pmatrix} h(0) \\ h \end{pmatrix}, \quad \tilde{D} = \begin{pmatrix} D \\ 0 \end{pmatrix} \in \mathscr{L}(K, H \times V),$$
and
$$\tilde{C} = (0, C).$$

It can be shown that $\tilde{C}$ and $\tilde{T}_t$ satisfy conditions similar to (7.32) and so the filtering problem for (7.42)–(7.43) and hence for the equivalent system (7.36)–(7.39) has a unique solution.

**Example 7.5** For the signal process take the stochastic delay system (5.25) of Example 5.2,
$$z(t) = \tilde{T}_t h + \int_0^t \tilde{T}_{t-s} \tilde{D} \, dw(s), \tag{7.44}$$
where $\tilde{T}_t$ is a semigroup on $\mathscr{M}^2 \cong R^n \times L_2(-b, 0; R^n)$ with generator $\tilde{A}$ of the form
$$(\tilde{A}h)(\theta) = \begin{cases} \tilde{A}_1 h, & \theta = 0 \\ \dfrac{dh}{d\theta}, & \theta \neq 0, \end{cases} \tag{7.45}$$

where $\tilde{A}_1$ is given by (2.12), and

$$\mathscr{D}(\tilde{A}) = \left\{ \begin{pmatrix} x \\ h \end{pmatrix} : h \in W^{1,2}(-b, 0; R^n); \ h(0) = x \right\}.$$

Furthermore,

$$\tilde{D} = \begin{pmatrix} D \\ 0 \end{pmatrix}$$

and

$$z(t) = \begin{pmatrix} x(t) \\ h(t, \theta) \end{pmatrix},$$

where $h(t, \theta) = h(t + \theta)$ and so (7.40) is already of the form (7.42). For our observation, we take (7.43) where $\tilde{C} = (0, C)$, and $C$ is given by (7.39) on $W^{1,2}(-b, 0; R^n)$. So there exists a unique optimal filter, which is given by

$$\hat{z}(t) = \int_0^t Y(t, s)(\tilde{C}P(r))^*(FVF^*)^{-1} \, dy(r),$$

where $Y(t, s)$ is the perturbation of $T_t$ by $-(\tilde{C}P(t))^*(FVF^*)^{-1}\tilde{C}$ and $P(t)$ satisfies a Riccati equation on $\mathscr{M}^2$, which again can be decomposed into three coupled matrix integro differential equations as for Example 5.2.

Equations (7.36)–(7.37) can also represent a distributed system and so we can solve the filtering problem for the noisy heat equation of Example 5.1 with delayed observations. The problems of smoothing and prediction with delayed observation can also be solved using this approach and it should be possible to prove a separation principle for the stochastic regulator problem with delayed control action and delayed observations. (The theory for the deterministic regulator problem with delayed control action already exists in [34].)

### F. Systems with State-Dependent Noise

So far we have limited our attention to linear stochastic evolution equation of the type

$$dz(t) = A(t)z(t) \, dt + B(t) \, dw(t), \qquad z(0) = z_0. \tag{7.46}$$

A more general class of stochastic evolution equations is one that allows for state-dependent noise:

$$dz(t) = A(t)z(t) \, dt + B[t, z(t)] \, dw(t). \tag{7.47}$$

Existence and uniqueness of partial differential equations of this type with nonlinear $B$ have been studied by Pardoux [41] using a Lions approach [36]. In the case where $A(t) = A$ generates a strongly continuous semigroup and $B \in \mathscr{L}(H)$, existence and stability results have been obtained [5, 30, 33, 49].

ESTIMATION AND STOCHASTIC CONTROL FOR LINEAR SYSTEMS

In [32], the linear stochastic regulator problem is studied for time-dependent abstract evolution equations with state- and control-dependent noise, but for simplicity, we only describe the results for the time-invariant system

$$z(t) = T_t z_0 + \int_0^t T_{t-s} Bu(s)\, ds + \int_0^t T_{t-s} Cu(s)\, dw_1(s)$$
$$+ \int_0^t T_{t-s} D(z(s))\, dw_2(s) + \int_0^t T_{t-s} F\, dw_3(s), \qquad (7.48)$$

where $T_t$ is a strongly continuous semigroup on $H$, $u \in L_2(0, t_1; U)$, $C \in \mathscr{L}(U, \mathscr{L}(K_1, H))$, $D \in \mathscr{L}(H, \mathscr{L}(K_2, H))$, $\mathscr{L}(F \in K_3, H)$, and $H, U, K$; $i = 1, 2, 3$ are real separable Hilbert spaces. $w_i(t)$; $i = 1, 2, 3$ are $K_i$-valued Wiener processes with incremental covariance operator $W_i$ and $z_0$, $w_1$, $w_2$, and $w_3$ are mutually independent. Equation (7.48) is motivated by the stochastic evolution equation

$$dz(t) = (Az - Bu)\, dt + C(u)\, dw_1(t) + D(z)\, dw_2(t) + F\, dw_3(t)$$
$$z(0) = z_0. \qquad (7.49)$$

This defines the finite-dimensional problem considered in [48] and [29], but in infinite dimensions (7.49) will not have strong solutions in general and so we define the system by (7.48).

The class of admissible controls is taken to be the linear feedback controls of the form $u(t) = -K(t)z(t)$, where $K \in \mathscr{B}_\infty(0, t_1; \mathscr{L}(H, U))$ and the problem is to minimize the following cost functional over all such $K$.

$$J(K) = E\left\{ \langle Gz(t_1), z(t_1) \rangle + \int_0^{t_1} \langle Mz(s), z(s) \rangle\, ds \right\}$$
$$+ E\left\{ \int_0^{t_1} \langle RK(s)z(s), K(s)z(s) \rangle\, ds \right\}, \qquad (7.50)$$

where $M, G \in \mathscr{L}(H)$, $R, R^{-1} \in \mathscr{L}(U)$ are self-adjoint and positive. The optimal control is given by

$$u_*(t) = [R + \Gamma(Q(t))]^{-1} B^*(t) Q(t) z(t),$$

where $Q(t)$ is the unique solution of

$$\frac{d}{dt} \langle Q(t)h, k \rangle + \langle Q(t)h, Ak \rangle + \langle Ah, Q(t)k \rangle$$
$$+ \langle Mh, k \rangle + \langle \Delta(Q)h, k \rangle = \langle (R + \Gamma(Q))^{-1} B^* Q(t)h, B^* Q(t)k \rangle, \qquad (7.51)$$

for $h, k \in \mathscr{D}(A)$ and $\Gamma(Q)$ and $\Delta(Q)$ are given by

$$\langle \Gamma(Q)u, v \rangle = \operatorname{tr}(C^* v) Q(t)(Cu) W_1 \qquad \forall u, v \in U,$$
$$\langle \Delta(Q)h, k \rangle = \operatorname{tr}(D^* k) Q(t)(Dh) W_2 \qquad \forall h, k \in H.$$

Similar results also hold for the time-dependent system defined in terms of a mild evolution operator $U(t,s)$, with $Q(t)$ satisfying an integrated version of (7.51). Together with sufficient conditions for the existence of a solution to the stochastic regulator on the infinite interval, these results [32] are the natural generalization of the finite-dimensional theory of the linear stochastic control problem with state- and control-dependent noise [48, 29].

## REFERENCES

1. Balakrishnan, A. V., "Applied Functional Analysis," Springer – Verlag, 1976.
2. Balakrishnan, A. V., Identification and control of a class of distributed systems with boundary noise, *Proc. Control Theory, Numerical Methods, and Computer Systems Modelling* Springer-Verlag, Berlin, and New York, 1974.
3. Bensoussan, A., "Filtrage Optimal des Systemes Lineaires." Dunod, Paris, 1971.
4. Bensoussan, A., and Viot, M., Optimal control of stochastic linear distributed parameter systems, *SIAM J. Control* 13 (1975), 904–926.
5. Chojnowska-Michalik, A., Stochastic differential equations in Hilbert spaces and their applications, Proceedings of the Banach Cent. Prob. Semester, Warsaw, 1976.
6. Curtain, R. F., and Falb, P. L., Stochastic differential equations in Hilbert space, *J. Differential Equations* 10 (1971), 412–430.
7. Curtain, R. F., Infinite dimensional filtering, *SIAM J. Control* 13 (1975), 89–104.
8. Curtain, Ruth F., A survey of infinite dimensional filtering, *SIAM Rev.*, 17 (1975), 395–411.
9. Curtain, Ruth F., The infinite dimensional Riccati equation with applications to affine hereditary differential systems, *SIAM J. Control* 13 (1975), 1130–1143.
10. Curtain, Ruth F., A Kalman–Bucy filtering theory for affine differential equations, *Int. Symp. on Control Theory, Numerical Methods, and Computer Systems Modelling* (June 1974), Springer-Verlag, Berlin, 1974.
11. Curtain, Ruth F., Infinite Dimensional Estimation Theory for Linear Systems. Control Theory Cent. Rep. No. 38, Univ. of Warwick, Coventry England, 1976.
12. Curtain, Ruth F., Estimation theory for abstract evolution equations excited by general white noise processes, *SIAM J. Control* 6 (1977), 1124–1150.
13. Curtain, Ruth F., Stochastic evolution equations with general white noise disturbance, *J. Math. Anal. Appl.* 60 (1977), 570–595.
14. Curtain, Ruth F., and Pritchard, A. J., The infinite dimensional Riccati equation for systems described by evolution operators, *SIAM J. Control* 14 (1976), 951–983.
15. Curtain, Ruth F., Infinite dimensional estimation theory applied to a water pollution problem, Optimization Techniques, *Proc. of the 7th Conf.*, fflice, 1975.
16. Curtain, Ruth F., and Ichikawa, A., The separation principle for stochastic evolution equations, *SIAM J. Control* (to appear).
17. Curtain Ruth F., and Pritchard, A. J., An abstract theory for unbounded control action for distributed parameter systems, *SIAM J. Control* 15 (1977).
18. Curtain, Ruth F., Linear Stochastic Control for Distributed Systems with Boundary Control, Boundary Noise and Point Observations Control Theory Cent. Rep. No. 46, 1976, Univ. of Warwick, Coventry, England.
19. Curtain, Ruth F., Ichikawa, Akira, and Ryan, E. G., Optimal location of point sensors, *Proc. IFIP Working Conf. on Modelling and Identification of Distributed Systems*, Rome,

June 1976.
20. Curtain, Ruth F., Ichikawa, A., and Ryan E. G., Modelling and control of river pollution, presented at the IFAC Symp. on Environmental Systems Planning, Disegn and Control, Kyoto, Japan, 1977.
21. Curtain, R. F., and Pritchard, A. J., "Infinite Dimensional Linear Systems Theory," Lecture Notes in Information and Control. Springer-Verlag (to appear).
22. Dakto, R., A linear control problem in abstract Hilbert space, *J. Differential Equations* **9** (1971), 346–359.
23. Davis, M. H., "Linear Estimation and Stochastic Control," Oxford Univ. Press, England, 1977.
24. Delfour, M. C., and Mitter, S. K., A class of affine systems and adjoint problems, *J. Differential Equations* **18** (1975), 18–28.
25. Delfour, M. C., and Mitter, S. K., Controllability, observability and optimal feedback control of affine hereditary differential systems, *SIAM J. Control* **10** (1972), 298–327.
26. Doob, L., "Stochastic Processes." Wiley, New York, 1972.
27. Dunford, N., and Schwartz, J., "Linear Operators Part I." Wiley (Interscience), New York, 1957.
28. Fattorini, H. O., On complete controllability of linear systems, *J. Differential Equations* **3** (1967), 391–402.
29. Haussmann, U., Optimal stationary control with state and control dependent noise, *SIAM J. Control* **9** (1971), 184–198.
30. Haussmann, U., Asymptotic stability of the linear Ito equation in infinite dimensions, *J. Differential Equations* (to appear).
31. Hille, E., and Phillips, R. S., Functional analysis and semigroups, *Colloq. Amer. Math. Soc.* **31** (1957).
32. Ichikawa, A., Optimal control of a linear stochastic evolution equation with state and control dependent noise, Proc. I.M.A. Conf. "Recent Theoretical Developments in Control," Leicester, September 1976.
33. Ichikawa, A., Linear stochastic evolution equations in Hilbert space, Control Theory Cent. Rep. No. 51, Univ. of Warwick, Coventry, England, 1976.
34. Ichikawa, A., Optimal quadratic control and filtering for evolution equations with delay in control and observation, Control Theory Cent. Rep. No. 53, Univ. of Warwick, Coventry, England, 1976.
35. Kato, T., Abstract Evolution equations of parabolic type in Banach and Hilbert spaces, *Nagoya Math. J.* **19** (1961), 93–125.
36. Lions, J. L., "Optimal Control of Systems Described by P.D.E's." Springer-Verlag, Berlin and New York, 1976.
37. Lukes, D. L., and Russel, D. L., The quadratic criterion for distributed systems, *SIAM J. Control* **7** (1969), 101–121.
38. Metivier, M., and Pistone, G., Une formule d'isometrie pour l'integrale stochastique hilbertienne et equations d'evolution lineares stochastiques. *Z. Wahrschein. revw.* **33** (1975), 1–18.
39. Manitius, A., Optimal control of hereditary systems, lecture notes for the course "Control Theory and Topics in Functional Analysis," IAEA, Vienna, 1976.
40. Mitter, S. K., and Vinter, R. B., Filtering for linear stochastic hereditary differential systems, *in Int. Symp. on Control Theory, Numerical Methods and Computer Systems Modelling*,1974, Lecture Notes in Economics and Mathematical Systems, 107, Springer-Verlag, 1974.
41. Pardoux, E., Nonlinear Stochastic Partial Differential Equations, Doctoral thesis, l'Universite de Paris Sud, Centre d'Orsay, 1975.

42. Parthasarathy, K. R., "Probability Measures on Metric Spaces." Academic Press, New York, 1967.
43. Russel, D. L., Boundary value control of the higher dimentional wave equation, *SIAM J. Control* **9** (1971), 29–42.
44. Triggiani, R., Extensions of rank conditions for controllability and observability to Banach spaces and unbounded operators, *SIAM J. Control* **14** (1976), 313–338.
45. Vinter, R. B., Filter stability for stochastic evolution equations, Technical Report, Imperial College London, Department of Computing and Control, 1976.
46. Vinter, R. B., and Johnson, T. L., Optimal control of nonsymmetric hyperbolic systems in $n$ variables on the half space, *SIAM J. Control* **5** (1977), 129–143.
47. Wonham, W. M., On the separation principle of stochastic control, *SIAM J. Control* **6** (1968), 312–326.
48. Wonham, W. M., Optimal stationary control of a linear system with state-dependent noise, *SIAM J. Control* **5** (1967), 486–500
49. Zabczyk, J., On stability of infinite dimensional linear stochastic systems, *Proc. Banach Cent. Prob. Semester*, Warsaw, 1976.

# Random Integrodifferential Equations

*D. KANNAN*

DEPARTMENT OF MATHEMATICS
UNIVERSITY OF GEORGIA
ATHENS, GEORGIA

and

DEPARTMENT OF MATHEMATICS AND STATISTICS
UNIVERSITY OF GUELPH
GUELPH, ONTARIO, CANADA

|      |                                                                  |     |
|------|------------------------------------------------------------------|-----|
| I.   | Introduction and Preliminaries                                   | 87  |
|      | A. General Introduction                                          | 87  |
|      | B. Random Integrodifferential Equations                          | 90  |
|      | C. Preliminaries                                                 | 94  |
| II.  | Existence and Uniqueness of Solutions                            | 103 |
|      | A. Linear Equations                                              | 103 |
|      | B. Nonlinear Equations                                           | 109 |
|      | C. Itô Integrodifferential Equation (Second-Order Itô Equations) | 119 |
| III. | Some Stochastic Properties of Solution Processes                 | 126 |
|      | A. Volterra-Type Random Integrodifferential Equation             | 127 |
|      | B. Itô Integrodifferential Equation                              | 140 |
| IV.  | Small Perturbations                                              | 148 |
| V.   | Vibrating String                                                 | 157 |
|      | References                                                       | 165 |

## I. Introduction and Preliminaries

### A. General Introduction

Mathematical equations are the lifeblood of all branches of applied mathematical sciences. They play a central role in the modeling, analyzing, and predicting of behaviors of various phenomena that arise in physical, biological, engineering, and social sciences. These equations involve several parameters and coefficients, which arise out of different aspects governing a particular phenomenon; for instance, the diffusion coefficient in heat conduction, refractive index in wave propagation, volume-scattering coefficient in underwater acoustics, growth rate, competition coefficient, and carrying capacity in a population of competing species, and the selection intensity in population genetics are few such coefficients. The magnitudes of these coefficients are experimentally determined and, hence, it is the mean value of a set of experimental values that is used for a particular coefficient. The averaged values may serve the purpose in certain instances; but in many cases the variance may be large enough to prevent the deterministic system from describing the realistic situation. Clearly, these coefficients are stochastic in nature. Because of the stochastic nature of the real environment, the environmental parameters in any ecological system are random. The situation in physical sciences is also similar. So, it becomes important to provide stochastic formulations in most of the scientific disciplines. By taking various random effects into consideration we see that the mathematical equation governing a scientific phenomenon is a random equation. Because of its applicatory value, the theory of random equations is a very active area of research.

There is an extensive literature dealing with the four basic classes of random equations, namely: (1) random algebraic equations (see the forthcoming book by Bharucha-Reid on this topic and also [1]), (2) random difference equations (cf. Bharucha-Reid [3]), (3) random differential equations (see Gihman and Skorohod [18] (for Itô equations), Soong [52], Srinivasan and Vasudevan [53]), and (4) random integral equations (see, Bharucha-Reid [2], Gihman and Skorohod [18], Tsokos and Padgett [54]). The random differential and random integral equations have been extensively studied. The corresponding random integrodifferential equations did not receive enough attention. Even in the deterministic case the complexity of the problems encountered in the study of differential and integral equations is not mitigated in the analysis of integrodifferential equations. Even the linear random integrodifferential equations are little known. Our purpose, in this chapter, is to present some elements of certain random integrodifferential equations.

Integrodifferential equations arise quite naturally in the mathematical formulation of many scientific phenomena. This class of mathematical equations arise, for example, in reactor dynamics, heat transfer by conduction and radiation, atomic scattering, fluctuations in the brightness of stars, automatic systems, fluctuations in the kinetic theory of gases, and prey–predator populations with historical actions, among others. A general form of a deterministic integrodifferential equation is

$$x^{(n)}(t) = \left( f(t, x(t), \ldots, x^{(n-1)}(t)), \lambda \int_{t_0}^{T} K(t, s, x(s), \ldots, x^{(m)}(s)) \, ds \right), \quad (1.1)$$

with initial conditions

$$x^{(k)}(t_0) \equiv x_k, \quad k = 0, \ldots, n-1, \quad (1.2)$$

where $m \leq n$ and $x^{(k)}(t) = d^k x / dt^k$. As remarked earlier, the function $f$ would involve experimentally determined parameters and coefficients influenced by the stochastic environment. Equation (1.1) becomes a *random integrodifferential equation* due to any possible combination of (i) random coefficients, (ii) random forcing functions, (iii) random initial conditions, and (iv) random kernels. Consider, for example, the following nonlinear integrodifferential system governing a prey–predator population with hereditary effects or historical actions:

$$\frac{dH(t)}{dt} = H(t) \left[ a - bP(t) - \int_{-\infty}^{t} \alpha_1(t-s) \theta_1(t-s) P(s) \, ds \right],$$

$$\frac{dP(t)}{dt} = P(t) \left[ -c + dH(t) + \int_{-\infty}^{t} \alpha_2(t-s) \theta_2(t-s) H(s) \, ds \right], \quad (1.3)$$

where $a, b, c, d > 0$, $H(t)$ (resp $P(t)$) is the prey size (resp predator size) at time $t$, $\alpha(\cdot)$ is the age distribution and $\theta(\cdot)$ is the food utilization function (cf. Volterra [57]). It is a well-known dictum of Ehrlich and Birch (see [13], and May [40]) that "models must be stochastic not deterministic." Instead of questioning its strength, we must accept the importance of this dictum. The birth and death rates, $\{a, c\}$, the competition rates $\{b, d\}$, and the initial population sizes are actually random (see Kannan [29] and, Gard and Kannan [17]). Thus, system (1.3) is an integrodifferential system with random coefficients and random initial conditions. There are certain stochastic elements in the food utilization function also (cf. May [40]), and this will make the system (1.3) a random integrodifferential system with random kernel. Next consider a second-order differential equation

$$\frac{d^2 x(t)}{dt^2} = f(x(t), \dot{x}(t)). \quad (1.4)$$

This is equivalent to the two-dimensional dynamical system

$$dx(t) = \dot{x}(t) \, dt \quad (1.5)$$

$$d\dot{x}(t) = f(x(t), \dot{x}(t)) \, dt. \quad (1.6)$$

Let $x(0) = x_0$ and $\dot{x}(0) = \dot{x}_0$ be the initial values. The function $f(x, \dot{x})$ can be interpreted as the force field of a flow. The dynamical systems often are subject to random fluctuations, and Eqs. (1.5)–(1.6) can be written, in such case, as

$$x(t) = x_0 + \int_0^t \dot{x}(s)\,ds, \tag{1.7}$$

$$\dot{x}(t) = \dot{x}_0 + \int_0^t f(x(s), \dot{x}(s))\,ds + R(t, \omega), \tag{1.8}$$

where $R(t, \omega)$ is the random force and $\omega$ is the sample variable. One can note that the integrodifferential equations with random forcing functions arise very naturally, for example, in the study of the motion of quantum mechanical particles, and wave propagation in one-dimensional random media (see Kannan [26]). At this point we would like to remark that there are many physical and biological phenomena which have been analytically formulated but not probabilistically analyzed, and which lead to interesting random integrodifferential equations when stochastically formulated.

## B. Random Integrodifferential Equations

In this chapter we will be treating some linear and nonlinear random integrodifferential equations. As remarked earlier, even the linear equations are little known. We will primarily analyze the Volterra equation,

$$\dot{x}(t, \omega) = \xi(t, \omega) + k(t)x(t, \omega) + \int_0^t K(t, s)x(s, \omega)\,ds$$

$$x(0, \omega) = x_0(\omega). \tag{1.9}$$

Equation (1.9) is a random integrodifferential equation with random forcing function and initial condition. To keep the mathematics simple we take $K(t, s)$ to be deterministic. Equations of the form (1.9) arise in many important physical phenomena. The generalized Langevin equation

$$\dot{v}(t) = -\int_{t_0}^t \gamma(t-s)v(s)\,ds + A(t) + R(t, \omega), \qquad t \geq t_0$$

$$v(t_0, \omega) = v_0(\omega) \tag{1.10}$$

governing the velocity of an impurity particle, say, is a special case of Eq. (1.9). Whenever possible we will work with this equation to illustrate the analysis of Eq. (1.9) (cf. Kannan [28] and Kannan and Bharucha-Reid [33]). Equation (1.10) is one of the basic equations in statistical mechanics, and, therefore, we shall outline, here, a derivation of Eq. (1.10) referring the reader to the articles of Kubo [35] and Mori [45] for the details.

Let us consider a lattice chain dynamical model of an impurity particle

$X_i = (x_i, v_i)$

with large but finite mass, see Figure 1. Here $M$ and $m$ are the masses of the particles, $x_n$ is the displacement from the equilibrium position of the particle at the $n$th coordinate, $v_n$ is the corresponding velocity and $F$ is the nearest-neighbor harmonic force constant. The Hamiltonian associated with this infinite chain is

$$H = \frac{M}{2}v_0^2 + \frac{m}{2}\sum_{n \neq 0} v_n^2 + \frac{F}{2}\sum_n (x_{n+1} - x_n)^2, \tag{1.11}$$

where $M \gg m$. The corresponding equations of motion are

$$\frac{d}{dt}v_0(t) = \frac{F}{M}(x_1(t) - 2x_0(t) + x_{-1}(t)),$$

$$\frac{d}{dt}v_n(t) = \frac{F}{m}(x_{n+1}(t) - 2x_n(t) + x_{n-1}(t)), \quad n \neq 0. \tag{1.12}$$

The solutions of these equations are linear combinations of initial displacements and velocities. (Here $w_n = x_{n+1} - x_n$, $n = 0, \pm 1, \pm 2, \ldots$ are Gaussian $N[0, (kT/F)]$, and the distributions of $v_n$ are in terms of the Maxwellian distribution, $k$ is the Boltzmann constant, and $T$ the absolute temperature.) The velocity of the impurity can be expressed as

$$v_0(t) = \alpha(t)v_0(0) + g(t),$$

$$g(t) = \sum_{n \neq 0} a_n(t)v_n(0) + \sum_n b_n(t)w_n(0), \tag{1.13}$$

where $\alpha(\cdot)$ is the reduced autocorrelation function of $v_0(t)$ and it also describes the mean decay of the velocity of the impurity from the initial value $v_0(0)$. Following Ford et al. [15] let us differentiate (1.13) with respect to $t$. Then,

$$\frac{d}{dt}v_0(t) = -a(t)v_0(t) + b(t),$$

$$a(t) = \frac{1}{\alpha(t)}\frac{d}{dt}\alpha(t),$$

$$b(t) = \frac{d}{dt}g(t) + a(t)g(t). \tag{1.14}$$

Taking limits as $t \to \infty$ and $(m/M) \to 0$ such that $mt/M$ is finite, one arrives at the Langevin equation

$$\frac{d}{dt} v_0(t) = -a v_0(t) + n(t), \tag{1.15}$$

where $a$ is a friction constant and $n(t)$ is the Gaussian noise with $En(t) = 0$, $En(t)n(s) = (2akT/M)\delta(t-s)$, $E$ is the mathematical expectation and $\delta(\cdot)$ is the Dirac delta function. If $m \ll M/2$, then the motion of impurity particle is equivalent to that of a free Brownian particle. Equation (1.14) is rather unsatisfactory as a generalization of the Langevin equation when the mass of the impurity is large but finite. It can be shown that $a(t)$ is not constant for $t \gg a^{-1}$, and that Eq. (1.13) cannot be solved for $v_0(0)$ to arrive at Eq. (1.14). So, what one can do is to solve the equation of motion with respect to the displacements $x_n(t)$, $n \neq 0$, treating $x_0(t)$ as an external force. Then,

$$x_n(t) = \sum_{i \neq 0} [a_{ni}(t) x_i(0) + b_{ni}(t) v_i(0)] + \int_0^t c_n(t-s) x_0(s) \, ds, \qquad n \neq 0$$

and

$$\frac{d}{dt} v_0(t) = \frac{F}{M} \int_0^t \{c_1(s) - 2\delta(s-0) + c_{-1}(s)\} x_0(t-s) \, ds$$

$$+ \frac{F}{M} \sum_{n=\pm 1} \sum_{i \neq 0} \{a_{ni}(t) x_i(0) + b_{ni}(t) v_i(0)\},$$

which reduces to the form

$$\frac{d}{dt} v_0(t) = -\int_0^t \gamma(t-s) v_0(s) \, ds + \theta(t), \tag{1.16}$$

where $\theta(t)$ is a linear combination of the initial values $\{v_n(0), n \neq 0; x_n(0)\}$. If an external force $A(t)$ acts on the impurity particle, then Eq. (1.16) takes the form of Eq. (1.10). See also Mori [45] and Kubo [35].

We will also study a nonlinear random integrodifferential equation of the form

$$\dot{x}(t, \omega) = h(t, x(t, \omega)) + \int_0^t K(t, s, \omega) f(x(s, \omega)) \, ds, \tag{1.17}$$

and the perturbed equation

$$x(t, \omega) = h(t, x(t, \omega)) + \int_0^t K(t, s, \omega) f(s, x(s, \omega)) \, ds$$

$$+ \int_0^t k(t, s, \omega) g(x(s, \omega)) \, dM(s, \omega), \tag{1.18}$$

where $M(t)$ is a martingale (see Padgett and Tsokos [49], Rao and Tsokos [51], and Tsokos and Padgett [54]). Equation (1.17) is a stochastic version of the equation

$$u'(t) = -b(t) - \int_0^t a(t-s)g(u(s))\,ds,$$

which arises in the analysis of the following system of equations occurring in reactor dynamics:

$$\frac{du}{dt} = -\int_{-\infty}^{\infty} \alpha(x)T(x,t)\,dx,$$

$$a\frac{\partial T}{\partial t} = b\frac{\partial^2 T}{\partial x^2} + \eta(x)g(u(t)),$$

where $x$ is the position along the reactor, $u(x)$ the logarithm of the total reactor power, $T(x,t)$ the deviation of the temperature from the equilibrium level, etc. (see Levin and Nohel [37]). By the same token, we remark that the equations of the form (1.9) also arise in reactor dynamics (see Gyftopulos [21]; with suitable transformations, his kinetic equations can be rewritten in the form of (1.9)).

Another interesting equation is the second-order Itô equation, which can be written as the following system:

$$x(t) = x(0) + \int_0^t \dot{x}(s)\,ds$$

$$\dot{x}(t) = \dot{x}(0) + \int_0^t m[s, x(s), \dot{x}(s)]\,ds + \int_0^t \sigma[s, x(s), \dot{x}(s)]\,d\beta(s),$$

(1.19)

where $\beta(t)$ is a Brownian motion. The Brownian oscillator is expressed by such an equation. The basic properties of a second-order Itô equation have been investigated by Borchers and Goldstein (cf. Borchers [5] and Goldstein [19]; see also Kannan [31]). We will present some of their results.

The basic questions concerning random equations are essentially the same as those for deterministic equations, viz., the existence and uniqueness of solutions, the explicit form of the solutions, and their properties. Both the qualitative and quantitative theories of random equations are important from a mathematical, as well as an applicatory, point of view. The study of random equations emphasizes probabilistic problems such as finding the finite-dimensional distributions of the solution process, computing higher moments, threshold-crossing problems, sample path properties, etc. The introduction of probabilistic aspects leads to several important problems as well as to specific mathematical difficulties. In this article we treat only a few types of random integrodifferential equations and present some representative properties of the solutions of these equations (at this point we simply remark that there is a lot more to be done).

## C. Preliminaries

Throughout, we work with a complete probability space $(\Omega, \mathscr{A}, P)$. Every sub-$\sigma$-algebra of $\mathscr{A}$ is assumed to be complete relative to $P$. The theory of stochastic processes is basic to our analysis. Take our linear equation,

$$\dot{x}(t, \omega) = \xi(t, \omega) + K(t) x(t, \omega) + \int_0^t K(t, s) x(s, \omega) \, ds,$$

for instance. This equation relates the unknown stochastic process

$$\{x(t, \omega), \, t \geqslant 0, \, \omega \in \Omega\},$$

its derived process $\{\dot{x}(t, \omega)\}$, and its weighted integral [with weight $K(\cdot, \cdot)$] with a given process $\{\xi(t, \omega)\}$. The solution of this equation defines a process $\{x(t, \omega), \, t \geqslant 0\}$, which is a function of the process $\{\xi(t)\}$. The derivative $\dot{x}(s, \omega)$ and the integral $\int_0^t K(t, s) x(s, \omega) \, ds$ are with respect to the time variable. It is not obvious whether the derivative and integral are random variables, for fixed $t > 0$. So one has to make clear in what sense these stochastic derivatives and integrals are defined, and what are their properties The problems of continuity, differentiability and integrability of stochastic processes now become the same problems as for the deterministic functions with values in abstract space.

$(\Omega, \mathscr{A}, P)$ is a (complete) probability space, $T = [0, \infty)$, $\mathscr{T}$ is the $\sigma$-algebra of Borel subsets of $T$, and $\lambda$ is the Lebesgue measure on $\mathscr{T}$. A stochastic process $X = \{x(t), \, t \in T\}$ is an indexed family of random variables $x_t(\cdot) \colon \Omega \to R^1$. $X$ is a function $x(\cdot, \cdot)$ of two variables $(t, \omega)$. If the $(t, \omega)$-function $x(\cdot, \cdot)$ on $T \times \Omega$ is $\mathscr{T} \times \mathscr{A}-$ measurable, then $X$ is called a *measurable process*. Without loss of generality we assume that all our processes are separable relative to closed sets (see Doob [9], Meyer [43]). (Separability assumption is really necessary to obtain a fruitful theory.) If the sample paths $x(\cdot, \omega) \colon T \to R$, for almost all $\omega$, are right continuous (with finite left limits), then $X$ is separable and measurable. Now we shall look, briefly, at certain stochastic calculi.

An important, but difficult, method of treating a random integrodifferential equation is the sample path approach. Stochastic processes describing a natural phenomenon are a family of deterministic functions and hence the random equations represent families of deterministic sample equations. When the collection of all solution trajectories becomes a stochastic process, they can be investigated with the tools of probability theory.

A stochastic process $\{x(t)\}$ is called *sample continuous*, or simply *continuous*, if almost all sample paths are continuous. (Here, the exceptional set can be taken to be the empty set.) A very useful condition in order that a process be continuous is given by Kolmogorov: *If a (separable) process $\{x(t,\omega)\}$ satisfies the condition $E|x(t)-x(s)|^{1+\alpha} \leq M|t-s|^{1+\beta}$, for some $\alpha, \beta, M > 0$ then the process $x(t)$ is continuous.* The absolutely continuous process and Hölder continuous process are similarly defined. Let $\{\pi_n\}$ be a sequence of successively finer partitions of $[0,\tau] \subset T$. Define $X_n = \sum_{\pi_n} |x(t_{k+1}) - x(t_k)|$. Clearly $\{X_n\}$ is an increasing sequence of random variables and hence $V_X = \sup_n X_n \leq \infty$. The process $\{x(t)\}$ is said to be of *bounded variation* on $[0,\tau]$ if $P\{V_X < \infty\} = 1$. (For the conditions for Hölder continuity, etc. we refer to Cramér and Leadbetter [8].)

A process $x(t, \omega)$ is said to be *continuously sample differentiable* if its $\omega$ sections $x_\omega(t)$ are continuously differentiable on $T$ for all $\omega \in \Omega \setminus N$ with $P(N) = 0$. Without loss of generality we take $N = \varnothing$. The sample derivative $x'(t, \omega)$ is a continuous stochastic process. If $x(t, \omega)$ is continuously sample differentiable with sample derivative $x'(t, \omega)$, then $x(t, \omega) - x(0, \omega) = \int_0^t x'(s, \omega)\, ds$. Here the sample integral $\int_0^t x'(s, \omega)\, ds$ is defined as follows. For each $\omega \in \Omega$, the Riemann integral $\int_0^t x'_\omega(s)\, ds$ exists for all $t \in T$. The collection of these integrals is denoted by $\int_0^t x'(s, \omega)\, ds$.

By $L_p(\Omega)$ we denote the collection of all (equivalence class of) random variables $X(\omega)$ such that $\|X\|_p = \{\int_\Omega |X(\omega)|^p\, dP(\omega)\}^{1/p} = E^{1/p}\{|X(\omega)|^p\} < \infty$, $1 \leq p < \infty$. Let $x(t, \omega)$ be a process such that $x_t \in L_p(\Omega)$ for every $t \in T$. Treating the process $X$ as the map $x: T \to L_p(\Omega)$ one can define the continuity, differentiability, and integrability of $x(t, \omega)$ as in the case of a function taking values in a Banach space (cf. Hille and Phillips [22]). We will mainly be considering the calculus in the $L_2(\Omega)$-space. The treatment of random equations by the $L_2$-calculus is attractive because of its simplicity. We remark that the techniques of the $L_2$-calculus are applicable whenever the stochastic sample solution is an $L_2$-process.

Let $\{x(t, \omega)\}$ be a centered [i.e., $Ex_t(\omega) = 0$, for every $t \in T$] $L_2$-process and $\Gamma(s, t) = E(x(s)x(t))$ denote the covariance function of $X$. The process $\{x(t)\}$ is called *strongly continuous at $t$* if $\lim_{h \to 0} \|x(t+h) - x(t)\|_2 = 0$. It has a *strong derivative* $\dot{x}(t, \omega)$ at $t \in T$ if $\{\dot{x}(t)\}$ is an $L_2$-process such that $\lim_{h \to 0} \|(1/h)[x(t+h) - x(t)] - \dot{x}(t)\|_2 = 0$. Let $\rho(t, s)$ be a jointly continuous function. Consider a sequence $\{\pi_n\}$ of successively finer partitions of $[0, t]$. Then the $L_2$-integral of $\rho(t, s)x(s)$ is defined as follows: Set $X_n(t) = \sum_{\pi_n} \rho(t, t_k) x(t_k)(t_{k+1} - t_k)$ and $\int_0^t \rho(t, s) x(s)\, ds = L_2 - \lim_n X_n(t) = X(t)$, where the mesh of the partition goes to zero as $n \to \infty$. The $L_2$-calculus has a well-developed theory (cf. Loéve [38]):

Let $x(t)$ be a centered $L_2$-process with covariance function $\Gamma(s, t)$. Then,
(1) $x(t)$ is strongly continuous if, and only if, $\Gamma(t, s)$ is continuous on the

diagonal of $T \times T$; (2) $x(t)$ is strongly differentiable at $t$ if, and only if, the second generalized derivative

$$\lim_{t,s \to 0} (1/ts)[\Gamma(t+s, u+s) - \Gamma(t+s, u) - \Gamma(t, u+s) + \Gamma(t, u)]$$

exists at $(t, t)$ and is finite; (3) the strong interval $\int_0^t \rho(t, s) x(s) \, ds$ exists if, and only if, $\int_0^t \int_0^t \rho(t, u) \rho(t, v) \Gamma(u, v) \, du \, dv$ exists and is finite; (4) Let $x(t)$ be strongly integrable, $\rho(t, s)$ be jointly continuous with finite first partial derivative $\partial \rho(t, s)/\partial t$, then the strong derivative of $\xi(t) = \int_0^t \rho(t, s) x(\omega) \, ds$ exists at all $t \in T$ and $\dot{\xi}(t) = \rho(t, t) x(t) + \int_0^t [\partial \rho(t, s)/\partial t] x(s) \, ds$; (5) let $x(t)$ be strongly differentiable on $T$ and let $\rho(t, s)$ be continuous on $T \times T$ whose partial derivative $\partial \rho(t, s)/\partial s$ exists. If $\eta(t) = \int_0^t \rho(t, s) \dot{x}(s) \, ds$, then $\eta(t) = \rho(t, s) x(s)|_0^t - \int_0^t [\partial \rho(t, s)/\partial s] x(s) \, ds$.

**Theorem 1.1** Let $I = [0, \tau]$ and, $X = \{x(t), T \in I\}$ and $Y = \{y(t), t \in I\}$ be any two real stochastic processes such that (a) $Y$ is a continuous process, (b) there is an $L_2$-random variable $\eta \geq 0$ with $|y_t| \leq \eta$, almost surely, and for all $t \in I$, and (c) $X$ is an absolutely continuous $L_2$-process with $L_2 - \lim_{h \to 0} [x(t+h) - x(t)/h] = y_t$, $0 \leq t < T$. Then,

(a) the strong derivative $\dot{x}_t$ exists on $I$, and $\{\dot{x}(t)\}$ is a continuous process satisfying $P\{\dot{x}(t) = y(t)\} = 1$, $t \in I$; and
(b) for almost all $\omega$, the $t$ functions $\dot{x}(\cdot, \omega)$, $x'(\cdot, \omega)$ (the sample derivative), and $y(\cdot, \omega)$ are equal for almost all $t \in I$.

*Proof* To prove (a), follow the method used to prove Lemma 2 in Edward and Moyal [12]. To prove (b) first observe that, for each $t \in I$, $x(t) = x(0) + \int_0^t y(u) \, du$ a.s., and hence $x'(\cdot, \omega) = y(\cdot, \omega)$ a.e. on $I$ and almost every $\omega$. Noting that $E|\dot{x}(t) - y(t)|^2 = 0$ and appealing to Fubini's theorem, we get $\int_0^T |\dot{x}(s, \omega) - y(s, \omega)|^2 \, ds = 0$ a.s. Hence, for almost every $\omega$, $\dot{x}(\cdot, \omega) = y(\cdot, \omega)$ a.e. on $I$. Hence the theorem.

In the study of stochastic motion of a particle the conditional forward and backward velocities play a basic role, (see Nelson [47]). Let us introduce this notion. $T = [0, \infty)$, $\{\mathscr{A}_t, t \in T\}$ (resp $\mathscr{B}_t, t \in T$) be an increasing (resp decreasing) family of sub-$\sigma$-algebras of $\mathscr{A}$. Let $\{x(t), t \in T\}$ be adapted to $\{\mathscr{A}_t\}$ as well as to $\{\mathscr{B}_t\}$. ($\mathscr{A}_t$ represents the past, $\mathscr{B}_t$ the future, and $\mathscr{A}_t \cap \mathscr{B}_t$ the present.) Let $x(t)$ be an $L_1$-process such that $t \to x(t)$ is continuous from $T$ into $L_1$. We say that the *conditional forward derivative* (or *velocity*) of $x(t)$ exists at $t$ if

$$D^+ x(t) = \lim_{h \downarrow 0} E\{(1/h)[x(t+h) - x(t)] | \mathscr{A}_t\}$$

exists as a limit in $L_1$ and $t \to D^+ x(t)$ is continuous from $T$ into $L_1$, where

$E\{\cdot\,|\,\mathscr{A}_t\}$ is the conditional expectation operation. The *conditional backward derivative* $D^-x(t)$ is defined as follows:

$$D^-x(t) = \lim_{h\downarrow 0} E\{(1/h)[x(t)-x(t-h)]\,|\,\mathscr{B}_t\}$$

whenever it exists as a limit in $L_1$ and $t \to D^-x(t)$ is continuous from $T$ into $L_1$. If $t \to x(t)$ is strongly differentiable in $L_1$, then $\dot{x}(t) = D^+x(t) = D^-x(t)$. Let $\{\xi(t)\}$ be a Markov process and $\mathscr{A}_t = \sigma\{\xi(s)\colon 0 \leq s \leq t\}$. If $f$ is in the domain of the infinitesimal generator $\mathbf{A}$ of $\xi(t)$, set $x(t) = f(\xi(t))$. Then, $D^+x(t) = D^+f(\xi(t)) = \mathbf{A}f(\xi(t))$.

**Theorem 1.2** *Let $\{x(t), t \in T\}$ be a strongly continuous $L_1$ process whose conditional forward derivative exists on $T$. Let $a \leq b$ and $a, b \in T$. Then,*

$$E\{x(b)-x(a)\,|\,\mathscr{A}_a\} = E\left\{\int_a^b D^+x(s)\,ds\,\Big|\,\mathscr{A}_a\right\},$$

*where the integral exists, by the continuity of $D^+x(s)$, as a Riemann integral in $L_1$. Similar result holds for conditional backward derivative. Consequently, $\{x(t), \mathscr{A}_t, t \in T\}$ is a martingale if, and only if, $D^+x(t) = 0$, $t \in T$; it is a submartingale (resp supermartingale) if, and only if, $D^+(x) \geq 0$ (resp $D^+x(t) \leq 0$), $t \in T$.*

*Proof* Let $\varepsilon > 0$. Define

$$I = \left\{ t \in [a,b]\colon \left\|E[x(s)-x(a)\,|\,\mathscr{A}_a] - E\left[\int_a^s D^+x(u)\,du\,\Big|\,\mathscr{A}_a\right]\right\|_1 \leq \varepsilon(s-a),\right.$$
$$\left. \text{for all } a \leq s \leq t \right\}.$$

Here $I \neq \emptyset$ since $a \in I$, and $I$ is a closed subinterval $[a,t] \subset [a,b]$. If $t = b$, we are done. If $t < b$, there exists a $\delta > 0$ such that $t + \delta \leq b$ and

$$\|E\{x(t+h)-x(t)\,|\,\mathscr{A}_t\} - D^+x(t)h\|_1 \leq \varepsilon h/2 \quad \text{for } 0 \leq h \leq \delta,$$

and hence

$$\|E\{x(t+h)-x(t)\,|\,\mathscr{A}_a\} - E\{D^+x(t)h\,|\,\mathscr{A}_a\}\| \leq \varepsilon h/2,$$

or

$$\left\|D^+x(t)h - \int_t^{t+h} D^+x(s)\,ds\right\| \leq \varepsilon h/2,$$

by the $L_1$ continuity of $D^+x(s)$. Therefore,

$$\left\|E\{D^+x(t)h\,|\,\mathscr{A}_a\} - E\left\{\int_t^{t+h} D^+x(s)\,ds\,\Big|\,\mathscr{A}_a\right\}\right\|_1 \leq \varepsilon h/2,$$

for $0 \leq h \leq \delta$. Thus $t+h \in I$, a contradiction, and hence $I = [a,b]$. Since $\varepsilon$ is arbitrary we get our theorem.

**Theorem 1.3** *Let $x(t)$ be a strongly continuous $L_1$ process such that $D^+x(t)$ and $D^-x(t)$ exist. Then, (a) $ED^+x(t) = ED^-x(t)$ for all $t \in T$, and (b) $x$ is a constant random variable for all $t$ if, and only if $D^+x = D^-x = 0$.*

*Proof* (a) From Theorem 1.2 we have

$$E \int_a^b D^+x(s)\,ds = E[x(b)-x(a)] = E \int_a^b D^-x(s)\,ds$$

for all $a,b \in T$. Now, (a) follows from the strong continuity of $D^+x(s)$ and $D^-x(s)$.

(b) Necessity is trivial. Now, let $D^+x = D^-x = 0$. By the martingale property given in Theorem 1.2, $E\{x(t_1)|x(t_2)\} = x(t_2)$ and $E\{x(t_2)|x(t_1)\} = x(t_1)$, $t_1 \neq t_2$. We claim that $x(t_1) = x(t_2)$ a.s. Let $\mu$ be the distribution of $(x(t_1), x(t_2))$ on the plane and $\nu$ be the distribution of $x(t_1)$. Set $x_i = x(t_i)$, $i = 1,2$. Then, there is a conditional probability $p(x_1, \cdot)$ such that for every positive Baire function $f$ on $R^2$, we have

$$\int f(x_1,x_2)\,d\mu(x_1,x_2) = \int\int f(x_1,x_2)\,p(x_1,dx_2)\,\nu(dx_1)$$

(see Doob [9] and Neveu [48]), and for every $\phi$ with $\phi(x_2) \in L_1$,

$$E\{\phi(x_2)|x_1\} = \int \phi(x_2)\,p(x_1,dx_2), \qquad \nu \text{ a.e.}$$

Choosing a strictly convex $\phi$ such that $|\phi(\xi)| \leq |\xi|$ for all $\xi \in R^1$, we get, for each $x_1$,

$$\phi\left\{\int x_2\,p(x_1,dx_2)\right\} < \int \phi(x_2)\,p(x_1,dx_2),$$

by Jensen's inequality, unless $\phi(x_2) = \int \phi(x_2)\,p(x_1,dx_2)$. Since

$$\int x_2\,p(x_1,dx_2) = x_1, \qquad \nu \text{ a.e.,}$$

we have

$$\phi(x_1) < \int \phi(x_2)\,p(x_1,dx_2),$$

unless $x_2 = x_1$, $\nu$ a.e., and consequently $E\phi(x_1) < E\phi(x_2)$. Arguing similarly, $E\phi(x_1) > E\phi(x_2)$ unless $x_1 = x_2$, a.e. Hence $x(t_1) = x(t_2)$, and the theorem.

Itô's calculus and equations have been extensively studied (see Gikhman and Skorokhod [18], McKean [42], Doob [9], Friedman [16], Dynkin [11], Itô and McKean [24], Kunita and Watanabe [36], and references given therein). We will give only a quick introduction to Itô calculus. Let

$\{\beta(t), \mathscr{A}_t, t \in T\}$ be an adapted family. The process $\{\beta(t), t \in T\}$ is called a *Brownian motion* if it has independent Gaussian increments such that $E\{\beta(t) - \beta(s)\} = 0$ and $E\{|\beta(t) - \beta(s)|^2\} = \sigma^2|t-s|$, $\sigma > 0$. For any $\alpha > \frac{1}{2}$, almost all sample paths of a Brownian motion are nowhere Hölder continuous with exponent $\alpha$; consequently, almost all sample paths of a Brownian motion are nowhere differentiable. A continuous process $\{\beta(t), t \in T\}$ adapted to $\{\mathscr{A}_t\}$ is a Brownian motion if and only if the relations

$$E\{[\beta(t) - \beta(s)] | \mathscr{A}_s\} = 0 \quad \text{and} \quad E\{[\beta(t) - \beta(s)]^2 | \mathscr{A}_s\} = \sigma^2(t-s),$$

hold a.s. for $0 \leqslant s \leqslant t$. These relations hold if and only if $\beta(t)$ and $\beta^2(t) - t$ are martingales.

Let $\{M(t), \mathscr{A}_t, t \in T\}$ be a martingale and $F(t)$ be a continuous nondecreasing function such that $E|M(t) - M(s)|^2 = E\{|M(t) - M(s)|^2 | \mathscr{A}_s\} = F(t) - F(s)$, for $0 \leqslant s \leqslant t$. Let $\Phi$ be the class of measurable processes $f(t, \omega)$ adapted to $\{\mathscr{A}_t\}$ and such that $\int_0^\infty E|f(t, \omega)|^2 \, dF(t) < \infty$. $\Phi_0$ will denote the step processes in $\Phi$. Let $I = [0, \tau]$ and $f \in \Phi_0$ such that $f(t, \omega) = f_j(\omega) = f(t_j, \omega)$ if $t_j \leqslant t < t_{j+1}$, where $\{0 = t_0 < t_1 < \cdots < t_n\}$ is a partition of $I$. Then, define

$$\int_0^\tau f(t, \omega) \, dM(t, \omega) = \sum_j f_j(\omega) [M(t_{j+1}, \omega) - M(t_j, \omega)].$$

Next define the following metrics: For $f, g \in \Phi$,

$$\|f - g\|^2 = \int_0^\infty E|f(t) - g(t)|^2 \, dF(t)$$

$$\left\|\int f \, dM - \int g \, dM\right\|^2 = E\left|\int f \, dM - \int g \, dM\right|^2.$$

Let $f \in \Phi$, $f_n \in \Phi_0$, $n \geqslant 1$, such that $\|f - f_n\| \to 0$. Then, the integral of $f$ is defined as the limit, in the norm $\|\|\cdot\|\|$, of the corresponding integrals of $f_n$. Set $y(t) = \int_0^t f(s, \omega) \, dM(s)$, $t \geqslant 0$. Let $M(s)$ be a continuous process. Then, $y(t)$ is a continuous martingale. For the details of this Itô–Doob integral we refer to Doob [9].

Let $\Omega$ be the space of all continuous paths $\boldsymbol{\beta}: R_+^1 \times R^1 \to R^1$ such that $\boldsymbol{\beta}(0, x) = 0 = \boldsymbol{\beta}(t, 0)$ for all $t \in R_+^1 = [0, \infty)$ and $x \in R^1 = (-\infty, \infty)$, $\mathscr{A}_\infty$ be the $\sigma$-algebra generated by the events $\{\boldsymbol{\beta} \in \Omega: a \leqslant \boldsymbol{\beta}(t, x) < b\}$, $a < b$, and, for $S = \{(t, x): a \leqslant t < b, c \leqslant x < d\}$, let

$$\boldsymbol{\beta}(s) = \boldsymbol{\beta}(b, d) - \boldsymbol{\beta}(a, d) - \boldsymbol{\beta}(b, c) + \boldsymbol{\beta}(a, c).$$

A probability space $(\Omega, \mathscr{A}_\infty, P)$ is called a *Brownian motion on* $R_+^1 \times R^1$ if the probability $P$ satisfies the following: (a) the variable $\boldsymbol{\beta}(S)$ is a centered Gaussian variable with variance $|S|$, the area of $S$, and (b) if $S_1$ and $S_2$ are two disjoint rectangles as above, then $\boldsymbol{\beta}(S_1)$ and $\boldsymbol{\beta}(S_2)$ are independent variables. Let $\mu$ be a *canonical measure* on $R^1$, that is, $\mu$ is a Borel measure which is finite on closed intervals and positive on open intervals. Using the same $\mu$ denote the distribution function

$$\mu(x) = \mu(x+) = \begin{cases} \mu\{(0,x]\} & \text{for } x \geq 0, \\ -\mu\{(x,0]\} & \text{for } x < 0. \end{cases}$$

Let $R_\mu$ be the range of the function $\mu$ and let $\beta_\mu$ be a Brownian motion on $R^1_+ \times R_\mu$. Then, the process $\beta$ on $R^1_+ \times R^1$ defined by $\beta(t,x) = \beta_\mu(t,\mu(x))$ is called a *$\mu$-Brownian motion* on $R^1_+ \times R^1$. If $\beta$ is a $\mu$-Brownian motion, then $\beta(S)$ is a centered Gaussian variable with variance $|S|$, the $dt \times d\mu$ measure of $S$.

Now we shall introduce a stochastic double integral following the standard ideas as presented, for example, in Gikhman and Skorokhod [18] and McKean [42]. Let $\{\mathscr{A}_t, t \in R_+\}$ be an increasing family of $\sigma$-algebras such that for each $t \in R_+$, $\mathscr{A}_t \supset \mathscr{B}_t = \sigma\{\beta(s,x): 0 \leq s \leq t, x \in R^1\}$ and is independent of $\mathscr{B}^1_+ = \{\beta(t+s,x) - \beta(t,x), s \geq 0\}$. Let $\mathscr{A}^0 = \sigma\{\bigcup_t \mathscr{A}_t\}$, $\mathscr{B} = \mathscr{B}_{R^1_+ \times R^1'}$ and $H$ be a separable Hilbert space. A function $f: R^1_+ \times R^1 \times \Omega \to H$ is said to be a *nonanticipating functional* if (a) $f(t,x)$ is $\mathscr{A}_t$-measurable for each $t \in R_+$ and $x \in R^1$, and (b) $f$ is $\mathscr{B} \otimes \mathscr{A}^0$-measurable. For simplicity, denote $\iint_{(0,\tau) \times R^1}$ by $\int_0^\tau \int$, where $\tau > 0$ (or $\tau = \infty$). When

$$P\left\{\int_0^\tau \int f^2(t,x)\, dt\, d\mu < \infty\right\} = 1,$$

define the integral

$$\int_0^t \int f\, d\beta = \int_0^t \int f(s,x)\, d\beta(s,x), \qquad 0 \leq t < \tau,$$

as a function of $t$ and of the Brownian path such that the following conditions are satisfied:

(a) if $f = I_S$, the indicator function of the rectangle $S \subset (0,\tau) \times R^1$, then $\int_0^\tau \int f\, d\beta = \beta(S)$;

(b) $\displaystyle\int_0^t \int (f+g)\, d\beta = \int_0^t \int f\, d\beta + \int_0^t \int g\, d\beta,$ for all $t \in (0,\tau)$;

(c) $\displaystyle\int_0^t \int Kf\, d\beta = K \int_0^t \int f\, d\beta,$ for any real constant $K$;

(d) $\int_0^t \int f\, d\beta$ is a continuous process;

(e) if $\tau^*$ is a stopping time with respect to $\{\mathscr{B}_t\}$ and $I_{\tau^*} = I_{\{t:t<\tau^*\}}$, then $\int_0^{\tau^*} \int f\, d\beta = \int_0^\tau \int f I_{\tau^*}\, d\beta$; and

(f) if $E\{\int_0^\tau \int \|f\|^2\, dt\, d\mu\} < \infty$, then

$$E\left\{\int_0^t \int f\, d\beta\right\} = 0, \quad E\left\{\int_0^t \int f\, d\beta \,\Big|\, \mathscr{A}_s\right\} = \int_0^s \int f\, d\beta, \qquad 0 \leq s < t < \tau.$$

and
$$E\left\{\left\|\int_0^\tau \int f\,d\boldsymbol{\beta}\right\|^2\right\} = E\left\{\int_0^\tau \int \|f\|^2\,dt\,d\mu\right\}.$$

**Theorem 1.4** (*Itô's formula*) *Let* $\phi: R_+^1 \times H \to R^1$ *be a function with continuous partial derivatives* $D_0\phi = \partial\phi/\partial t$, $D_x\phi: R_+^1 \times H \to H$, *and* $D_{xx}^2\phi: R_+^1 \times H \to \mathscr{E}(H)$, *the space of endomorphisms of* $H$, *such that* $D_{xx}^2$ *is symmetric and continuous in the uniform topology, and*

$$\phi(t_0+t, z_0+z) = \phi(t_0, z_0) + D_0\phi(t_0, z_0)t + \langle D_x\phi(t_0, z_0), z\rangle$$
$$+ \tfrac{1}{2}\langle D_{xx}^2\phi(t_0, z_0)z, z\rangle + o_1(t, z) + o_2(t, z),$$

*where* $\lim_{t,\|z\|\to 0} t^{-1}o_1(t,z) = 0 = \lim_{t,\|z\|\to 0} \|z\|^{-2}o_2(t,z)$. *Let* $\boldsymbol{\beta}$ *be a* $\mu$-*Brownian motion on* $R_+^1 \times R^1$ *and let* $g(t), f(t,x)$ *be nonanticipating* $H$-*valued processes such that* $E\int_0^\tau \|g(t)\|^2\,dt < \infty$, *and* $E\int_0^\tau \int \|f(t,x)\|^2\,dt\,d\mu < \infty$. *If*

$$\eta(t) = a + \int_0^t g(s)\,ds + \int_0^t \int f(s,x)\,d\boldsymbol{\beta}(s,x),$$

*then the following Itô's formula holds*:

$$\phi(t, \eta(t)) = \phi(0, a) + \int_0^t D_0\phi[s, \eta(s)]\,ds + \int_0^t \langle D_x\phi(s, \eta(s)), g(s)\rangle\,ds$$
$$+ \int_0^t \int \langle D_x\phi(s, \eta(s)), f(s,x)\rangle\,d\boldsymbol{\beta}(s,x)$$
$$+ \tfrac{1}{2}\int_0^t \int \langle D_{xx}^2\phi(s, \eta(s))f(s,x), \ f(s,x)\rangle\,ds\,d\mu,$$

*a.s., for all* $0 \leqslant t < \tau$.

The proof of this formula is standard, as in the scalar case, (see Cabaña [6], Friedman [16], Kannan [25], Kunita and Watanabe [36], and McKean [42]).

We will introduce further preliminaries as we need them. Before closing this Section I, we shall give a brief summary of the sections to follow. In the study of mathematical equations it is natural to treat, first, the existence and uniqueness theorems, and that is what we do in Section II. In Section II.A we present conditions for the existence and uniqueness of strong and mild solutions of the linear integrodifferential equation (1.9). In Sections II.B and II.C we treat some nonlinear equations. All these equations contain an Itô–Doob integral term. Section II.B treats a nonlinear version of the Generalized Langevian equation and a stochastic version of an integrodifferential equation that arises in nuclear reactor dynamics, and in Section II.C we consider the so-called Itô integrodifferential equation. In proving these theorems one extends the methods used in the deterministic theory.

The stochasticity involved in these equations leads to probabilistic problems, which are not always easy to work with. The methods we chose here are illustrative and extensions of applications of the methods of polygonal approximations, successive approximations, contraction principles, resolvent operators, etc.

After establishing the existence of a unique solution of an equation, it becomes necessary to analyze the behavior of the trajectories of the solution. Our equations, being random equations, yield random solutions and they require a careful stochastic analysis. Sections III and IV treat several stochastic properties of random integrodifferential equations of Volterra and Itô types. Section III begins with a derivation of an integrodifferential equation governing the moments of the bounded solutions of the Volterra type equation. Because the Volterra equation has one degree of randomness, it is important to find the one-dimensional distributions of its solution. We derive a diffusion-type equation satisfied by these one-dimensional distributions, and establish a Liouville–Gibbs theorem. Theorems of this type are basic in hydrodynamics, among other areas. In the remainder of Section III.A we work with the example of generalized Langevin equations. We establish a comparison theorem (for nonlinear equations), and derive the fluctuation–dissipation relations and the force correlation, which are important in the statistical mechanics of irreversible processes. If a free particle is driven by a continuous time Markov chain $\eta(t, \omega)$, then $(x(t, \omega), \eta(t, \omega))$ becomes a Markov process. We briefly study this process and present a necessary and sufficient condition for the existence of a steady-state (stationary) distribution. Section III.B treats the so-called Itô integrodifferential equation on the second-order Itô equation. We selected to present here several results from Borchers [5] and Goldstein [19]. These results are parallel to those in the corresponding first-order case. We show that the position $x(t)$ and the velocity $\dot{x}(t)$ jointly form a Markov process $\mathbf{X}(t) = (x(t), \dot{x}(t))$, and compute an absolutely continuous probability which is used to estimate certain exponential moments. A result on the asymptotic behavior of the sample paths of $x(t)$ and $\dot{x}(t)$, as $t \uparrow \infty$, and a result on the variation of the sample paths are presented.

The penetration of cosmic particles into matter and several other physical phenomena lead to the study of small perturbations of dynamical systems. The asymptotic behavior of the solution, as the fluctuation decays, is the theme of Section IV. First we illustrate the method of asymptotic expansion applied to the Volterra-type random integrodifferential equation, and show that the solution, up to the first order of approximation, is a Gaussian process. Next we present Dubrovskii's result on the asymptotic behavior, as the perturbation decays, of the stationary distributions of the solutions of a deterministic integrodifferential system perturbed by small noise-type impulses.

The last section treats Feller's generalized equation of vibrating string. By rewriting the equation as an integrodifferential system, Cabaña has investigated certain energy properties, such as the conservation of energy, decay of energy under damping, and bounds for energy, when vibrating string is driven by a plane white noise $\partial^2 \beta / \partial t \, \partial \mu$. We present, in Section V, Cabaña's results (but, also see Feller [14]).

In recent years, several researchers have treated deterministic equations by adjoining to it a random external force term. The ideas in establishing the path properties essentially depend on introducing a suitable stochastic calculus. These equations involve random functions. But, there are also a number of articles where deterministic integrodifferential operators are studied by probabilistic methods. These operators arise, for example, in the kinetic theory of gases (Boltzmann type equations) and in diffusions with Lévy generators. Though important, we have avoided including these topics here; but the interested reader should read the relevant recent articles of Komatsu, McKean, and Stroock, among others.

## II. Existence and Uniqueness of Solutions

The first order of business regarding any random equation is the existence and uniqueness of a solution of the equation. The existence and uniqueness theorem in its general form has very limited applicability. It is desirable if one can express the solution in a suitable explicit form which will enable us to study the analytical behavior of the solution process. This section is devoted to the existence and uniqueness theorems corresponding to the random integrodifferential equations mentioned in the Introduction (Section I.B). In establishing these theorems one mostly adapts the methods used in the corresponding deterministic cases, but not without probabilistic analytical difficulties—one has to work with suitable function spaces and stochastic calculus, and the convergence problems are more involved. We will mostly work with $L_2$-random variables and stochastic processes (equivalence classes), and the necessary function spaces will be introduced as we need them.

### A. Linear Equations

The linear random integrodifferential equations are easier, compared to the nonlinear case, to work with; nevertheless, they arise in important physical models such as the physical motion of impurity particles and in reactor dynamics. The linear equation that we treat first is

$$\dot{x}(t,\omega) = \xi(t,\omega) + k(t)x(t,\omega) + \int_0^t K(t,s)x(s,\omega)\,ds,$$
$$x(0,\omega) = x_0(\omega). \tag{2.1}$$

Let $\mathscr{C} = \mathscr{C}[R^+, B]$ be the locally convex space of all continuous functions from $R_+$ into a Banach space $B$ with the topology of uniform convergence on $[0, T]$ for every $T > 0$. Let $\mathscr{L}_p = \mathscr{L}_p[R_+, B]$, $1 \leq p < \infty$, be the space of all functions $f$ such that $\|f\|^p$ is locally integrable; $\mathscr{L}_\infty = \mathscr{L}_\infty[R^+, B]$ is defined similarly. An $L_2$ process $\{x(t, \omega), t \geq 0\}$ is called an $L_2$ *solution* (or a *strong solution*) on an interval $[0, T]$, of the initial-value problem (2.1) if $x(0, \omega) = x_0(\omega) \in L_2(\Omega)$, and $\dot{x}(t, \omega)$ exists as a strong derivative on $[0, T]$ and satisfies Eq. (2.1), $T > 0$. (Here, one sometimes requires that $x: [0, T] \to L_2(\Omega)$ is strongly absolutely continuous.) A continuous process $\{x(t, \omega), t \geq 0\}$ is called a *sample solution* of Eq. (2.1) if almost all trajectories of the process $x(t, \omega)$ are defined on the interval $[0, T]$, at each $t \in [0, T]$ the process $x(t, \omega)$ is a Borel function of $\{k(s), K(t, s), \xi(s, \omega)\}$, $0 \leq s < t \leq T$, and $\dot{x}$ exists as a continuous sample derivative and satisfies Eq. (2.1). For the most part we will be working with the strong solution case, and, here, we assume that

$$k(t) \in \mathscr{L}_\infty[R_+, R], \quad K(t, s) \in \mathscr{L}_\infty(R_+ \times R_+, R), \quad 0 \leq s \leq t.$$

**Theorem 2.1** (a) *If $x_0 \in L_2(\Omega)$ and $\xi \in \mathscr{L}_1[R_+, L_2(\Omega)]$, then there exists a unique (up to equivalence) strong solution $x(t, \omega)$ to Eq. (2.1) on $[0, T]$, for each $T > 0$. If $\dot{x}(t, \omega)$ exists as a bounded strong derivative on $[0, T]$, then almost all sample paths $x(\cdot, \omega)$ are of bounded variation on $[0, T]$.*

(b) *If*

$$k(t) \in \mathscr{C}[R_+], \quad K(t, s) \in \mathscr{C}[R_+ \times R_+], \quad 0 \leq s \leq t,$$

*and*

$$\xi \in \mathscr{C}[R_+, L_2(\Omega)],$$

*then there exists a unique strong solution, on $[0, T]$ for every $T > 0$, of Eq. (2.1) such that almost all sample paths of the solution $x(t, \omega)$ are continuous on $[0, T]$.*

*Proof* For every $t \geq 0$, define

$$(Ax)(t) = k(t)x(t) + \int_0^t K(t, s)x(s)\,ds. \tag{2.2}$$

Then, Eq. (2.1) takes the form

$$\dot{x}(t) - A(t)x(t) = \xi(t), \quad x(0) = x_0. \tag{2.3}$$

From (2.2) and the hypotheses in (a) [or (b)] it is easy to see that $A(t)$ is a uniformly measurable and locally Bochner integrable mapping of $L_2(\Omega)$ into itself (see Hille and Phillips [22]). Hence by a well-known theorem (cf. Massera and Schäffer [39]), Eq. (2.3), and hence Eq. (2.1) has a unique strong solution on $[0, T]$ for every $T > 0$.

Having taken care of the existence and uniqueness of the solution of (2.1) in both cases (a) and (b), we shall first complete the proof of (b). It is clear that $\dot{x}$ exists boundedly on $[0, T]$, for every $T > 0$. Now,

$$E|x(t) - x(s)|^2 = 2E|x(t) - x(s) - (t-s)\dot{x}(t)|^2 + 2E|(t-s)\dot{x}(t)|^2$$
$$\leq 2(M^2 + 1)|t-s|^2,$$

where $M$ is the bound for $\dot{x}$ in $[0, T]$. Noting that finite number of intervals of the form $|t-s| < \delta$ cover $[0, T]$, Kolmogorov's condition for the sample continuity of stochastic process is satisfied and hence we obtain part (b) of the theorem.

As above, there is a constant $c > 0$ such that $E|x(t+h) - x(t)|^2 \leq ch^2$. Thus, if $\{\pi_n\}$ is a sequence of successively finer partitions of $[0, T]$ such that $\|\pi_n\| \to 0$ as $n \to \infty$,

$$\sup_n \sum_{\pi_n} E|x(t_{k+1}) - x(t_k)|^2 \leq c^2 \sup_n \sum_{\pi_n} (t_{k+1} - t_k)^2 \leq 2c^2 T^2 < \infty.$$

Hence, from the Schwarz inequality, there is a $C > 0$, such that

$$\sup_n \sum_{\pi_n} E|x(t_{k+1}) - x(t_k)| = C < \infty,$$

or

$$E[V_x] = E\left[\sup_n \sum_{\pi_n} |x(t_{k+1}) - x(t_k)|\right] < \infty.$$

Let $N$ be a positive integer. From the Markov inequality,

$$P[V_x > N] \leq C/N \to 0, \quad \text{as} \quad N \to \infty.$$

Thus, $V_x(\omega) < \infty$ for almost all $\omega$. This proves the theorem.

In order to define the so-called mild solution let us introduce the following functions:

$$R(t, s) = k(t) + \int_s^t K(t, u)\, du, \qquad 0 \leq s \leq t < \infty, \qquad (2.4)$$

$$\rho(t, s) = 1 + \int_s^t \rho(t, u) R(u, s)\, du, \qquad 0 \leq s \leq t < \infty, \qquad (2.5)$$

where $K(t, s) = R(t, s) = \rho(t, s) \equiv 0$, for $s > t \geq 0$. Let $k(\cdot) \in \mathscr{L}_\infty[R_+, R^1]$ and $K(\cdot, \cdot) \in \mathscr{L}_\infty[R_+ \times R_+, R^1]$. Then, we note the following:

(a) For fixed $t$, $R(t, \cdot)$ is continuous in $s$.
(b) From (a) it is trivial that $\rho(t, s)$ exists on $0 \leq s \leq t$.
(c) $R(\cdot, \cdot) \in \mathscr{L}_\infty[R_+ \times R_+, R]$ from the definition.

(d) Using the Gronwall lemma it is easy to see that

$$|\rho(t,s)| \leq 1 + \left[\int_0^t \left\{|k(u)| + \int_0^u |K(u,v)|\,dv\right\} du\right]$$
$$\times \exp\left[\int_0^t \left\{|k(u)| + \int_0^u |K(u,v)|\,dv\right\} du\right], \qquad (2.6)$$

and $\rho \in \mathscr{L}_\infty[R_+ \times R_+, R^1]$.

(e) $\rho(t,s)$ is the formal solution of

$$\frac{\partial}{\partial s}\rho(t,s) = -\rho(t,s)k(s) - \int_s^t \rho(t,u)K(u,s)\,du, \qquad \rho(t,t) = 0, \quad (2.7)$$

on $0 \leq s \leq t$.

(f) $(\partial/\partial s)\rho(t,s) \in \mathscr{L}_\infty$.

(g) From the estimate (2.6) it follows that $\rho(t,s)$ is continuous in $t$, for fixed $s$.

(h) If $\phi(t)$ denotes the right-hand side of (2.6), then

$$\left|\frac{\partial}{\partial s}\rho(t,s)\right| \leq \left\{k(s) + \int_s^\tau K(u,s)\,du\right\}\phi(\tau), \qquad (2.8)$$

for $0 \leq s \leq t \leq \tau$.

(i) From (2.8), $\rho(t,s)$ is continuous in $s$ uniformly for $0 \leq s \leq t \leq \tau$, and hence it follows from (g) that $\rho(t,s)$ is jointly continuous.

An $L_2$-process $x(t,\omega)$ given by the relation

$$x(t,\omega) = \rho(t,0)x_0(\omega) + \int_0^t \rho(t,u)\xi(u,\omega)\,du \qquad (2.9)$$

is called a *mild solution* of Eq. (2.1).

**Theorem 2.2** *Let $x_0 \in L_2(\Omega)$ and $\xi \in \mathscr{L}_2(R_+, L_2(\Omega))$. Then, any strong solution $x(t,\omega)$ on $[0,\tau]$ of the initial value problem (2.1) is also a mild solution. If a strongly differentiable function $x: [0,\tau] \to L_2(\Omega)$ is given by (2.9), then $x(t,\omega)$ is a strong solution on $[0,\tau]$ of Eq. (2.1).*

*Proof* Let $x(t,\omega)$ be a strong solution of Eq. (2.1) on $[0,\tau]$. Fix $t \in [0,\tau]$. Then,

$$x(t,\omega) - \rho(t,0)x_0(\omega) - \int_0^t x(u,\omega)\frac{\partial}{\partial u}\rho(t,u)\,du$$
$$= \int_0^t \rho(t,u)x(u,\omega)\,du \qquad \text{(integration by parts)}$$
$$= \int_0^t \rho(t,u)\xi(u,\omega)\,du + \int_0^t \rho(t,s)k(s)x(s,\omega)\,ds$$

$$+ \int_0^t \rho(t,u) \int_0^u K(u,s) x(s,\omega) \, ds \, du,$$

$$= \int_0^t \rho(t,u) \xi(u,\omega)$$
$$+ \int_0^t \left\{ \rho(t,s) k(s) \, du + \int_s^t K(u,s) \rho(t,u) \, du \right\} x(s,\omega) \, ds$$

$$= \int_0^t \rho(t,u) \xi(u,\omega) \, du - \int_0^t \left\{ \frac{\partial}{\partial u} \rho(t,u) \right\} x(u,\omega) \, du,$$

where we have used (2.1), (2.6), Fubini's theorem, and (2.7). Hence, from Note (i), it follows that $x(t,\omega)$ satisfies relation (2.9).

Now let $x(t,\omega)$ be a strongly differentiable mild solution of Eq. (2.1) on $[0,T]$, for each $\tau > 0$. Then, for any $t \in (0,T]$,

$$\int_0^t \rho(t,u) \left[ \dot{x}(u) - \xi(u) - k(u) x(u) - \int_0^u k(u,v) x(v) \, dv \right] du$$

$$= \left\{ x(t) - \rho(t,0) x_0 - \int_0^t \rho(t,u) \xi(u) \, du \right\}$$

$$- \left\{ \int_0^t x(u) \frac{\partial}{\partial u} \rho(t,u) \, du + \int_0^t \rho(t,u) k(u) x(u) \, du \right.$$

$$+ \left. \int_0^t \int_0^u \rho(t,u) K(u,v) x(v) \, dv \, du \right\}$$

$$= \mathcal{B}_1 - \mathcal{B}_2, \quad \text{say}.$$

Here $\mathcal{B}_1 = 0$ by (2.9). From (2.6), Fubini's theorem, and (2.7), it follows that $\mathcal{B}_2 = 0$. Hence, $x(t,\omega)$ is a strong solution.

**Corollary 2.3** *Let $x_0 \in L_2(\Omega)$, $\xi \in \mathcal{L}_2[R_+, L_2(\Omega)]$, and $\rho_t(t,s)$ exist. Then, for every $\tau > 0$, a mild solution $x(t,\omega)$ is also a strong solution on $[0,\tau]$ such that almost all trajectories of $x(t,\omega)$ are (1) continuous, (2) of bounded variation, and (3) Hölder continuous of order $\alpha$, for $0 < \alpha < \frac{1}{2}$.*

*Proof* Clear.

We will, for the most part, be working with $L_2$-solutions. We pause here to point out how the corresponding sample solution case can be treated.

**Theorem 2.4** *Let $k \in \mathcal{L}_1[R_+, R^1]$, $K \in \mathcal{L}_1[R_+ \times R_+, R^1]$ and $\xi(t,\omega)$ be a continuous process. Then, the initial value problem (2.1) has a unique sample solution on $[0,\tau]$, for every $\tau > 0$, given by (2.9), where the integral is a sample path integral.*

*Proof* Let

$$\{y_0(\omega), \eta(t,\omega), t \geq 0\}$$

be a representation of

$$\{x_0(\omega), \xi(t,\omega), t \geq 0\}$$

such that $\eta(t,\omega)$ is a continuous process. Fix an $\omega \in \Omega$ and consider the sample equation

$$\dot{y}_\omega(t) = \eta_\omega(t) + k(t) y_\omega(t) + \int_0^t K(t,s) y_\omega(s) \, ds,$$

$$y_\omega(0) = y_{\omega,0}. \tag{2.10}$$

From the deterministic theory (cf. Grossman [20], and Miller [44]), the sample solution of Eq. (2.10) is given by

$$y_\omega(t) = \rho(t,0) y_{\omega,0} + \int_0^t \rho(t,s) \eta_\omega(s) \, ds,$$

$0 \leq s \leq t \leq \tau$. Noting that $\rho(t,s)$ is independent of $\omega$, we see that the totality of all the sample solutions can be written as

$$y(t,\omega) = \rho(t,0) y_0(\omega) + \int_0^t \rho(t,s) \eta(s,\omega) \, ds, \tag{2.11}$$

$(\omega, t) \in \Omega \times [0, \tau]$, $0 \leq s \leq t \leq \tau$, where the integral is a sample Riemann integral. Being the limit of a sequence of Riemann sums, the above integral is a Borel function of the random variables $\eta_s(\omega)$, $0 \leq s \leq t$. But, $y(t,\omega)$ is a representation of

$$x(t,\omega) = \rho(t,0) x_0(\omega) + \int_0^t \rho(t,s) \xi(s,\omega) \, ds,$$

and hence $x(t,\omega)$, given by (2.9), solves Eq. (2.1), a.s. From the properties of $\rho(t,s)$, $x(t,\omega)$ is sample differentiable and satisfies the initial-value problem (2.1).

*Example* We saw (Section I.B) that the generalized Langevin equation (1.10) is a special case of the random integrodifferential equation (2.1). We treat, in the next section, a nonlinear extension of (1.10). Here we consider the following physical motion (cf. Kannan [28]). Let $\xi(s,\omega)$ be the stochastic process describing the motion of an impurity particle during $0 \leq s \leq t_0$. Beginning time $t_0$ the motion is governed by the equation

$$\dot{x}(t,\omega) = -\int_{t_0}^t \gamma(t-s) x(s,\omega) \, ds + \zeta(t,\omega), \qquad t \geq t_0. \tag{2.12}$$

For instance, the particle might undergo a heavy bombardment at time $t_0$ and lose some mass, and the resulting mass undergoes a motion described by (2.12). Let $x(t, \omega; t_0, \xi)$ denote the solution of

$$\dot{x}(t, \omega) = -\int_{t_0}^{t} \gamma(t-s) x(s, \omega) \, ds + \zeta(t, \omega), \qquad t \geq t_0$$

$$x(s, \omega) = \xi(s, \omega), \qquad 0 \leq s \leq t_0. \qquad (2.13)$$

Then, Eq. (2.13) can be rewritten as

$$\dot{x}(t, \omega; t_0, \xi) = \zeta(t + t_0, \omega) - \int_{t_0}^{t} \gamma(t-s) x(s + t_0, \omega; t_0, \xi) \, ds$$

$$- \int_{0}^{t} \gamma(t + t_0 - s) \xi(s, \omega) \, ds, \qquad t \geq 0,$$

$$x(t_0, \omega; t_0, \xi) = \xi(t_0, \omega). \qquad (2.14)$$

Let $\gamma(\cdot) \in \mathscr{L}_\infty[R_+, R^1]$ and $\zeta \in \mathscr{L}_1[R_+, L_2(\Omega)]$. Then, the mild solution of Eq. (2.14) is given by

$$x(t + t_0, \omega; t_0, \xi) = \rho(t) \xi(t_0, \omega) + \int_0^t \rho(t-s)$$

$$\times \left[ \zeta(s + t_0, \omega) + \int_0^{t_0} \gamma(s + t_0 - u) \xi(u, \omega) \, du \right] ds, \qquad (2.15)$$

for all $t \geq 0$, where

$$\rho(t) = 1 - \int_0^t \rho(t-s) \int_0^s \gamma(u) \, du \, ds.$$

### B. Nonlinear Equations

For a Brownian particle under an external force field (e.g., a harmonic oscillator), the Langevin equation takes the form

$$\dot{x}(t) = -\gamma x(t) + A(t) + R(t, \omega),$$

where $A(t)$ is the external force and $R(t, \omega)$ is a centered Gaussian random force such that

$$E\{R(s) R(t)\} = \sigma \delta(s-t), \qquad (2.16)$$

where $\delta$ is the Dirac delta function. As noted in the derivation of the generalized Langevin equation, the time scale of the molecular motion need not be shorter than that of the impurity particle, and this forces us to drop assumption (2.16). In place of the generalized equation (1.10) we consider a nonlinear extension, which is given in the integrated form as follows (see Kannan and Bharucha-Reid [33]):

$$x(t,\omega) = x(t_0,\omega) + \int_{t_0}^{t}\int_{t_0}^{s} \gamma(s-u)x(u,\omega)\,du\,ds$$
$$+ \int_{t_0}^{t} m(u,x(u,\omega))\,du + \int_{t_0}^{t} \sigma(u,x(u,\omega))\,dM(u,\omega), \quad (2.17)$$

where $M(t,\omega)$ is a suitable martingale and the last integral in (2.17) is the Itô–Doob integral. Equation (2.17) is to be solved for $t \in (t_0, T]$. Toward this we assume the following.

($A_1$) $\{M(t), \mathscr{A}_t, t \in [t_0, T]\}$ is a continuous martingale such that there exists a nondecreasing function $F(t)$, $t \in [t_0, T]$, with the property that, for $s < t$, $E\{|M(t) - M(s)|^2 \,|\, \mathscr{A}_s\} = F(t) - F(s)$.

($A_2$) The functions $m(t,x)$ and $\sigma(t,x)$ are measurable in the pair $(t,x)$ for $t_0 \leq t \leq T$ and $x \in R^1$.

($A_3$) For each $t \in (t_0, T)$, we have

$$\int_{t_0}^{t} E\{|m(s,x)|^2 + |\sigma(s,x)|^2\}\,dF(s) < \infty.$$

Let $\gamma \in L_1([t_0, T])$ and $x \in L_\infty([t_0, T])$. Then, for $s \in [t_0, T]$, define the convolution $\gamma * x$ by $(\gamma * x)(s) = \int_{t_0}^{s} \gamma(s-u)x(u)\,du$. Now Eq. (2.17) becomes

$$x(t) = x_0 + \int_{t_0}^{t} (\gamma * x)(u)\,du + \int_{t_0}^{t} m(u,x(u))\,du + \int_{t_0}^{t} \sigma(u,x(u))\,dM(u).$$
(2.18)

Through the integral equation (2.18) we establish the existence and uniqueness of the solution of the generalized Langevin equation

$$dx(t) = \left\{\int_{t_0}^{t} \gamma(t-s)x(s)\,ds\right\}dt + m(t,x(t))\,dt$$
$$+ \sigma(t,x(t))\,dM(t).$$

**Lemma 2.5** Let assumptions ($A_1$)–($A_3$) hold. Define a mapping $\phi$ on $L_\infty = L_\infty([t_0, T], L_2(\Omega))$ by

$$\phi x(t) = x_0 + \int_{t_0}^{t} \gamma_x(u)\,du + \int_{t_0}^{t} m(u,x(u))\,du$$
$$+ \int_{t_0}^{t} \sigma(u,x(u))\,dM(u),$$

where $\gamma_x = \gamma * x$. Assume the uniform growth condition:

($A_4$) there is a constant $K$ such that for $t \in [t_0, T]$ and every $x$,
$$|m(t,x)|^2 + |\sigma(t,x)|^2 \leq K^2(1 + |x|^2).$$

Then, $\phi$ maps $L_\infty$ into $L_\infty$.

*Proof* From the growth condition ($A_4$),
$$\|m(\cdot, x(\cdot))\|_\infty \leqslant K^2(1+\|x\|_\infty^2) < \infty.$$
Thus, $m(\cdot, x(\cdot)) \in L_\infty$, and similarly $\sigma(\cdot, x(\cdot)) \in L_\infty$. From the uniform continuity of $\gamma_x$ on $[t_0, T]$, there is a constant $K_0$ such that $|\gamma_x| \leqslant K_0$, and
$$\left\|\int_s^t \gamma_x(u)\, du\right\|_2^2 \leqslant K_0^2(t-s)^2.$$
Now,
$$\left\|\int_s^t m(u, x(u))\, du\right\|_2^2 \leqslant (t-s)^2 \|m(\cdot, x(\cdot))\|_\infty^2$$
$$\leqslant K^2(t-s)^2(1+\|x\|_\infty^2) < \infty;$$
and
$$\left\|\int_s^t \sigma(u, x(u))\, dM(u)\right\|_2^2 = \int_s^t \|\sigma(u, x(u))\|_2^2\, dF(u)$$
$$\leqslant \|\sigma(\cdot, x(\cdot))\|_\infty^2 (F(t) - F(s))$$
$$\leqslant K^2(1+\|x\|_\infty^2)\mathscr{V}F < \infty,$$
where $\mathscr{V}F$ is the total variation of $F$ in $[t_0, T]$. Hence,
$$\|\phi x\|_\infty^2 = 4\|x_0\|_\infty^2 + 4K_0^2(T-t_0)^2$$
$$+ 4K^2(1+\|x\|_\infty^2)\{(T-t_0)^2 + \mathscr{V}F\} < \infty,$$
and $\phi: L_\infty \to L_\infty$, which proves the lemma.

**Lemma 2.6** *In addition to the hypotheses made in Lemma 2.5 let us assume also that*

($A_5$) *the uniform Lipschitz condition,*
$$|m(t, x) - m(t, y)| + |\sigma(t, x) - \sigma(t, y)| \leqslant K|x - y|,$$
*holds, for a certain (Lipschitz) constant $K$. Then, some power of $\phi: L_\infty \to L_\infty$ is a contraction.*

*Proof* For $\xi, \eta \in L_\infty$, define
$$\Delta_i(\cdot) = \phi^i \xi(\cdot) - \phi^i \eta(\cdot), \quad \Delta m_i(\cdot) = m(\cdot, \phi^i \xi(\cdot)) - m(\cdot, \phi^i \eta(\cdot)),$$
and
$$\Delta \sigma_i(\cdot) = \sigma(\cdot, \phi^i \xi(\cdot)) - \sigma(\cdot, \phi^i \eta(\cdot)).$$
Then
$$\|\Delta_i(t)\|_2^2 \leqslant 3\left\|\int_{t_0}^t [\gamma * (\phi^{i-1}\xi - \phi^{i-1}\eta)](u)\, du\right\|_2^2$$

$$+ 3\left\|\int_{t_0}^{t}\Delta m_i(u)\,du\right\|_2^2 + 3\left\|\int_{t_0}^{t}\Delta\sigma_i(u)\,dM(u)\right\|_2^2$$

$$\leq 3(T-t_0)^2 K^2 \int_{t_0}^{t}\|\Delta_{i-1}\|_2^2\,du + 3K^2\int_{t_0}^{t}\|\Delta_{i-1}\|_2^2\,dF(\omega)$$

$$\leq A\int_{t_0}^{t}\|\Delta_{i-1}\|_2^2\,du + B\int_{t_0}^{t}\|\Delta_{i-1}\|_2^2\,dF(u),$$

where $L = (T-t_0)$, $A = 3L\|\gamma\|_1^2 + 3k^2$, and $B = 3k^2$. In particular, for $i = 1$ and 2, we have

$$\|\Delta_1(t)\|_2^2 \leq A\|\Delta_0\|_\infty^2\int_{t_0}^{t}du + B\|\Delta_0\|_\infty^2\int_{t_0}^{t}dF(u),$$

and

$$\|\Delta_2(t)\|_2^2 \leq A\int_{t_0}^{u}\left\{A\|\Delta_0\|_\infty^2\int_{t_0}^{t}du + B\|\Delta_0\|_\infty\int_{t_0}^{t}dF(u)\right\}du$$

$$+ B\int_{t_0}^{u}\left\{A\|\Delta_0\|_\infty^2\int_{t_0}^{t}du + B\|\Delta_0\|_\infty^2\int_{t_0}^{t}dF(u)\right\}du$$

$$= \|\Delta_0\|_\infty^2\left\{[A^2(t-t_0)^2/2!] + AB\left[(u-t_0)F(u)\Big|_{t_0}^{t}\right.\right.$$

$$\left.\left. - \int_{t_0}^{t}F(u)\,du + \int_{t_0}^{t}(F(u)-F(t_0))\,du\right] + (B^2/2!)(\mathscr{V}F)^2\right\}$$

$$\leq \|\Delta_0\|_\infty^2\{(A^2L^2/2!) + AB(t-t_0)(F(t)-F(t_0)) + [(BV)^2/2!]\}$$

$$\leq \|\Delta_0\|_\infty^2(AL+BV)^2/2!,$$

where $V = \mathscr{V}F$. Thus,

$$\|\Delta_1\|_\infty^2 \leq \|\Delta_0\|_\infty^2(AL+BV)$$

and

$$\|\Delta_0\|_\infty^2 \leq \|\Delta_0\|_\infty^2(AL+BV)^2/2!,$$

and by induction one can show that

$$\|\Delta_m\|_\infty^2 \leq \|\Delta_0\|_\infty^2(AL+BV)^m/m!.$$

Now, given a $\lambda$, $0 < \lambda < 1$, we can choose a sufficiently large $n$ such that $(AL+BV)^n/n! \leq \lambda^2$, and therefore,

$$\|\phi^n\xi - \phi^n\eta\|_\infty \leq \lambda\|\Delta_0\|_\infty = \lambda\|\xi-\eta\|_\infty.$$

Hence, $\phi^n$ is a contraction, and this proves the lemma.

**Theorem 2.7** *Let conditions* $(A_1)$–$(A_5)$ *hold and $\phi$ be as defined in Lemma 2.5. Then, there exists a unique solution $x(t)$ solving the nonlinear generalized Langevin equation (2.17).*

*Proof* By Lemma 2.6, there is a positive integer $n$ such that $\phi^n$ is a contraction. Therefore, by contraction principle the mapping $\phi$ has a unique fixed point; that is, there is a unique (up to equivalence) process $x(t)$ such that $\phi x(t) = x(t)$. Thus, $x(t)$ satisfies Eq. (2.18) and hence Eq. (2.17). This proves the theorem.

By replacing $L_\infty$ with the space of continuous processes one can similarly obtain continuous solutions of the Eq. (2.17). Next we illustrate the method of approximating the solution process $x(t)$ by a Cauchy polygon.

**Cauchy Polygonal Approximation** Let $m(t, x)$ and $\sigma(t, x)$ be jointly continuous in $(t, x) \in [t_0, T] \times (-\infty, \infty)$ and satisfy the usual Lipschitz continuity in $x$. Consider a bounded solution $x(t)$: $\|x(t)\|_2 \leq A$.

From the uniform continuity of $(\gamma * x)$ on $I = [t_0, T]$, there is a constant $K_0 > 0$ such that $\|\gamma * x\|_2 \leq K_0$. For fixed $x$, $m(t, x)$ and $\sigma(t, x)$ are continuous on $I$ and hence are bounded there. By Lipschitz continuity, there exists a constant $L \geq K_0$ such that $\|m(t, x)\|_2 \leq L$ and $\|\sigma(t, x)\|_2 \leq L$. Define

$$\Gamma(\delta) = \sup_{|t-s|<\delta} \|\gamma_x(t) - \gamma_x(s)\|_2,$$

$$m(\delta) = \sup_{|t-s|<\delta} \|m(t, x(t)) - m(s, x(s))\|_2,$$

$$\sigma(\delta) = \sup_{|t-s|<\delta} \|\sigma(t, x(t)) - \sigma(s, x(s))\|_2,$$

and

$$g(\delta) = \max[\Gamma(\delta), m(\delta), \sigma(\delta)].$$

Under the assumptions made above, we have

$$g(\delta) \to 0 \quad \text{as} \quad \delta \to 0. \tag{2.19}$$

Let $F(t)$ be differentiable everywhere, (it is so a.e.), and have a bounded derivative $f(t)$ with $|f(t)| \leq B$. Let $\pi: t_0 < t_1 < \cdots < t_n = T$ be a partition of $I$, and $\|\pi\| = \max_i(t_i - t_{i-1})$. Define the Cauchy polygon as follows:

(i) $x_\pi(t_0) = 0$ and
(ii) if $x_\pi$ has been defined for $t_0 \leq t \leq t_i$,

then we define

$$x_\pi(t_{i+1}) = x_\pi(t_i) + \gamma_{x_\pi}(t_i)(t_{i+1} - t_i)$$
$$+ m(t_i, x_\pi)(t_{i+1} - t_i) + \sigma(t_i, x_\pi)[M(t_{i+1}) - M(t_i)],$$

and define $x_\pi(t)$, $t_i < t < t_{i+1}$, by linear interpolation between $x_\pi(t_i)$ and $x_\pi(t_{i+1})$. We claim that *for each $\varepsilon > 0$, there is a $\delta > 0$ with the following* *property*: *If $x(t)$ is a bounded solution process of Eq.* (2.17) *and $x_\pi(t)$ is the Cauchy polygon defined above corresponding to a partition $\pi$ with $\|\pi\| < \delta$, then*

$$\|x(t) - x_\pi(t)\|_2 < \varepsilon, \qquad t \in I. \tag{2.20}$$

First, for $t_0 \leq s \leq t \leq T$,

$$\|x(t) - x(s)\|_2^2 = E[x(t) - x(s)]^2$$

$$\leq 3 \left\{ \left\| \int_s^t \gamma_x(u) \, du \right\|_2^2 + \left\| \int_s^t m(u, x(u)) \, du \right\|_2^2 \right.$$

$$\left. + \left\| \int_s^t \sigma(u, x(u)) \, dM(u) \right\|_2^2 \right\}$$

$$\leq 3L^2 \{ (T - t_0)(t - s) + (T - t_0)(t - s) + B(t - s) \}.$$

Thus, there is a constant $C > 0$ such that

$$\|x(t) - x(s)\|_2 \leq C(t - s)^{\frac{1}{2}}.$$

Next we formalize the definition of $x_\pi$ as follows. For any $t \in I$, there is an $i$, $0 \leq i \leq n-1$, such that $t \in [t_i, t_{i+1})$ and define

$$x_\pi(t) = x_0 + \int_{t_0}^t \gamma_{x_\pi}(T_\pi s) \, ds + \int_{t_0}^t m(T_\pi s, x_\pi) \, ds$$

$$+ \int_t^{t_i} \sigma(T_\pi s, x) \, dM(s) + q(t) \int_{t_i}^{t_{i+1}} \sigma(t_i, x_\pi) \, dM(s), \tag{2.21}$$

where $T_\pi t = t_i$ if $t_i \leq t < t_{i+1}$, $0 \leq i \leq n-1$, and $q(t) = (t - t_i)/(t_{i+1} - t_i)$. Then,

$$x_\pi(t) - x(t) = \int_{t_0}^t \{x_\pi(T_\pi s) - \gamma_x(T_\pi s)\} \, ds$$

$$+ \int_{t_0}^t \{\gamma_x(T_\pi s) - \gamma_x(s)\} \, ds$$

$$+ \int_{t_0}^t \{m(T_\pi s, x_\pi) - m(T_\pi s, x)\} \, ds$$

$$+ \int_{t_0}^t \{m(T_\pi s, x) - m(s, x)\} \, ds$$

$$+ \int_{t_0}^{t_i} \{\sigma(T_\pi s, x_\pi) - \sigma(T_\pi s, x)\} \, dM(s)$$

$$+ \int_{t_0}^{t_i} \{\sigma(T_\pi s, x) - \sigma(s, x)\} \, dM(s)$$

$$+ q(t) \int_{t_i}^{t_{i+1}} \{\sigma(t_i, x_\pi) - \sigma(t_i, x)\} \, dM(s)$$

$$+ q(t) \int_{t_i}^{t_{i+1}} \sigma(t_i, x) \, dM(s)$$

$$+ \int_{t_i}^{t} -\sigma(s, x(s)) \, dM(s)$$

$$= \sum_{k=1}^{9} I_k, \quad \text{say}.$$

Define $\psi(t)$ as follows: for $t \in I = [t_0, T]$,

$$\psi(t) = \sup_{t_0 \leq s \leq t} E\{[x_\pi(s) - x(s)]^2\}.$$

Noting that $(\gamma^2 * 1)$, where 1 is the function identically equal to 1, can be majorized by a constant $D > 0$, we get

$$\|I_1\|_2^2 \leq D \int_{t_0}^{t} \psi(s) \, ds \quad \text{and} \quad \|I_2\|_2^2 \leq (t - t_0^2) \, \Gamma^2(\delta).$$

By the Lipschitz continuity and Cauchy–Schwarz inequality,

$$\|I_3\|_2^2 \leq (t - t_0) K^2 \int_{t_0}^{t} \|x_\pi - x\|_2^2 \, ds \leq K^2 (t - t_0) \int_{t_0}^{t} \psi(s) \, ds;$$

$$\|I_4\|_2^2 \leq (t - t_0)^2 [m(\delta)]^2.$$

From the Lipschitz continuity and stochastic integral isometry, we obtain

$$\|I_5\|_2^2 \leq K^2 B \int_{t_0}^{t_i} \psi(s) \, ds;$$

$$\|I_6\|_2^2 \leq (t_i - t_0) B [\sigma(\delta)]^2;$$

$$\|I_7\|_2^2 \leq [q(t)]^2 \int_{t_0}^{t_{i+1}} \|\sigma(t_i, x_\pi) - \sigma(t_i, x)\|_2^2 \, dF(s)$$

$$\leq [q(t)]^2 K^2 \int_{t_i}^{t_{i+1}} \psi(t_i) \, dF(s) \leq K^2 B \int_{t_i}^{t_{i+1}} \psi(t_i) \, ds$$

$$\|I_8\|_2^2 \leq L^2 B \|\pi\|, \quad \text{and} \quad \|I_9\|_2^2 \leq L^2 B \|\pi\|.$$

Let $N = \max\{K, L\}$. From the above estimates, we get

$$\psi(t) \leq D \int_{t_0}^{t} \psi(s) \, ds + (t - t_0)^2 [g(\delta)]^2$$

$$+ N^2 (t - t_0) \int_{t_0}^{t} \psi(s) \, ds + (t - t_0)^2 [g(\delta)]^2$$

$$+ N^2 B \int_{t_0}^{t} \psi(s) \, ds + (t - t_0) B [g(\delta)]^2 + 2BN^2 \|\pi\|$$

$$= \xi(t) + \eta(t) \int_{t_0}^{t} \psi(s) \, ds,$$

where
$$\xi(t) = [g(\delta)]^2[2(t-t_0)^2 + B(t-t_0)] + 2BN^2\|\pi\|,$$
and
$$\eta(t) = \{D + N^2[B+(t-t_0)]\}.$$
Hence, from Gronwall's lemma,
$$\psi(t) \leq \xi(t) + \int_{t_0}^{t} \eta(t)\xi(s)\left[\exp\int_s^t \eta(u)\,du\right]ds.$$
Noting that $\max_t \xi(t) \to 0$, as $\|\pi\| \to 0$, we establish our claim.

*Note* For a given admissible error $\varepsilon^* > 0$ and the Cauchy polygonal approximation $x_\pi$, one can see that $\|x(s)\|_2$ is majorized by the sum $\varepsilon^* + \sup_{t_0 \leq s \leq t} \|x_\pi(s)\|_2$. This enables us to find $\xi(t)$ and $\eta(t)$. Hence, $\psi(t)$ can be estimated, which gives the accuracy of the approximation.

Certain special cases and deterministic forms of the following random integrodifferential equation

$$x(t,\omega) = h(t,x(t,\omega)) + \int_0^t K_1(t,s,\omega)f_1(s,x(s,\omega))\,ds$$
$$+ \int_0^t K_2(t,s,\omega)f_2(s,x(s,\omega))\,dM(s,\omega)$$
$$x(0,\omega) = x_0(\omega) \qquad (2.22)$$

arise in automatic systems, reactor dynamics, and ecosystems with memory. Equations of the form (2.22) have been studied by Padgett and Rao and Tsokos (see [49, 51]). In Eq. (2.22), $x(t,\omega)$ is the unknown process, $h(t,x)$ is a scalar function, $K_i(t,s,\omega)$, $i = 1, 2$, are scalar stochastic kernels defined for $0 \leq s \leq t < \infty$, $\omega \in \Omega$, $f_i(t,x)$, $i = 1, 2$, are scalar functions, and $M(t,\omega)$ is a real martingale adapted to an increasing family $\{\mathscr{A}_t, t \geq 0\}$ of sub-$\sigma$-algebras of $\mathscr{A}$ such that there is a continuous nondecreasing function $F(t)$ with the property that

$$E|M(t)-M(s)|^2 = E\{|M(t)-M(s)|^2\,|\,\mathscr{A}_s\} = F(t) - F(s), \qquad \text{a.s.}$$

We now establish the existence and uniqueness of the solution of Eq. (2.22). Toward this we introduce certain function spaces.

Let $\mathbf{C} = \mathbf{C}[R_+, L_2(\Omega)]$ be the Fréchet space of all continuous mappings $x: R_+ \to L_2(\Omega)$ such that $\{x(t), \mathscr{A}_t, t \geq 0\}$ is an adapted family and the topology in $\mathbf{C}$ is defined by the seminorms $\|x\|_n = \sup_{0 \leq t \leq n} \|x(t)\|_{L_2}$, $n \geq 1$. Let $\mathbf{C}_g = \mathbf{C}_g[R_+, L_2(\Omega)]$ be the Banach space of all $x \in \mathbf{C}$ such that $\|x(t)\|_2 \leq Ag(t)$, where $A > 0$ and $g(t)$ is a positive continuous function on $R_+$, and the topology in $\mathbf{C}_g$ is defined by the norm

$$\|x\|_g = \sup_{t \geq 0} \{\|x(t)\|_{L_2}/g(t)\}.$$

By $\mathbf{C}^b = \mathbf{C}^b[R_+, L_2(\Omega)]$ we shall denote the Banach space of all $x \in \mathbf{C}$ that are bounded maps from $R_+$ into $L_2(\Omega)$, where the norm in $\mathbf{C}^b$ is given by $\|x\| = \sup_{t \geq 0} \|x(t)\|_{L_2}$.

Define the linear operators $T_1$, $T_2$, and $T_3$ on $\mathbf{C}$ by

$$T_1 x(t) = \int_0^t x(s)\,ds, \quad T_2 x(t) = \int_0^t G_1(t,s) x(s)\,ds,$$

$$T_3 x(t) = \int_0^t G_2(t,s) x(s)\,dM(s)\,ds,$$

where $G_i(t,s) = \int_0^t K_i(u,s)\,du$, $i = 1, 2$. We make the following assumptions: ($H_1$) for $0 \leq s \leq t < \infty$, the kernels $K_i(t,s,\omega)$ are $\mathscr{A}_s$-measurable and $P$ essentially bounded, and ($H_2$) $K_i$ are continuous maps from $\mathscr{D} = \{(t,s): 0 \leq s \leq t < \infty\}$ into $L_\infty(\Omega)$. Under these assumptions one can easily show that $T_i$ maps $\mathbf{C}$ continuously into itself, $i = 1, 2, 3$. In order to make the next assumption we need the notion of admissibility of a pair of Banach space relative to an operator; for the details we refer to Massers and Schaffer [39]. A Banach subspace $\mathbf{B} \subset \mathbf{C}$ is said to be *stronger* than $\mathbf{C}$, if every sequence $\{x_n\}$ that converges in the norm $\|\cdot\|_\mathbf{B}$ also converges in the topology of $\mathbf{C}$. Let $\mathbf{B}, \mathbf{D} \subset \mathbf{C}$ be two Banach subspaces of $\mathbf{C}$ and $T: \mathbf{C} \to \mathbf{C}$ be a linear operator. The pair $(\mathbf{B}, \mathbf{D})$ is said to be *admissible* relative to $T$ is $T\mathbf{B} \subset \mathbf{D}$. Let $T$ be an endomorphism of $\mathbf{C}$, $\mathbf{B}$ and $\mathbf{D}$ be two Banach subspaces of $\mathbf{C}$ stronger than $\mathbf{C}$, and $(\mathbf{B}, \mathbf{D})$ be admissible relative to $T$. Then, it is a consequence of closed graph theorem (see, Yosida [60]) that $T$ is a continuous linear operator from $\mathbf{B}$ into $\mathbf{D}$. Now the next assumption follows.

($H_3$) $\mathbf{B}$ and $\mathbf{D}$ are Banach subspaces stronger than $\mathbf{C}$ such that the pair $(\mathbf{B}, \mathbf{D})$ is admissible relative to each endomorphism $T_i$ of $\mathbf{C}$, $i = 1, 2, 3$. Then, by the continuity of $T_i$ from $\mathbf{B}$ into $\mathbf{D}$, there are constants $N_i > 0$ such that

$$\|T_i x\|_\mathbf{D} \leq N_i \|x\|_\mathbf{B}.$$

($H_4$) $f_i: R_+^1 \times \mathbf{C} \to \mathbf{C}$, $i = 1, 2$.

By integrating Eq. (2.22) one can see that it is equivalent to

$$x(t, \omega) = x_0(\omega) + \int_0^t h(s, x(s,\omega))\,ds + \int_0^t G_1(t,s,\omega) f_1(s, x(s,\omega))\,ds$$

$$+ \int_0^t G_2(t,s,\omega) f_2(s, x(s,\omega))\,dM(s,\omega), \quad (2.23)$$

where $G_i(t,s) = \int_0^t K_i(u,s)\,du$. So, we establish the existence and uniqueness of solution of Eq. (2.23) (see Rao and Tsokos [51]).

**Theorem 2.8** *Let assumptions* $(H_1)$–$(H_4)$ *hold. Let*

$$S \equiv \{x \in \mathbf{D}: \|x\|_{\mathbf{D}} \leq \rho\},$$

*where $\rho > 0$ is a constant, and the mappings $x \to h(t,x)$, $x \to f_i(t,x)$, $i = 1, 2$, map $S$ into $\mathbf{B}$ such that there are constants $\lambda_i > 0$, $i = 1, 2, 3$, satisfying the following Lipschitz continuity:*

$$\|h(t,x(t,\omega)) - h(t,y(t,\omega))\|_{\mathbf{B}} \leq \lambda_1 \|x(t,\omega) - y(t,\omega)\|_{\mathbf{D}}$$

$$\|f_i(t,x(t,\omega)) - f_i(t,y(t,\omega))\|_{\mathbf{B}} \leq \lambda_{i+1} \|x(t,\omega) - y(t,\omega)\|_{\mathbf{D}},$$

*for $x, y \in \mathbf{D}$ and $i = 1, 2$. As an $L_2$-random variable independent of time, let the initial value $x_0 \in \mathbf{D}$. Then, there is a unique solution to Eq. (2.23) in $S$, provided that*

$$\alpha = \lambda_1 N_1 + \lambda_2 N_2 + \lambda_3 N_3 < 1 \tag{2.24}$$

*and*

$$\|x_0\|_{\mathbf{D}} + N_1 \|h(t,0)\|_{\mathbf{B}} + N_2 \|f_1(t,0)\|_{\mathbf{B}} + N_3 \|f_2(t,0)\|_{\mathbf{B}}$$
$$\leq \rho(1 - \lambda_1 N_1 - \lambda_2 N_2 - \lambda_3 N_3). \tag{2.25}$$

*Proof* Define the mapping $\Phi: S \to \mathbf{D}$ by

$$(\Phi x)(t,\omega) = x_0(\omega) + \int_0^t h(s,x(s,\omega))\,ds + \int_0^t G_1(t,s,\omega)f_1(s,x(s,\omega))\,ds$$
$$+ \int_0^t G_2(t,s,\omega)f_2(s,x(s,\omega))\,dM(s,\omega).$$

By the admissibility hypothesis, $T_i: \mathbf{B} \to \mathbf{D}$ continuously, and hence $\Phi S \subset \mathbf{D}$. Using the Lipschitz continuity of $h$ and $f_i$, the existence of constant $N_i$ due to the continuity of the operator $T_i: \mathbf{B} \to \mathbf{D}$, and condition (2.24), we get, for $x, y \in \mathbf{D}$, that

$$\|\Phi x(t,\omega) - \Phi y(t,\omega)\|_{\mathbf{D}} \leq \alpha \|x(t,\omega) - y(t,\omega)\|_{\mathbf{D}},$$

and hence $\Phi: S \to D$ is a contraction. Again using the Lipschitz continuity of $h$ and $f_i$ for the elements $x, 0 \in S$, and the condition (2.25), we see that $\Phi S \subset S$. Now, Banach's fixed point theorem completes the proof of this theorem.

**Theorem 2.9** *In addition to hypotheses $(H_1)$ and $(H_2)$ let us also assume that*

(i) *there are a constant $\alpha > 0$ and a positive continuous function $g(t)$ on*

$R_+$ such that

$$\left\| \int_s^t K_1(s,u,\omega)\, du \right\|_{L_\infty(\Omega)} g(s)\, ds + \int_0^t \left\| \int_s^t K_2(s,u,\omega)\, du \right\|_{L_\infty(\Omega)} g(s)\, dF(s) \leq \alpha;$$

(ii) $h$ and $f_i$, $i = 1, 2$, are continuous functions for $(t, x) \in R_+^1 \times R^1$ such that

$$|h(t,x) - h(t,y)| + \sum_{i=1}^{2} |f_i(t,x) - f_i(t,y)| \leq \lambda g(t)|x-y|,$$

for some $\lambda > 0$, and $h(t,0)$, $f_i(t,0) \in \mathbf{C}_g$; and

(iii) $x_0(\omega) \in \mathbf{C}^b$.

Then, there exists a solution to Eq. (2.22) such that

$$\sup_{t \geq 0} \|x(t)\|_{L_2} < \rho,$$

for some $\rho > 0$, provided $\lambda$ is small enough.

*Proof* If the pair $(\mathbf{C}_g, \mathbf{C}^b)$ is admissible relative to each of the operators $T_i$, $i = 1, 2, 3$, then the conditions of this theorem would imply the conditions of Theorem 2.8 and that will complete the proof. Now we claim that $(\mathbf{C}_g, \mathbf{C}^b)$ is admissible relative to $T_i$, $i = 1, 2, 3$. First,

$$\|(T_2 x)(t,\omega)\|_{L_2} \leq \int_0^t \|G_1(t,s,\omega)\|_{L_\infty} \|x(s,\omega)\|_{L_2}\, ds$$

$$\leq \sup_{0 \leq s} \frac{\|x(s)\|_{L_2}}{g(s)} \int_0^t \left\| \int_s^t K_1(u,s,\omega)\, ds \right\|_{L_\infty} g(s)$$

$$\leq \alpha \|x\|_g,$$

and hence $(\mathbf{C}_g, \mathbf{C}^b)$ is admissible relative to $T_2$. Similarly $(\mathbf{C}_g, \mathbf{C}^b)$ is admissible relative to $T_1$ and $T_3$. This completes the proof.

## C. Itô Integrodifferential Equation (Second-Order Itô Equations)

The first-order stochastic differential equation has been extensively studied (cf. [2, 9, 11, 16, 18, 23, 42]). Important physical phenomena, such as a Brownian oscillator, lead to second order stochastic differential equations (cf. [5, 7, 12, 19, 31, 58]). Consider the deterministic equation $\ddot{x}(t) = m(t, x(t), \dot{x}(t))$, which describes a dynamical system. Due to the inherent stochasticity in the system, the random impulses perturb the system. As a result of random fluctuations, the above equation takes the form

$$d\dot{x}(t) = m(t, x(t), \dot{x}(t))\, dt + \sigma(t, x(t), \dot{x}(t))\, d\beta(t), \quad (2.26)$$

or the symbolic stochastic integrodifferential equation form,

$$\dot{x}(t) = \eta_0 + \int_0^t m(s, x(s), \dot{x}(s))\, ds + \int_0^t \sigma(t, x(t), \dot{x}(t))\, d\beta(t), \quad (2.27)$$

The stochastic equation of this form has been introduced and investigated by Borchers [5] and Goldstein [19]. In this section we present the existence and uniqueness theorem of Borchers; here, the proof illustrates the classical method of successive approximation extended to the present equation (see also Itô [23], Dynkin [11], Gihman and Skorokhod [18], Friedman [16], and McKean [42]).

The initial-value problem we study in this section is that of Eq. (2.26) with initial values $x(0) = \xi_0$ and $\dot{x}(0) = \eta_0$. Equivalently we consider either the random integrodifferential equation (2.27) or the equation

$$x(t) = \xi_0 + \eta_0 t + \int_0^t \left\{ \int_0^s m(u, x(u), \dot{x}(u))\, du + \int_0^s \sigma(u, x(u), \dot{x}(u))\, d\beta(\omega) \right\} ds. \quad (2.28)$$

Before presenting the meaning of these integrals certain assumptions are in order.

($B_1$) $m$ and $\sigma$ are Borel functions of $R_+^1 \times R^1 \times R^1$ into $R^1$ satisfying local Lipschitz and growth conditions: for each compact interval $I \subset R_+^1$ there is a positive constant $K = K(I)$ such that

$$|m(t, x_1, y_1) - m(t, x_2, y_2)| + |\sigma(t, x_1, y_1) - \sigma(t, x_2, y_2)|$$
$$\leqslant K(|x_1 - x_2|^2 + |y_1 - y_2|^2)^{\frac{1}{2}},$$
$$|m(t, x, y)|^2 + |\sigma(t, x, y)|^2 \leqslant K(1 + |x|^2 + |y|^2),$$

with $t \in I$, $x_i, y_i \in R^1$, $i = 1, 2$.

($B_2$) $\{\beta(t), t \geqslant 0\}$ is a continuous $L_2$-martingale with independent increments such that there is a continuous, nondecreasing function $F(t)$ on $R_+^1$ with the property that $E\{[\beta(t) - \beta(s)]^2 | \mathscr{A}_s\} = F(t) - F(s)$, a.s., $(s < t)$, where $\mathscr{A}_s = \sigma\{\xi_0, \eta_0, \beta(v) - \beta(u), 0 \leqslant u \leqslant v \leqslant s\}$, $\xi_0, \eta_0 \in L_2(\Omega)$ and $\xi_0, \eta_0$ are independent of the increments $\{\beta(v) - \beta(u)\}$, $0 \leqslant u \leqslant v$.

To discuss the meaning of the right-hand side (rhs) of Eq. (2.28) let us assume, in addition to ($B_1$) and ($B_2$), that the adapted family $\{x(t), \mathscr{A}_t, t \geqslant 0\}$ is an absolutely continuous process satisfying the following conditions:

(i) $\dot{x}(t)$ exists for every $t \geqslant 0$, and

(ii) $\int_0^t E\{|x(s)|^2\}\, dG(s) < \infty$, $t \geqslant 0$, where $G(t) = t + F(t)$.

Then, the $(t, \omega)$-function $x^*(t, \omega)$ defined by the rhs of Eq. (2.28) is well defined. From Hölder's inequality, the growth condition, absolute continuity of $x(t)$, and condition (ii) above, it is easy to see that the first integral on the rhs of (2.28) exists. Regarding the second integral, first note that $\sigma(\cdot, x(\cdot, \cdot))$ is $\lambda \times P$-measurable as a consequence of Borel measurability of $\sigma(\cdot, \cdot, \cdot)$ and the measurability of $x$ and $\dot{x}$. Also, $\sigma(\cdot, x(\cdot, \cdot), \dot{x}(\cdot, \cdot))$ is adapted to $\{\mathcal{A}_t, t \geq 0\}$. Noting that $x(t, \omega) = \xi_0(\omega) + \int_0^t \dot{x}(s, \omega)\, ds$, (see Section I.C), we have, for $0 \leq t \leq T_0$,

$$\int_0^t E[\sigma^2(s, x(s), \dot{x}(s))]\, dF(s)$$

$$\leq K^2 \int_0^t \{1 + E|x(s)|^2 + E|\dot{x}(s)|^2\}\, dF(s)$$

$$\leq c_1 + c_2 + 2T_0 K^2 \int_0^t \int_0^s E|\dot{x}(u)|^2\, du\, dF(s) + K^2 \int_0^t E|\dot{x}(s)|^2\, dF(s)$$

$$\leq C_1 + C_2 + C_3 \int_0^t E|\dot{x}(s)|^2\, dG(s), \qquad 0 \leq t \leq T_0,$$

$$< \infty,$$

where $C_1 = K^2[G(T_0) - G(0)]$, $C_2 = 2K^2[G(T_0) - G(0)]E(\xi_0^2)$ and $C_3 = 2T_0 K^2[G(T_0) - G(0)] + K^2$. The Itô–Doob integral

$$J(t) = \int_0^t \sigma(s, x(s), \dot{x}(s))\, d\beta(s)$$

exists and can be taken as a continuous martingale (cf. Doob [9]). It follows that almost all sample functions $J(\cdot, \omega)$ are Borel functions bounded on compact intervals $[0, T_0]$. Hence $\int_0^t J(s, \omega)\, ds$ exists, is finite a.s., and is an absolutely continuous process. It is not hard to show that $x^*(t, \omega)$ satisfies the properties of $x(t, \omega)$ and solves Eq. (2.27).

**Theorem 2.10** *Let $\xi_0$ and $\eta_0$ be any two random variables and $\{\beta(t), t \geq 0\}$ be a martingale satisfying hypotheses $(B_1)$ and $(B_2)$. Then, there exists a unique process $\{x(t), t \geq 0\}$ with absolutely continuous sample paths such that*

(i) $\{x(t)\}$ *solves Eq.* (2.28), *a.s.,*
(ii) $\{x(t)\}$ *is adapted to* $\{\mathcal{A}_t\}$,
(iii) $\dot{x}(t)$ *exists as an $L_2$-derivative,*
(iv) $x(0) = \xi_0$, $\dot{x}(0) = \eta_0$, *a.s.,*
(v) $\int_0^t E|\dot{x}(s)|^2\, dG(s) < \infty$,
(vi) $\dot{x}(t)$ *solves, a.s., Eq.* (2.27), *and*
(vii) $E\{\sup\limits_{s \leq t \leq u} |x(t)|^2\} < \infty$, $E\{\sup\limits_{s \leq t \leq u} |\dot{x}(t)|^2\} < \infty$.

*Proof Existence* Define, successively, the processes $x^{(n)}(t, \omega)$, $n = 0, 1, 2, \ldots$, as

$$x^{(0)}(t, \omega) = \xi_0(\omega) + \eta_0(\omega) t, \quad n = 0,$$

$$x^{(n+1)}(t, \omega) = \xi_0(\omega) + \int_0^t \left\{ \eta_0(\omega) + \int_0^s m(u, x^{(n)}(u, \omega), \dot{x}^{(n)}(u, \omega)) \, du \right.$$

$$\left. + \int_0^s \sigma(u, x^{(n)}(u, \omega), \dot{x}^{(n)}(u, \omega)) \, d\beta(u) \right\} ds, \quad (2.29)$$

for $n \geq 1$. These processes $x^{(n)}(t, \omega)$, $n \geq 1$, are well-defined, as seen above, and the derived processes $\dot{x}^{(n)}(t, \omega)$ exist and satisfy Eq. (2.27). Next define

$$\Delta_{n+1} x(t) = x^{(n+1)}(t) - x^{(n)}(t)$$

$$\Delta_{n+1} \dot{x}(t) = \dot{x}^{(n+1)}(t) - \dot{x}^{(n)}(t)$$

$$\Delta_{n+1} m(t) = m(t, x^{(n+1)}(t), \dot{x}^{(n+1)}(t)) - m(t, x^{(n)}(t), \dot{x}^{(n)}(t))$$

$$\Delta_{n+1} \sigma(t) = \sigma(t, x^{(n+1)}(t), \dot{x}^{(n+1)}(t)) - \sigma(t, x^{(n)}(t), \dot{x}^{(n)}(t))$$

$$\Delta G(T_0) = G(T_0) - G(0). \quad (2.30)$$

Then,

$$E |\Delta_{n+1} \dot{x}(t)|^2 \leq 2E \left\{ \left| \int_0^t \Delta_n m(s) \, ds \right|^2 + \left| \int_0^t \Delta_n \sigma(s) \, d\beta(s) \right|^2 \right\}. \quad (2.31)$$

The estimations of the two expectations in (2.31) are analogous; so, we shall estimate the second expectation.

Noting that $x^{(n)}(0) = \xi_0$ and $x^{(n)}(t) = \xi_0 + \int_0^t \dot{x}^{(n)}(s) \, ds$, from stochastic integral isometry and Lipschitz continuity we get

$$E \left| \int_0^t \Delta_n \sigma(s) \, d\beta(s) \right|^2$$

$$= \int_0^t E |\Delta_n \sigma(s)|^2 \, dF(s)$$

$$\leq \int_0^t E |\Delta_n \sigma(s)|^2 \, dG(s)$$

$$\leq \int_0^t \{ K^2 [E |\Delta_n x(s)|^2 + E |\Delta_n \dot{x}(s)|^2] \} \, dG(s)$$

$$\leq T_0 K^2 \int_0^t \int_0^s E |\Delta_n \dot{x}(u)|^2 \, du \, dG(s) + K^2 \int_0^t E |\Delta_n \dot{x}(s)|^2 \, dG(s)$$

$$\leq C_0 \int_0^t E |\Delta_n \dot{x}(s)|^2 \, dG(s), \quad (2.32)$$

where $C_0 \geq (T_0 \Delta G(T_0)+1)K^2$ and $0 \leq t \leq T_0$. Using (2.32) and a similar estimate for $E|\int_0^t \Delta_n m(s)\,ds|^2$ in (2.31), we get

$$E|\Delta_{n+1}\dot{x}(t)|^2 \leq C_1 \int_0^t E|\Delta_n \dot{x}(s)|^2\,dG(s), \tag{2.33}$$

for $0 \leq t \leq T_0$. From the hypothesis, the martingale property of the stochastic integral $J(t) = \int_0^t \sigma\,d\beta$ and Doob's submartingale inequality applied to $J^2$, one can easily see that

$$\sup_{0 \leq s \leq T_0} E|\Delta_1 \dot{x}(s)|^2 \leq 2E\{\sup_{0 \leq s \leq T_0}|\dot{x}^{(1)}(s)|^2 + \sup_{0 \leq s \leq T_0}|\dot{x}^{(0)}(s)|^2\} < \infty,$$

and hence,

$$E|\Delta_{n+1}\dot{x}(t)|^2 \leq C^n/n!, \quad 0 \leq t \leq T_0, \tag{2.34}$$

where $C$ is a constant depending only on $T_0$. We claim that the sequence $\{\dot{x}^{(n)}(t), n \geq 1\}$ converges uniformly on $[0, T_0]$, a.s.

Define $I_n(t) = |\int_0^t \Delta_n m(s)\,ds|$ and $J_n(t) = |\int_0^t \Delta_n \sigma(s)\,d\beta(s)|$. Using Lipschitz continuity, Chebyshev's inequality, the idea used to derive (2.32), and inequality (2.34), one can see that

$$P\{\sup_{0 \leq t \leq T_0} I_n(t) \geq 2^{-n}\} \leq C_1 \frac{4^n C^{n-1}}{(n-1)!}. \tag{2.35}$$

Similarly, using Doob's inequality to $\{J_n^2(t), t \geq 0\}$ and (2.32),

$$P\{\sup_{0 \leq t \leq T_0} J_n(t) \geq 2^{-n}\} \leq 4^{-n} E[J_n^2(T_0)]$$

$$\leq 4^n C_0 \int_0^{T_0} E|\Delta_n \dot{x}(s)|^2\,dG(s)$$

$$\leq C_0 \frac{4^n C^{n-1}}{(n-1)!}. \tag{2.36}$$

From (2.35), (2.36) and Borel–Cantelli lemma it follows, for large $n$, that

$$\sup_{t \leq 0 \leq T_0} |\Delta_{n+1}\dot{x}(t)| < 2^{-(n-1)}, \quad \text{a.s.}$$

Thus, for large $m, n, m > n$,

$$\sup_{0 \leq t \leq T_0} |\dot{x}^{(m)}(t) - \dot{x}^{(n)}(t)| \leq 2\sum_{k \geq n} 2^{-k}, \quad \text{a.s.,} \tag{2.37}$$

and hence, for each $T_0$, $\dot{x}^{(n)} \to x^*$ uniformly on $[0, T_0] \times (\Omega \setminus N_{T_0})$ where $P(N_{T_0}) = 0$.

Take $x^*(t, \omega) = 0$ wherever it is undefined. Clearly, $x^*(\cdot, \cdot)$ is product measurable, and

$$\int_0^t \dot{x}^{(n)}(s)\,ds \to \int_0^t x^*(s)\,ds, \tag{2.38}$$

uniformly on $[0, T_0]$, as $n \to \infty$, with probability 1. Next define a process $X = \{x(t), t \geq 0\}$ by

$$x(t, \omega) = \xi_0(\omega) + \int_0^t x^*(s, \omega) \, ds. \tag{2.39}$$

Note that $x^{(n)}(t) = \xi_0 + \int_0^t \dot{x}^{(n)}(s) \, ds \to \xi_0 + \int_0^t x^*(s) \, ds \equiv x(t)$, uniformly on $[0, T_0]$, a.s., and that $x(t, \omega)$ is a unique (up to equivalence), absolutely continuous process. Now we claim that $\dot{x}(t, \omega)$ exists for $x(t, \omega)$ and is equivalent to $x^*(t, \omega)$.

To prove this claim we will appeal to Theorem 1.1, and toward this we have to show that $x(t, \omega)$ and $x^*(t, \omega)$ satisfy the hypothesis of that theorem. First, we observe that $x^*(t)$ is a continuous process. Next, we shall show that there is a constant $C_{T_0}$ such that

$$E|x^*(t)|^2 \leq C_{T_0}, \quad 0 \leq t \leq T_0, \quad T_0 > 0. \tag{2.40}$$

To see this, note that

$$E|\dot{x}^{(m)}(t) - \dot{x}^{(n)}(t)|^2$$

$$\leq E\left|\sum_{k=n}^{m-1} \Delta_{k+1}\dot{x}(t)\right|^2 = E\left|\sum_{k=n}^{m-1} 2^{-k} 2^k \Delta_{k+1}\dot{x}(t)\right|^2$$

$$\leq \sum_{k=n}^{m-1} 4^{-k} \sum_{k=n}^{m-1} 4^k E|\Delta_{k+1}\dot{x}(t)|^2$$

$$\leq 4^{-n} \sum_{k=n}^{m-1} (4C)^k/k! \leq 4^{-n} e^{4C} \to 0, \quad \text{as } m > n \to \infty$$

This and inequality (2.37) imply, as $m \to \infty$, that $E|x^*(t) - \dot{x}^{(n)}(t)|^2 \leq 4^{-n} e^{4C}$, $0 \leq t \leq T_0$, and hence

$$\dot{x}^{(n)}(t) \xrightarrow{L_2} x^*(t)$$

uniformly in $[0, T_0]$. Now, the inequality (2.40) follows upon observing that

$$E|x^*(t)|^2 \leq 2E|x^*(t) - \dot{x}^{(n)}(t)|^2 + 2E|\dot{x}^{(n)}(s)|^2$$

$$\leq 2[4^{-n} e^{4C} + E\{\sup_{0 \leq t \leq T_0} |\dot{x}^{(n)}(t)|^2\}].$$

Finally, it is easy to see that

$$\lim_{h \downarrow 0} E\left|\frac{x(t+h) - x(t)}{h} - x^*(t)\right| = 0, \quad t \geq 0. \tag{2.41}$$

Hence $\dot{x} \equiv x^*$ a.s.

To complete the proof of the existence, it remains to prove that $x$ and

$\dot{x}$ solve Eqs. (2.28) and (2.27), respectively. Clearly,

$$\sup_{0 \leq t \leq T_0} \left| \int_0^t [m(s, x^{(n)}(s), \dot{x}^{(n)}(s)) - m(s, x(s), \dot{x}(s))] \, ds \right|^2$$

$$\leq \Delta G(T_0) K \sup_{0 \leq t \leq T_0} |\dot{x}^{(n)}(s) - \dot{x}(s)|^2 \to 0, \quad \text{as } n \to \infty, \text{ a.s.}$$

From (2.40) and $x^* = \dot{x}$ it follows that $\int_0^t E|\dot{x}(s)|^2 \, dG(s) < \infty$, $t \geq 0$, and hence that the stochastic integral $\int_0^t \sigma(s, x(s), \dot{x}(s)) \, d\beta(s)$ exists. Now,

$$P \left\{ \sup_{0 \leq t \leq T_0} \left| \int_0^t [\sigma(s, x(s), \dot{x}(s)) - \sigma(s, x^{(n)}(s), \dot{x}^{(n)}(s))] \, d\beta(s) \right| \geq \frac{1}{n} \right\}$$

$$\leq n^2 E \left\{ \int_0^{T_0} [\sigma(t, x(t), \dot{x}(t)) - \sigma(t, x^{(n)}(t), \dot{x}^{(n)}(t))] \, d\beta(t) \right\}^2$$

$$= n^2 \int_0^{T_0} E|\sigma(t, x(t), \dot{x}(t)) - \sigma(t, x^{(n)}(t), \dot{x}^{(n)}(t))|^2 \, dF(t)$$

$$\leq n^2 K^2 \int_0^{T_0} E\{|x(t) - x^{(n)}(t)|^2 + |\dot{x}(t) - \dot{x}^{(n)}(t)|^2\} \, dF(t)$$

$$\leq n^2 K^2 T_0 \int_0^{T_0} \int_0^t E|\dot{x}(s) - \dot{x}^{(n)}(s)|^2 \, ds \, dG(t)$$

$$+ n^2 K^2 \int_0^{T_0} E|\dot{x}(t) - \dot{x}^{(n)}(t)|^2 \, dG(t)$$

$$\leq n^2 K^2 (T_0 \Delta G(T_0) + 1) \int_0^{T_0} E|\dot{x}(t) - \dot{x}^{(n)}(t)|^2 \, dG(t)$$

$$\leq n^2 K^2 (T_0 \Delta G(T_0) + 1) 4^{-n} e^{4C} \Delta G(T_0),$$

where we have used Doob's inequality, stochastic integral isometry, and the hypotheses. Hence by Borel–Cantelli lemma we have that

$$\sup_{0 \leq t \leq T_0} \left| \int_0^t [\sigma(t, x(t), \dot{x}(t)) - \sigma(t, x^{(n)}(t), \dot{x}^{(n)}(t))] \, d\beta(u) \right|^2 < \frac{1}{n},$$

a.s., for large $n$.

Now let $n \to \infty$ in (2.29) and the corresponding equation for $\dot{x}^{(n+1)}(t)$ to obtain (2.28) and (2.27), respectively. Hence the existence.

*Uniqueness* Let $x(t, \omega)$ and $\tilde{x}(t, \omega)$ be any two solutions of Eq. (2.28) as guaranteed by the existence part of the theorem. Then, the corresponding $\dot{x}(t, \omega)$ and $\dot{\tilde{x}}(t, \omega)$ satisfy Eq. (2.27). Now define

$$\Delta \dot{x}(t) = \dot{x}(t) - \dot{\tilde{x}}(t), \quad \Delta m(t) = m(t, x(t), \dot{x}(t)) - m(t, \tilde{x}(t), \dot{\tilde{x}}(t)),$$

and

$$\Delta \sigma(t) = \sigma(t, x(t), \dot{x}(t)) - \sigma(t, \tilde{x}(t), \dot{\tilde{x}}(t)).$$

Then,
$$\Delta \dot{x}(t) = \int_0^t \Delta m(s)\,ds + \int_0^t \Delta \sigma(s)\,d\beta(s),$$
and
$$E|\Delta \dot{x}(t)|^2 \leq \alpha \int_0^t E|\Delta \dot{x}(s)|^2\,dG(s), \qquad 0 \leq t \leq T_0,$$
where $\alpha > 0$, depending only on $T_0$. Define
$$\phi(t) = E|\Delta \dot{x}(t)|^2.$$
Then, $\phi(t) \geq 0$, and
$$\phi(t) \leq 2\left[\sup_{0 \leq t \leq T_0} E|\dot{x}(t)|^2 + \sup_{0 \leq t \leq T_0} E|\dot{x}^\sim(t)|^2\right]$$
$$\leq 2E\{\sup_{0 \leq t \leq T_0} |\dot{x}(t)|^2\} + 2E\{\sup_{0 \leq t \leq T_0} |\dot{x}^\sim(t)|^2\}$$
$$\leq M < \infty.$$

Hence, by iteration,
$$0 \leq \phi(t) \leq \alpha \int_0^t \phi(s)\,dG(s) \leq \alpha^{n+1} M \frac{[G(t)-G(0)]^n}{n!}.$$
Let $n \to \infty$. Then, since $G(t) \uparrow G(T_0) < \infty$,
$$E|\Delta \dot{x}(t)|^2 = 0, \qquad 0 \leq t \leq T_0,$$
and $\Delta \dot{x}(t) = 0$, a.s., for each $t \geq 0$. Hence,
$$P[\dot{x}(t) = \dot{x}^\sim(t), \ t \geq 0] = 1,$$
and consequently,
$$x(t,\omega) = \xi_0(\omega) + \int_0^t \dot{x}(s,\omega)\,ds = \xi_0(\omega) + \int_0^t \dot{x}^\sim(s,\omega)\,ds = x^\sim(t,\omega),$$
for all $t \geq 0$, with probability 1. This completes the proof.

## III. Some Stochastic Properties of Solution Processes

The last section illustrates several methods to establish the existence and uniqueness of random solutions of different random integrodifferential equations which govern certain physical or biological phenomena. Under relevant hypotheses we obtained random solutions (sample path or $L_2$-solutions) that are suitable stochastic processes representing the appropriate phenomena. As in the deterministic case, it becomes necessary to (probabilistically) analyze the solution process. Many of the problems in such an analysis are challenging and difficult and still await solutions. In this section we present a few representative properties of solutions of random integrodifferential equations of Volterra type and Itô type (cf. Kannan [27, 28, 30, 31], Borchers [5], Goldstein [19] and, Kannan and Bharucha-Reid [33]). A reasonable amount of work on these two types of equations has been done, but even more remains to be done.

## A. *Volterra-Type Random Integrodifferential Equation*

In this section we consider the initial-value problem

$$\dot{x}(t,\omega) = \xi(t,\omega) + k(t)x(t,\omega) + \int_0^t K(t,s)x(s,\omega)\,ds, \qquad x(0,\omega) = x_0(\omega) \tag{3.1}$$

and the corresponding mild $L_2$-solution process

$$x(t,\omega) = \rho(t,0)x_0(\omega) + \int_0^t \rho(t,s)\xi(s,\omega)\,ds. \tag{3.2}$$

Throughout this section we assume that $\rho_t(t,s) = \partial \rho(t,s)/\partial t$ exists boundedly, but one will observe that this condition is not needed in establishing some of the results.

Several pieces of important information about the physical systems governed by the random equations are contained in the moments of the solution process. It is very useful to know the asymptotic behavior of the variance erned by the random equations are contained in the moments of the solution process. It is very useful to know the asymptotic behavior of the variance as $t \uparrow \infty$. So, first we consider the moments of the solution. Towards this, we shall derive an integro-partial differential equation governing the moments of $x(t,\omega)$, where it is implicitly assumed that we treat only the bounded solutions of Eq. (3.1). Let $f: R_+ \times R \to R$ be a function $f(t,x)$ with bounded continuous derivatives $f_x, f_{xx}$, etc. Using Taylor expansion up to the second order, we get $\Delta f = f_x \Delta x + \tfrac{1}{2} f_{xx}(\Delta x)^2 + o(\Delta x)^2$. But,

$$\Delta x = \rho_t(t,0)x_0\Delta t + \xi(t)\Delta t + \Delta t \int_0^t \rho_t(t,s)\xi(s)\,ds + o(\Delta t)^2$$

and $(\Delta x)^2$ consists of terms involving the factor $(\Delta t)^2$. Noting that

$$E[\Delta x \mid x(t)] = E[x_0 \mid x(t)]\rho_t(t,0)\Delta t + E[\xi(t) \mid x(t)]\Delta t$$
$$+ \Delta t \int_0^t \rho_t(t,s)E[\xi(s) \mid x(t)]\,ds + o(\Delta t),$$

we see that

$$E[\Delta f] = E[E[\Delta f \mid x]]$$
$$= E[f_x \Delta x] + \tfrac{1}{2}E[f_{xx}\{(\cdots)(\Delta t)^2\}] + o(\Delta t).$$

Dividing this expression throughout by $\Delta t$, taking the limit as $\Delta t \to 0$, and permuting the expectation and differentiation operations we get

$$\frac{d}{dt} E[f(t, x(t, \omega))] = \rho_t(t, 0) E[x_0(\omega) f_x(t, \omega)] + E[\xi(t, \omega) f_x(t, x(t, \omega))]$$

$$+ \int_0^t \rho_t(t, s) E[\xi(s, \omega) f_x(t, x(t, \omega))] \, ds. \quad (3.3)$$

For proper choice of $f(t, x)$ on suitable domain we get the integrodifferential equations governing the moments of bounded solution $x(t, \omega)$ of Eq. (3.1).

In the case of the classical Langevin equation, it is well known that, if the infinitesimal variability of the particle motion is always above a positive level then, the variance $\mathcal{D}^2[x(t, \omega)]$ of the solution $x(t, \omega)$ of the Langevin equation satisfies the following property.

$$\mathcal{D}^2[x(t, \omega)] \to \infty, \quad \text{as} \quad t \to \infty. \quad (3.4)$$

Let us verify this property for the generalized Langevin equation

$$x(t) = x_0 - \int_{t_0}^t \int_{t_0}^s \gamma(s-u) x(u) \, du \, ds + \int_{t_0}^t \sigma[s, x(s)] \, dM(s), \quad (3.5)$$

(See Section II.B). Let $\sigma(t, x(t)) \geq \theta > 0$, and $F(t)$ be monotonically increasing unboundedly as $t \uparrow \infty$. Let also $x_0$ be independent of the martingale $M(t)$. Then,

$$\mathcal{D}^2[x(t, \omega)] \geq \mathcal{D}^2[x(t, \omega) - x_0(\omega)]$$

$$= \mathcal{D}^2 \left[ \int_{t_0}^t \int_{t_0}^s \gamma(s-u) x(u) \, du \, ds \right] + \int_{t_0}^t E[\sigma(s, x(s, \omega))]^2 \, dF(s)$$

$$\geq \int_{t_0}^t E[\sigma(s, x(s, \omega))]^2 \, dF(s) \geq \theta^2 [F(t) - F(t_0)]$$

$$\to \infty, \quad \text{as} \quad \to \infty.$$

This motivates us to find conditions such that relation (3.4) is obtained for the solution $x(t, \omega)$ of Eq. (3.1). For another illustration, we consider the mild mean square solution $x(t, \omega)$ of Eq. (2.14). The process $x(t, \omega)$ is given by (2.15), and here we assume that $\xi(t, \omega)$ is a white noise uncorrelated to $\zeta(\cdot, \omega), \gamma(\cdot)$, is continuous such that $\gamma(\cdot) \geq a > 0$, and that $\rho(t) \geq \alpha$. Define $G(s+t_0) = \int_0^{t_0} \gamma(s+t_0-u) \xi(u) \, du$. Then,

$$E[x(t+t_0, \omega; t_0, \xi)]^2$$

$$= \rho^2(t) E[\xi(t_0)]^2 + 2\rho(t) \int_0^t \int_0^{t_0} \rho(t-s) \gamma(s+t_0-u) \delta(t_0-u) \, du \, ds$$

$$+ \int_0^t \int_0^t \rho(t-s)\rho(t-u)\, \Gamma_\zeta(s+t_0, u+t_0)\, ds\, du$$

$$+ \int_0^t \int_0^t \rho(t-s)\rho(t-u)\, E\{G(s+t_0)G(u+t_0)\}\, ds\, du$$

$$\geq 2\alpha^2 a \int_0^t \int_0^{t_0} \delta(t_0-u)\, du\, ds + \alpha^2 t^2$$

$$\geq \alpha^2 t^2 \uparrow \infty, \quad \text{as} \quad t \uparrow \infty,$$

where $\Gamma_\zeta$ is the covariance function of $\zeta$. Similarly it is easy to see the following proposition regarding the mean square solution $x(t, \omega)$ of Eq. (3.1).

**Proposition 3.1** *Let $x_0$ and $\xi(t, \omega)$ be centered and uncorrelated. If $\rho(t,s) \geq \alpha$ and $\Gamma_\xi(s,u) \geq a > 0$, for $0 \leq s, u \leq t$, then $\mathscr{D}^2[x(t)] \uparrow \infty$, as $t \uparrow \infty$.*

Let $x(t, \omega)$ be the $L_2$ solution of Eq. (3.1). The stochastic properties of $x(t, \omega)$ are described in terms of the $n$-dimensional, $(n \geq 1)$, distributions of $x(\cdot, \cdot)$. In particular, the one-dimensional distributions of $x(t, \omega)$ are of great interest. Since our equation is of first order, and will have one degree of randomness, it is important to find the one-dimensional distributions of the solution process $x(t, \omega)$. We will find these one-dimensional distributions in the course of the proof of the following theorem of Liouville–Gibbs type for dynamical systems. This kind of theorem finds applications in hydrodynamics and kinetic theory.

**Theorem 3.2** *Let $x(t, \omega)$ be the mean square solution of the random integrodifferential equation (3.1), where it is assumed that there is a second-order separable process $\eta(t, \omega)$ with independent Gaussian increments such that $x_0$ is independent of these increments and $\xi(t, \omega)\, dt = d\eta(t, \omega)$. Let $x_0 \in N(0, 1)$. Denote by $f(t, \theta)$ the density function of $x(\omega)$ and by $g(t; \theta)$ the conditional expectation $E[\dot{x}(t) | x(t) = \theta]$. Assume that $f(t, \theta)\, g(t, \theta)$ is continuously differentiable. Then, for all $t \geq 0$, and $\theta \in R^1$,*

$$\frac{\partial}{\partial \theta}[f(t,\theta)g(t,\theta)] - M(t)\frac{\partial}{\partial \theta}f(t,\theta) + \tfrac{1}{2}S(t)\frac{\partial^2}{\partial \theta^2}f(t,\theta) = 0, \quad (3.6)$$

*where*

$$M(t) = \int_0^t \frac{\partial}{\partial t}\rho(t,s)\, d\mu(s), \quad (3.7)$$

$$S(t) = \frac{\partial}{\partial t}\rho^2(t,0) + \int_0^t \frac{\partial}{\partial t}\rho^2(t,s)\, dv(s) \geq 0, \quad (3.8)$$

*$\mu$ and $v$ are real functions vanishing with $t$, and $v(t)$ is monotonically increasing.*

*Proof* By Theorem 2.2, the solution process $x(t,\omega)$ is given by (3.2). Let, for the moment, $\eta(t,\omega)$ be an arbitrary second-order process with independent increments (without the Gaussian qualification). Let $\psi(t,u)$ denote the characteristic function of $x(\omega)$. Then, from the $L_2$-convergence and Lévy–Cramér theorem,

$$\psi(t,u) = E\left\{\exp\left[iu\left\{\rho(t,0)x_0 + \int_0^t \rho(t,s)\,d\eta(s)\right\}\right]\right\}$$

$$= E\left\{\exp[iu\rho(t,0)x_0]\lim_n \exp\left[iu\sum_{k=0}^{n-1}\rho(t,s_k)(\eta(s_{k+1})-\eta(s_k))\right]\right\} \tag{3.9}$$

where $0 = s_0 < s_1 < \cdots < s_n = t$ is a partition of $[0,t]$ such that the mesh of the partition goes to zero as $n \to \infty$. Let $\psi^*(t,u) = \ln\psi(t,u)$. Then, from (3.9),

$$\psi^*(t,u) = -\frac{u^2}{2}\rho^2(t,0) + iu\int_0^t \rho(t,s)\,d\mu(s) - \frac{u^2}{2}\int_0^t \rho^2(t,s)\,d\nu(s)$$

$$+ \int_0^t \int_{R^1\setminus\{0\}} \left[\exp\{iu\rho(t,s)v\} - 1 - \frac{iu\rho(t,s)v}{1+v^2}\right]\lambda(ds,dv) \tag{3.10}$$

where $\mu(t)$, $\nu(t)$, and $\lambda(t,v)$ are real functions vanishing with $t$ and $\nu(t)$ is monotonically increasing. Now, let $\eta(t)$ have Gaussian increments. Noting that $(\partial/\partial t)\psi(t,u) = \psi(t,u)(\partial/\partial t)\psi^*(t,u)$, we get

$$\frac{\partial}{\partial t}f(t,\theta) = \tfrac{1}{2}S(t)\frac{\partial^2}{\partial\theta^2}f(t,\theta) - M(t)\frac{\partial}{\partial\theta}f(t,\theta), \tag{3.11}$$

where $M$ and $S$ are given by (3.7) and (3.8), respectively. [So, the Eq. (3.11) governs the density function of $x_t(\omega)$.] Thus, it remains to show that

$$\frac{\partial}{\partial\theta}[f(t,\theta)g(t,\theta)] + \frac{\partial}{\partial t}f(t,\theta) = 0. \tag{3.12}$$

Let $\alpha(\theta)$ be a nonnegative continuous function with compact support such that $d\alpha/d\theta$ and $d^2\alpha/d\theta^2$ exist continuously. Then, from the Borel–Lebesgue measurability of $f(t,\theta)$ in $R^1$, the integral $\int_a^b \alpha(\theta)f(t,\theta)\,d\theta$ exists and is finite, where $[a,b]$ is a compact interval containing the support of $\alpha(\cdot)$. Now,

$$\int_a^b \alpha(\theta)\frac{\partial f}{\partial t}\,d\theta = \frac{\partial}{\partial t}\int_a^b \alpha(\theta)f(t,\theta)\,d\theta$$

$$= \lim_{h\to 0}\frac{1}{h}\{E[\alpha(x(t+h,\omega)) - \alpha(x(t,\omega))]\}$$

$$= \lim_{h \to 0} \frac{1}{h} \int_{R^2} [\alpha(\theta') - \alpha(\theta)] f(t, \theta; t+h, \theta') \, d\theta \, d\theta'$$

$$= \lim_{h \to 0} \frac{1}{h} \int_{R^2} \left\{ (\theta' - \theta) \frac{d\alpha(\theta)}{d\theta} \right.$$

$$\left. + \frac{(\theta' - \theta)^2}{2} \frac{d^2\alpha}{d\theta^2} \bigg|_{\theta = \sigma} \right\} f(t, \theta; t+h, \theta') \, d\theta \, d\theta',$$

where $f(t, \theta; t+h, \theta')$ is the joint density of $x_t(\omega)$ and $x_{t+h}(\omega)$, and $\sigma = \theta + \tau(\theta' - \theta)$, $0 < \tau < 1$. Noting that $\alpha''(\theta)$ is bounded, it is easy to see that

$$\left| \frac{1}{h} \int_{R^2} \frac{(\theta' - \theta)^2}{2} \frac{d^2\alpha(\theta)}{d\theta^2} f(t, \theta; t+h, \theta') \, d\theta \, d\theta' \right| \to 0.$$

as $h \to 0$. Noting that $|\alpha'(\theta)| \leq C < \infty$,

$$\left| E\left\{ \alpha'(x(t)) \left[ \frac{x(t+h) - x(t)}{h} - \dot{x}(t) \right] \right\} \right|$$

$$\leq C \left\{ E\left[ \frac{x(t+h) - x(t)}{h} - \dot{x}(t) \right]^2 \right\}^{\frac{1}{2}} \to 0,$$

as $h \to 0$, and consequently

$$\lim_{h \to 0} E\{\alpha'(x(t, \omega))[x(t+h, \omega) - x(t, \omega)]/h\}$$

$$= E\{\alpha'(x(t, \omega)\dot{x}(t, \omega))\} = E\{\alpha'(x(t, \omega)) E[\dot{x}(t, \omega) | x(t, \omega)]\}$$

$$= \int_{R^1} \alpha'(\theta) g(t, \theta) f(t, \theta) \, d\theta = -\int_a^b \alpha(\theta) \frac{\partial}{\partial \theta} [g(t, \theta) f(t, \theta)].$$

Hence,

$$\int_a^b \alpha(\theta) \left\{ \frac{\partial f}{\partial t} + \frac{\partial (fg)}{\partial \theta} \right\} d\theta = 0,$$

and this gives (3.12) due to our choice of $\alpha(\cdot)$. This completes the proof.

For the remainder of the subsection we shall study some special properties of certain generalized Langevin equations. First we establish a comparison theorem for the solution processes of the nonlinear generalized Langevin equation

$$dx(t) = -\left\{ \int_{t_0}^t \gamma(t-s) x(s) \, ds \right\} dt + \sigma(t, x(t)) \, dM(t). \tag{3.13}$$

For each $i = 1, 2$, let $\{x_i(t, \omega), t \in [t_0, T]\}$ be the continuous solution process of the Eq. (3.13) corresponding to the initial value $x_{i0}(\omega)$. As in Section

II.B we will work with the integrated version of (3.13)

$$x_i(t) = x_{i0} - \int_{t_0}^{t} (\gamma_i^* x_i)(s)\, ds + \int_{t_0}^{t} \sigma(s, x_i(s))\, dM(s), \qquad (3.14)$$

with the same $\sigma(\cdot, \cdot)$ and $M(\cdot)$. We shall make the following assumptions.

($C_1$) $x_1(t)$ and $x_2(t)$ are the continuous solutions of the Eqs. (3.14) under the hypotheses of Theorem 2.7 (see the remark following this theorem);

($C_2$) $\sigma(t, x)$ is jointly continuous on $[t_0, T] \times R^1$, $\sigma(t, x) > 0$, and for every $c > 0$, there exist an $\alpha > \frac{1}{2}$ and $K > 0$ such that for $|x| \leq c$, $|y| \leq c$, we have $|\sigma(t, x) - \sigma(t, y)| \leq K|x - y|^\alpha$; and

($C_3$) $\gamma_1 * x_1 < \gamma_2 * x_2$ for every $t \in [t_0, T]$.

**Theorem 3.3** *Let conditions* ($C_1$)–($C_3$) *hold. If* $x_1(\tau) \geq x_2(\tau)$ *a.s., for some stopping time $\tau$ in $[t_0, T]$, then there is a $\tau_1 > \tau$ a.s., such that for $s \in (\tau, \tau_1)$,*

$$x_1(s) > x_2(s), \qquad \text{a.s.}$$

*Proof* For $s \in [t_0, T]$, define $\psi(s)$ as follows: (i) $\psi(s) = 1$ if and only if $\tau \leq s$, (ii)

$$\inf_{\tau \leq u \leq s} [(\gamma_2 * x_2)(u) - (\gamma_1 * x_1)(u)] > \tfrac{1}{2}[(\gamma_2 * x_2)(\tau) - (\gamma_1 * x_1)(\tau)],$$

and (ii) $\psi(s) = 0$ if and only if $s < \tau$ and, for $s > \tau$,

$$\inf_{\tau \leq u \leq s} [(\gamma_2 * x_2)(u) - (\gamma_1 * x_1)(u)] \leq \tfrac{1}{2}[(\gamma_2 * x_2)(\tau) - (\gamma_1 * x_1)(\tau)].$$

Also, define for $k > 0$, $c > 0$ and $s \in [t_0, T]$

$$\psi_k^c(s) = I_{[0, c]}\{\sup_{t_0 \leq u \leq s} [|x_1(u)| + |x_2(u)|]\} I_{[\tau, \tau + k]} \psi(s).$$

We claim that

$$\lim_{k \to 0} k^{-1} \int_{t_0}^{T} \psi_k^c(s)[\sigma(s, x_1(s)) - \sigma(s, x_2(s))]\, dM(s) = 0, \qquad \text{a.s.} \quad (3.15)$$

Suppose that we have established this claim. From the definition of $\psi_k^c$ we get

$$\psi_k^c(\tau + k) \int_{t_0}^{T} \psi_k^c(u)[(\gamma_2 * x_2)(u) - (\gamma_1 * x_1)(u)]\, du$$

$$\geq \psi_k^c(\tau + k) \int_{t_0}^{T} \psi_k^c(u)\, 2^{-1}[(\gamma_2 * x_2)(\tau) - (\gamma_1 * x_1)(\tau)]\, du$$

$$= 2^{-1} k \psi_k^c(\tau + k)[(\gamma_2 * x_2)(\tau) - (\gamma_1 * x_1)(\tau)]$$

and hence,

$$\psi_k^c(\tau + k)[x_1(\tau + k) - x_2(\tau + k)]$$

$$= \psi_k^c(\tau+k) \int_{t_0}^T \psi_k^c(u) [(\gamma_2 * x_2)(u) - (\gamma_1 * x_1)(u)] \, du$$

$$+ \psi_k^c(\tau+k) \int_{t_0}^T \psi_k^c(u) [\sigma(u, x_1(u)) - \sigma(u, x_2(u))] \, dM(u)$$

$$\geq k\psi_k^c(\tau+k) \Big\{ 2^{-1}[(\gamma_2 * x_2)(\tau) - (\gamma_1 * x_1)(\tau)]$$

$$+ k^{-1} \int_{t_0}^T \psi_k^c(u) [\sigma(u, x_1(u)) - \sigma(u, x_2(u))] \, dM(u) \Big\}.$$

From our claim (3.15), there exists an $h > 0$ such that, for $0 < k < h$,

$$k^{-1} \Big| \int_{t_0}^T \psi_k^c(u) [\sigma(u, x_1(u)) - \sigma(u, x_2(u))] \, dM(u) \Big|$$
$$< 4^{-1}[(\gamma_2 * x_2)(\tau) - (\gamma_1 * x_1)(\tau)],$$

and hence, for $0 < k < h$,

$$\psi_k^c(\tau+k)[x_1(\tau+k) - x_2(\tau+k)]$$
$$\geq 4^{-1} \psi_k^c(\tau+h)[(\gamma_2 * x_2)(\tau) - (\gamma_1 * x_1)(\tau)].$$

But for almost all $\omega$, there are positive constants $c$ and $k_0$ such that $\psi_k^c(\tau+k) = 1$, for $0 < k < k_0$. Hence, we get the conclusion of the theorem once we establish the claim (3.15). So, it remains to prove (3.15). Here we make an additional assumption that the increasing function $F(t)$ associated with the martingale $M(t)$ is continuously differentiable on $[t_0, T]$. One can avoid the above assumption by a random time change on the continuous martingale.

From the stochastic integral isometry and, repeated use of Hölder's inequality and Fubini's theorem we get the following estimate:

$$E\Big\{ \int_{t_0}^T \psi_k^c(s) [\sigma(s, x_1(s)) - \sigma(s, x_2(s))] \, dM(s) \Big\}^2$$

$$\leq \int_{t_0}^T E\{\psi_k^c(s) [\sigma(s, x_1(s)) - \sigma(s, x_2(s))]^2\} \, dF(s)$$

$$\leq K^2 \int_{t_0}^T E\{\psi_k^c(s) |x_1(s) - x_2(s)|^{2\alpha}\} \, dF(s)$$

$$\leq K^2 E\Big\{ \Big[\int_{t_0}^T \psi_k^c(s) \, dF(s)\Big]^{1-\alpha} \Big[\int_{t_0}^T \psi_k^c(s) |x_1(s) - x_2(s)|^2 \, dF(s)\Big]^\alpha \Big\}$$

$$\leq K^2 \Big\{\int_{t_0}^T E[\psi_k^c(s)] \, dF(s)\Big\}^{1-\alpha} \Big\{\int_{t_0}^T E[\psi_k^c(s) |x_1(s) - x_2(s)|^2] \, dF(s)\Big\}^\alpha$$

$$\leq K^2 \{E[F(\tau+k) - F(\tau)]\}^{1-\alpha} \Big\{\int_{t_0}^T E[\psi_k^c(s) |x_1(s) - x_2(s)|^2] \, dF(s)\Big\}^\alpha.$$

(3.16)

We note that $\psi_k^c(s) = 1$ if and only if

$$\sup_{t_0 \leq u \leq s} \{|x_1(u)| + |x_2(u)|\} \leq c, \quad s \in [t_0, T] \cap [\tau, \tau+k], \quad \text{and} \quad \psi(s) = 1.$$

For $u \leq s$, $\psi_k^c(s) = 1$ implies that $\psi_k^c(u) = I_{[\tau, s]}(u)$. Now

$$\psi_k^c(s)[x_1(s) - x_2(s)]$$

$$= \psi_k^c(s) \int_\tau^s \psi_k^c\{(\gamma_2 * x_2)(u) - (\gamma_1 * x_1)(u)\} \, du$$

$$+ \psi_k^c(s) \int_\tau^s \psi_k^c(u) \{\sigma(u, x_1(u)) - \sigma(u, x_2(u))\} \, dM(u). \quad (3.17)$$

Let us denote by $\Psi_k^c[\sigma_{12}(\cdot)]$ and $\Delta_k F(\tau)$ the following:

$$\Psi_k^c[\sigma_{12}(s)] = \psi_k^c(s)[\sigma(s, x_1(s)) - \sigma(s, x_2(s))],$$

$$\Delta_k F(\tau) = F(\tau+k) - F(\tau).$$

From (3.16), (3.17), and that there is a constant $C > 0$, due to the uniform continuity of $\gamma * x$ on $[t_0, T]$, such that $I_{[0, c]}(x)\{|\gamma_1 * x_1| + |\gamma_2 * x_2|\} \leq C$, we have

$$E\left\{\int_{t_0}^T \Psi_k^c[\sigma_{12}(s)] \, dM(s)\right\}^2$$

$$\leq K^2 \{E\Delta_k F(\tau)\}^{1-\alpha}$$

$$\times \left\{\int_{t_0}^T E\left[\psi_k^c(s)\left\{8C^2k^2 + 2\left(\int_{t_0}^s \Psi_k^c[\sigma_{12}(u)] \, dM(u)\right)^2\right\}\right] dF(s)\right\}^\alpha$$

$$\leq K^2 \{E\Delta_k F(\tau)\}^{1-\alpha}$$

$$\times \left\{8C^2k^2[E\Delta_k F(\tau)] + 2\int_{t_0}^T E\left[\psi_k^c(s)\left(\int_{t_0}^s \Psi_k^c[\sigma_{12}(u)]\right)^2\right] dF(s)\right\}^\alpha$$

$$\leq Uk^{2\alpha} E\Delta_k F(\tau) + V\{E\Delta_k F(\tau)\}^{1-\alpha}$$

$$\times \left\{\int_{t_0}^T E\left[\psi_k^c(s)\left(\int_{t_0}^s \Psi_k^c[\sigma_{12}(u)] \, dM(u)\right)^2\right] dF(s)\right\}^\alpha, \quad (3.18)$$

where $U$ and $V$ are certain constants. From Doob's inequality,

$$E\left\{\int_{t_0}^T \psi_k^c(s)\left[\int_{t_0}^s \Psi_k^c[\sigma_{12}(u)] \, dM(u)\right]^2 dF(s)\right\}$$

$$\leq 4E\Delta_k F(\tau) \int_{t_0}^T E\{\psi_k^c(u)[\sigma(u, x_1(u)) - \sigma(u, x_2(u))]^2\} \, dF(s). \quad (3.19)$$

Put $\Phi(k) = E\{\int_{t_0}^{T} \Psi_k^c[\sigma_{12}(u)]\, dM(u)\}^2$. Using (3.18), (3.19), and the stochastic integral isometry we obtain, for some constants $c_0$ and $c_1$, the inequality

$$\Phi(k) \leqslant c_0 k^{2\alpha}\{E\Delta_k F(\tau)\} + c_1 E\Delta_k F(\tau)[\Phi(k)]^\alpha.$$

Hence,

$$\Phi(k)[E\Delta_k F(\tau)]^{-1}k^{-2\alpha}$$
$$\leqslant c_0 + c_1 k^{-2\alpha + 2\alpha^2}[E\Delta_k F(\tau)]^\alpha\{\Phi(k)[E\Delta_k F(\tau)]^{-1}k^{-2\alpha}\}^\alpha. \qquad (3.20)$$

Here we claim that there are constants $D > 0$ and $\delta > 0$ such that

$$\Phi(k)[E\Delta_k F(\tau)]^{-1}k^{-2\alpha} \leqslant D, \quad \text{if } 0 < k \leqslant \delta. \qquad (3.21)$$

Suppose that the bound in (3.21) does not exist and that there is a sequence $\{k_i\}$ such that, as $i \to \infty$,

$$k_i \to \infty \quad \text{and} \quad \Phi(k_i)[E\Delta_{k_i} F(\tau)]^{-1}k_i^{-2\alpha} \to \infty. \qquad (3.22)$$

Let $A = [t_0, T] \times [-c, c]$ and $\frac{1}{2} < \beta < \alpha < 1$. If $(t, x)$ and $(t, y)$ are any two points in $A$, then by condition $(C_2)$ we have

$$|\sigma(t, x) - \sigma(t, y)| \leqslant K|x-y|^\alpha \leqslant K|x-y|^\beta, \quad \text{if } |x-y| \leqslant 1,$$

and

$$|\sigma(t, x) - \sigma(t, y)| \leqslant 2R \leqslant 2R|x-y|^\beta, \quad \text{if } |x-y| > 1,$$

for some $R > 0$. Let $N = \max\{K, 2R\}$. Then,

$$|\sigma(t, x) - \sigma(t, y)| \leqslant N|x-y|^\beta.$$

From (3.20) and (3.22) we have, since $\alpha < 1$,

$$\{\Phi(k)[E\Delta_{k_i} F(\tau)]^{-1}k_i^{-2\alpha}\}^{1-\alpha}$$
$$\leqslant c_0\{\Phi(k_i)[E\Delta_{k_i} F(\tau)]^{-1}k_i^{-2\alpha}\}^{-\alpha} + c_1[E\Delta_{k_i} F(\tau)k_i^{-1}]^\alpha k_i^{\alpha(2\alpha-1)}$$
$$\to \infty, \quad \text{as } i \to \infty. \qquad (3.23)$$

But $\alpha > \frac{1}{2}$. Hence, the first term on (3.23) goes to zero; and the second term goes to zero due to continuous differentiability of $F$. Thus, we get the contradiction $\infty \leqslant 0$. Thus, we get (3.21) and hence,

$$E\left\{\int_{t_0}^{T} \Psi_k^c[\sigma_{12}(s)]\, dM(s)\right\}^2 \leqslant DE[\Delta_k F(\tau)]k^{-2\alpha}.$$

Let

$$\eta(s) = I_{[0, c]}\{\sup_{t_0 \leqslant u \leqslant s} (|x_1(u)| + |x_2(u)|)\psi(s)[\sigma(u, x_1(u)) - \sigma(u, x_2(u))]\}$$

and

$$\zeta(t) = \int_{t_0}^{t} \eta(s)\, dM(s).$$

Noting that $\{z(k) = [\zeta(\tau+k) - \zeta(\tau)],\ k \geqslant 0\}$ is a martingale we get, from Doob's inequality, that for $0 < k_0 < \delta$,

$$E\left\{\sup_{0<k<k_0}\left[\int_{t_0}^{T}\Psi_k^c[\sigma_{12}(s)]\,dM(s)\right]^2\right\} \leqslant 4DE[\Delta_{k_0}F(\tau)]\,k^{-2\alpha}.$$

Consequently,

$$P\left\{\sup_{2^{-n-1}\leqslant k\leqslant 2^{-2}}\left|k^{-1}\int_{t_0}^{T}\Psi_k^c[\sigma_{12}(s)]\,dM(s)\right| > n^{-1}\right\}$$

$$\leqslant P\left\{\sup_{0\leqslant k\leqslant 2^{-2}}\left|\int_{t_0}^{T}\Psi_k^c[\sigma_{12}(s)]\,dM(s)\right| > n^{-1}2^{-n-1}\right\}$$

$$\leqslant n^2\,2^{2(n+1)}E\left\{\sup_{0\leqslant k\leqslant 2^{-2}}\left|\int_{t_0}^{T}\Psi_k^c[\sigma_{12}(s)]\,dM(s)\right|^2\right\}$$

$$\leqslant n^2\,2^{2n+4}DE[\Delta_{2^{-n}}F(\tau)]\,2^{-2\alpha n}$$

$$\leqslant 16D\,2^{-n(2\alpha-2)}n^2 E[\Delta_{2^{-n}}F(\tau)]$$

$$\leqslant 16DN_0\,n^2\,2^{-n(2\alpha-1)},$$

for some $N_0$ which obtains from the assumption on $F(\cdot)$. Since $\alpha > 2^{-1}$, $\sum_{n=1}^{\infty} n^2\,2^{-n(2\alpha-1)} < \infty$. Now, the Borel–Cantelli lemma yields our claim (3.15). This completes the proof.

**Theorem 3.4** Let conditions $(C_1)$–$(C_3)$ hold. If $P\{x_1(t_0) = x_2(t_0)\} = 1$, then $P\{x_1(t) \geqslant x_2(t)\} = 1$, for all $t \in [t_0, T]$.

*Proof* Let $\tau = \inf\{t \geqslant t_0 : x_1(t,\omega) \geqslant x_2(t,\omega)\}$. $\tau$ is a stopping time. From the continuity of the processes $x_i(t,\omega)$, we have $x_1(\tau(\omega),\omega) = x_2(\tau(\omega),\omega)$. By Theorem 3.3, there is a $k_0$ such that $x_1(t) > x_2(t)$ for $\tau < t < \tau + k_0$. So, if $\tau_1$ is the first zero of $[x_1(t,\omega) - x_2(t,\omega)]$, then there is another random time $\tau_2$ such that $x_1(t) > x_2(t)$ for $t \in (\tau_1, \tau_2)$. Continue this process. Arrange these zeros in an increasing sequence $\{\tau_i(\omega)\}$. If there is a maximal zero $\tau^*$, then Theorem 3.3 gives an $h > 0$ such that $x_1(t) = x_2(t)$ for $t \in (\tau^*, \tau^* + h)$. The sign of $(x_1(t) - x_2(t))$ remains the same in $(\tau^*, T)$ so that $x_1(t) > x_2(t)$ for $(\tau^*, T)$. Thus, for all $t \in [t_0, T]$, $x_1(t) > x_2(t)$ a.s. This proves the theorem.

Finally, consider the linear generalized Langevin equation

$$\dot{x}(t) = -\int_0^t \gamma(t-s)x(s)\,ds + \eta(t,\omega), \qquad x(0) = x_0. \tag{3.24}$$

Assume that both $x_0(\omega)$ and $\eta(t, \omega)$ are centered and of second order, and that $x_0$ and $\eta(\cdot)$ are independent. Let

$$x(t) = \rho(t)x_0 + \int_0^t \rho(t,s)\eta(s)\,ds, \tag{3.25}$$

where $\rho(t) = 1 - \int_0^t \rho(t-s)\{\int_0^s \gamma(u)\,du\}\,ds$, be the mild mean-square solution of the Eq. (3.24). We will present here some important physical properties of a particle whose motion is governed by the Eq. (3.24), (cf. Kannan [28]).

First let us recall (see Section I.B) that the random driving force on the particle and the frictional force arise out of random collisions between the particle and the media molecules. Because these forces arise out of a common source, it is natural to expect certain relations between them. These relations are generally known, in physics, as the fluctuation–dissipation theorems. These theorems are of fundamental importance in the statistical mechanics of irreversible processes; for instance, they are used in characterizing the fluctuations in the system, and to derive the admittance from thermal fluctuations. The Fourier-transformed version of these theorems were obtained by Kubo and Mori (cf. [35, 45]). But we shall derive the fluctuation–dissipation relation using the mild solution (3.25).

**Theorem 3.5** *Let $x(t, \omega)$ be the mild mean square solution of Eq. (3.24). Assume that $x_0$ and $\eta(\cdot)$ are independent, centered and are of second order. Also, let $\dot{\rho}(\cdot)$ exist continuously. Then,*

(1) *First fluctuation–dissipation theorem*:

$$\rho(t) = \Gamma_x(0,t)/\sigma^2; \tag{3.26}$$

(2) *second fluctuation–dissipation theorem*:

$$\Gamma_\eta(0,t) = \sigma^2 \gamma(t); \tag{3.27}$$

(3) *force correlation*:

$$\Gamma_{\dot{x}}(0,t) = \sigma^2\left\{\gamma(t) + \int_0^t \dot{\rho}(t-s)\gamma(s)\,ds\right\}, \tag{3.28}$$

*where $\sigma^2 = \mathrm{Var}(x_0)$ and $\Gamma_z(\cdot,\cdot)$ denotes the covariance function of a process $z(t, \omega)$.*

*Proof* (1) Clear.
(2) Since the system is in equilibrium, we have $E[\dot{x}(0)x(0)] = 0$, and $E[\dot{x}(0)x(t)] = -E[\dot{x}(t)x(0)]$. Now,

$$E[\eta(0)\eta(t)] = E\left\{\dot{x}(0)\left[\dot{x}(t) + \int_0^t \gamma(t-s)x(s)\,ds\right]\right\}$$

$$= E[\dot{x}(0)\dot{x}(t)] - \int_0^t \gamma(t-s) E[\dot{x}(s) x(0)] \, ds$$

$$= E[\dot{x}(0)\dot{x}(t)] - \sigma^2 \int_0^t \dot{\gamma}(t-s) \rho(s) \, ds$$

$$= E[\eta(0)\eta(t)] - \int_0^t \dot{\rho}(t-s) \Gamma_\eta(0,s) \, ds$$

$$+ \sigma^2 \int_0^t \gamma(t-s) \dot{\rho}(s) \, ds,$$

from which we obtain (3.27).

(3) Since $\Gamma_{\dot{x}}(0,t) = \Gamma_\eta(0,t) + \int_0^t \dot{\rho}(t-u) \Gamma_\eta(0,u) \, du$, we obtain (3.28) from (3.27).

Before closing this section we treat Eq. (3.24) as follows. Let us consider a free particle described by the deterministic integrodifferential equation

$$\dot{x}(t) = -\int_0^t \gamma(t-s) x(s) \, ds, \qquad x(0) = x_0. \tag{3.29}$$

On this free particle let a jump Markov process $\eta(t,\omega)$ act as a driving force. Then, Eq. (3.29) describes a particle motion under the action of a jump Markov process. Next, we point out some of the properties of the generalized Langevin equation in the said context. Here, we omit all the details, referring the reader to Kannan [27] for the methods to be adapted to this situation. To fix ideas, we shall take $\eta(t,\omega)$ as a continuous-time finite-state Markov chain with state space $S = \{\eta_1, \eta_2, \ldots, \eta_N\}$, (in [27], we considered a birth–death process). Denote the mild solution $x(t,\omega)$ by $x^*(t, x_0; \eta)$ to represent the particle motion driven by the action $\eta$. Let $\eta(0) = \eta_i \in S$. For this initial constant external force, the particle moves according to

$$x^*(t, x_0; \eta_i) = \rho(t) x_0 + \int_0^t \rho(t-s) \eta_i \, ds$$

until a new external force $\eta_j \in S$ is prescribed, where the prescription of the new action $\eta_j$ occurs randomly. That is, until a random time $\tau_1$, the particle moves according to $x^*(t, x_0; \eta_i)$, $t < \tau_1$, and at time $\tau_1$ the new action $\eta_j$ is prescribed as a new driving force. Now the particle evolves according to $x^*(t, x(\tau_1); \eta_j)$, $t \geq \tau_1$, until another $\eta_k$ is prescribed at a random time $\tau_2 > \tau_1$. The motion continues in this fashion. In other words, if $0 \equiv \tau_0 < \tau_1(\omega) < \tau_2(\omega) < \cdots$ form a sequence of random times, then between any two consecutive random times $\tau_n$ and $\tau_{n+1}$ the particle motion is described by $x^*(t, x(\tau_n); \eta_{\tau_n})$, $\tau_n \leq t < \tau_{n+1}$. The evolution of the particle motion is obtained by piecing together the bits $x^*(t, x(\tau_n); \tau_{\eta_n})$, $n \geq 0$.

Let $p_{ij}(t)$ be the transition probability of the (homogeneous) Markov chain $\eta(t,\omega)$ with infinitesimal transition matrix denoted by

$$\mathbf{Q} = \begin{bmatrix} -q_0 & q_{01} & \cdots & q_{0N} \\ q_{10} & -q_1 & \cdots & q_{1N} \\ \vdots & \vdots & \cdots & \vdots \\ q_{N0} & q_{N1} & \cdots & -q_N \end{bmatrix},$$

where $q_k = \lim_{t \to 0} t^{-1}(1-p_{kk}(t))$, $q_{ij} = \lim_{t \to 0} t^{-1} p_{ij}(t)$, $i \neq j$, and $\sum_{j \neq i} q_{ij} = q_i$, (see, for example, Doob [9]). Extending the ideas presented in Blumenthal and Getoor [4] and following the details in Kannan [27, Section 2] one can establish.

**Theorem 3.6** *Let $x^*(t, \xi; \eta)$ be the solution of the generalized Langevin equation (3.24), where $\eta$ is a Markov chain as described above. Then, the process*

$$\mathbf{x}(t,\omega) = (x(t), \eta(t)) = (x^*(t-\tau_n, x_{\tau_n}; \eta_{\tau_n}), \eta_{\tau_n}), \qquad \tau_n \leq t < \tau_{n+1}, \tag{3.30}$$

*governing the evolution of the particle is a Markov process, and*

$$\mathbf{x}_n(\omega) = (\tau_n(\omega), x_{\tau_n(\omega)}, \eta_{\tau_n(\omega)})$$

The transition probability of $\mathbf{x}(t,\omega)$ is given as follows:

$$P(t,(x,\eta_i),(A,\eta_j)) = P_{x,\eta_i}[x(t) \in A, \eta(t) = \eta_j]$$
$$= P_{\eta_i}[x^*(t,x;\eta(\cdot)) \in A, \eta(t) = \eta_j]$$
$$= p_{ij}(t) P[x^*(t,x;\eta(\cdot)) \in A \mid \eta(0) = \eta_i, \eta(t) = \eta_j].$$

It is not hard to prove the following.

**Theorem 3.7** *The transition probability $P(t,(x,\eta_i),(A,\eta_j))$ is a stochastically continuous Feller transition function, and hence $\mathbf{x}(t,\omega)$ is a strong Markov process.*

Let $\mathbf{D}$ denote collection of all functions $f(x,\eta_i)$ on $R^1 \times S$ such that $f$, $\partial f/\partial x$ and $(\gamma * x)(\partial f/\partial x)$ are all continuous functions vanishing at $\infty$, (here $S$ is given the discrete topology). Then, the infinitesimal generator $\mathbf{A}$ of $\mathbf{x}(t,\omega)$ is defined on $\mathbf{D}$ and is given by

$$\mathbf{A}f(x,\eta_i) = \sum_{j \neq i} q_{ij} f(x,\eta_j) - q_i f(x,\eta_i) - (\gamma * x) \frac{\partial f}{\partial x}.$$

Using the theory of dual semigroup of transition semigroup of operators of the strong Markov process $\mathbf{x}$ one can easily see the following theorem.

**Theorem 3.8** *Let $\mu(A,\eta_i)$ denote the initial distribution of the evolution* process $\mathbf{x}(t, \omega)$ and $\psi(x, \eta_i)$ be the corresponding density function. Then, a steady-state distribution of the evolution process $\mathbf{x}(t, \omega)$ exists if and only if

$$\frac{\partial}{\partial x}[\psi \cdot (\gamma * x)] = q_i \psi - \sum_{j \neq i} q_{ij} \psi.$$

### B. Itô Integrodifferential Equation

The purpose of this section is to investigate the properties of the Itô integrodifferential system

$$\dot{x}(t) = \eta_0 + \int_0^t m(s, x(s), \dot{x}(s))\, ds + \int_0^t \sigma(s, x(s), \dot{x}(s))\, d\beta(s),$$

$$x(t) = \xi_0 + \int_0^t \dot{x}(s)\, ds. \tag{3.31}$$

The results presented here are due to Borchers or Goldstein; but, for further results, see Kannan [31]. Throughout this section we assume that $B_1$ and $B_2$ of Section II.C hold. Under these conditions we have seen (Theorem 2.10) that the Itô integrodifferential system (3.31) has a unique solution. Next, we study some stochastic properties of this unique solution process. If $x(t, \omega)$ denotes the solution of Eq. (2.28), then we denote $\mathbf{X}(t, \omega) = (x(t, \omega), \dot{x}(t, \omega))$, where $\dot{x}(t)$ solves Eq. (2.27). For any Borel set $A$ in $R^2$ and any $t \geq s$, define

$$p(s, \mathbf{x}, t, A) = P\{\mathbf{X}_{s,\mathbf{x}}(t, \omega) \in A\}. \tag{3.32}$$

**Theorem 3.9** *The process $\mathbf{X}(t, \omega)$ is a Markov process*:

$$P\{\mathbf{X}(t) \in A \mid \mathscr{A}_s\} = P\{\mathbf{X}(t) \in A \mid \mathbf{X}(s)\} = p(s, \mathbf{X}(s), t, A), \quad \text{a.s.,} \tag{3.33}$$

*for all $t > s$ and any Borel set $A \subset R^2$. Moreover, $p(s, \mathbf{x}, t, A)$ is a transition probability function.*

*Proof* Let $0 \leq s < t < \infty$. Then, from (3.31),

$$\dot{x}(t) - \dot{x}(s) = \int_s^t m(u, \mathbf{X}(u))\, du + \int_s^t \sigma(u, \mathbf{X}(u))\, d\beta(u),$$

$$x(t) - x(s) = \int_s^t \dot{x}(u)\, du.$$

Thus, the random vector $\mathbf{X}_t = (x_t, \dot{x}_t)$ depends only on $x_s$, $x(u)$ with $s \leq u \leq t$, and the $\beta$-increments in $[s, t]$. One easily sees that $\{\mathbf{X}(t), \mathscr{A}_t, t \geq 0\}$ is an adapted family. Now, for any bounded measurable function

$$f(\mathbf{x}_0, x_1, \ldots, x_m) = f_0(\mathbf{x}_0) f_1(x_1, \ldots, x_m),$$

and any $0 < t_1 < t_2 < \cdots < t_m$,

$$E\{f(\mathbf{X}(s), \beta(t_1+s) - \beta(s), \ldots, \beta(t_m+s) - \beta(s)) \mid \mathscr{A}_s\}$$
$$= f_0(\mathbf{X}(s)) E\{f_1(\beta(t_1+s) - \beta(s), \ldots, \beta(t_m+s) - \beta(s)) \mid \mathscr{A}_s\}$$
$$= f_0(\mathbf{X}(s)) E\{f_1(\beta(t_1+s) - \beta(s), \ldots, \beta(t_m+s) - \beta(s))\},$$

and hence,

$$E\{f(\mathbf{X}(s), \beta(t_1+s) - \beta(s), \ldots, \beta(t_m+s) - \beta(s)) \mid \mathscr{A}_s\}$$
$$= [E\{f(\mathbf{x}_0, \beta(t_1+s) - \beta(s), \ldots, \beta(t_m+s) - \beta(s))\}]_{\mathbf{X}(s)=\mathbf{x}_0}. \quad (3.34)$$

If $f$ is a bounded continuous function, it is not hard to see that

$$E\{f(\mathbf{X}(t)) \mid \mathscr{A}_s\} = E\{f(\mathbf{X}(t)) \mid \mathbf{X}(s)\}.$$

If $A$ is an open set in $R^2$, then we see that (3.33) is obtained by taking an increasing sequence $f_n$ of bounded continuous functions that approach $I_A$. Now, by a theorem on monotone class of sets, we see that (3.33) holds for all Borel sets $A$. It remains to prove that $p(\cdot, \cdot, \cdot, \cdot)$ is a transition function.

From (3.32) it is clear that $P(s, \mathbf{x}, t, A)$ is a probability measure in $A$. Noting that each function in the successive approximation that converges to $\mathbf{X}_{s,\mathbf{x}}$ is Borel measurable in $\mathbf{x} = (\xi, \eta)$, say, we see that $P(s, \mathbf{x}, t, A)$ is Borel measurable in $\mathbf{x}$, for fixed $s$, $t$, and $A$. Finally, if $s < u < t$, we have, by the Markovian property of $\mathbf{X}(\cdot)$,

$$P_{s,\mathbf{x}}\{\mathbf{X}(t) \in A\} = E_{s,\mathbf{x}}\{P_{s,\mathbf{x}}\{\mathbf{X}(t) \in A \mid \mathbf{X}(u) = \mathbf{y}\}\}$$
$$= E_{s,\mathbf{x}}\{P_{s,\mathbf{y}}\{\mathbf{X}(t) \in A\}\},$$

and hence the Chapman–Kolmogorov relation holds. This completes the proof.

The following theorem describes the continuous dependence of the solution $\mathbf{X}_{s,\mathbf{x}}(t)$ on the initial condition $\mathbf{x}$; we omit the proof of this.

**Theorem 3.10** *For each interval* $[s, t] \subset [s, \infty)$ *there exists a constant* $c = c_{s,t}$ *such that*

$$E\{\sup_{s \leq u \leq t} |\mathbf{X}_{s,\mathbf{x}}(t) - \mathbf{X}_{s,\mathbf{y}}(t)|\}^2 \leq C|\mathbf{x} - \mathbf{y}|^2,$$

*where* $|\cdot|$ *is the usual Euclidean norm.*

Next, we study a class of absolutely continuous probabilities. First we state a theorem which computes asymptotic estimates for the infinitesimal means and variances connected with $\mathbf{X}(t)$, $t \geq 0$. We omit the proof because of lengthy computations (cf. Borchers [5]). Recall that we assumed that conditions $B_1$ and $B_2$ hold throughout this subsection.

**Theorem 3.11** *Let* $\mathbf{X}(t) = (x(t), \dot{x}(t))$ *be the Markovian solution of system* (3.31). *Then*

($a_1$) $E\{\sup_{s \leq t \leq s+h} |x(t)-x(s)|^2\} = O(h^2)[1+Ex^2(s)+E\dot{x}^2(s)];$

($a_2$) $E\{\sup_{s \leq t \leq s+h} |\dot{x}(t)-\dot{x}(s)|^2\} = O(h)[1+Ex^2(s)+E\dot{x}^2(s)];$

($b_1$) $E_{s,\mathbf{x}}\{\sup_{s \leq t \leq s+h} |x(t)-\xi_0|^2\} = O(h^2)[1+|\mathbf{x}|^2];$

($b_2$) $E_{s,\mathbf{x}}\{\sup_{s \leq t \leq s+h} |\dot{x}(t)-\eta_0|^2\} = O(h)[1+|\mathbf{x}|^2];$

($c_1$) $E_{s,\mathbf{x}}\{x(s+h) - \xi_0\} = \eta_0 h + O(h^{\frac{3}{2}})[1+|\mathbf{x}|^2]^{\frac{1}{2}};$

($c_2$) $E_{s,\mathbf{x}}\{\dot{x}(s+h) - \eta_0\} = \int_s^{s+h} m(u, \mathbf{X}) \, du + O(h^2)(1+|\mathbf{x}|^2)^{\frac{1}{2}};$

($d_1$) $E_{s,\mathbf{x}}\{|x(s+h) - \xi_0|^2\} = O(h^2)[1+|\mathbf{x}|^2];$

($d_2$) $E\mathbf{x}_{s,\mathbf{x}}|\dot{x}(s+h) - \eta_0|^2\} = \int_s^{s+h} (\sigma[t,\mathbf{X}])^2 \, dF(t) + O(h^{\frac{3}{2}})[1+|\mathbf{x}|^2];$

($d_3$) $E_{s,\mathbf{x}}\{(x(s+h) - \xi_0)(\dot{x}(s+h) - \eta_0)\} = O(h^{\frac{3}{2}})[1+|\mathbf{x}|^2];$

*where the order terms are uniform, for* $(s, \mathbf{x}) \in [0, T] \times R^2$, *on each compact interval* $[0, T]$. *Now, let* $\beta(t, \omega)$ *be a standard Brownian motion. Then,*

($e_1$) $P_{s,\mathbf{x}}\{\sup_{s \leq t \leq s+h} |x(t)-\xi_0| > \varepsilon\} = O(h^2)[1+|\mathbf{x}|^2];$

($e_2$) $P_{s,\mathbf{x}}\{\sup_{s \leq t \leq s+h} |\dot{x}(t)-\eta_0| > \varepsilon\} = O(h^{\frac{3}{2}})[1+|\mathbf{x}|^2]^{\frac{3}{2}};$

($e_3$) $P_{s,\mathbf{x}}\{\sup_{s \leq t \leq s+h} |\mathbf{X}(t)-\mathbf{x}| > \varepsilon\} = O(h^{\frac{3}{2}})[1+|\mathbf{x}|^2]^{\frac{1}{2}};$

*where for each* $\varepsilon > 0$, *the order terms are uniform on* $[0, T] \times R^2$, *provided* $h$ *is sufficiently small.*

From now on, we assume that, $\beta(t, \omega)$ is a standard Brownian motion.

**Theorem 3.12** *Let* $f(t, x)$ *be any Borel function that is bounded on* $[0, T] \times R^2$, $T \geq 0$. *Then, for every real* $\lambda$, (a)

$$E\left\{\exp\left[\lambda \int_0^t f(s, x(s), \dot{x}(s)) \, d\beta(s) - 2^{-1}\lambda^2 \int_0^t f(s, x(s), \dot{x}(s))^2 \, ds\right]\right\} = 1,$$

*and* (b)

$$E\left\{\exp\left[\lambda \int_0^t \int_0^s f(u, \mathbf{X}(u)) \, d\beta(u) \, ds - 2^{-1}\lambda^2 \int_0^t (t-s)^2 f(s, \mathbf{X}(s))^2 \, ds\right]\right\} = 1.$$

*Proof* Define $g(s, \omega) = f(s, x(s, \omega) \dot{x}(s, \omega))$, $s \in [0, T]$, $\omega \in \Omega$. Since

$X(t, \omega)$ is a continuous process, $\{g(s), \mathscr{A}_s, s \in [0, T]\}$ is an adapted family, almost all trajectories $g_\omega(s)$ are Borel functions of $s$, and $g(s, \omega)$ is bounded and measurable on $[0, T] \times R^2$.

Treating $\int_0^t \int_0^s g(u, \omega) \, d\beta(u, \omega) \, ds$ as the sample path integral of the (separable) martingale $\int_0^s g(u, \omega) \, d\beta(u, \omega)$, $0 \le s \le t$, and using the standard approximation of the Itô integral by finite sums, it is not hard to see that

$$\int_0^t \int_0^s g(u) \, d\beta(u) \, ds = \int_0^t (t-s) g(s) \, d\beta(s), \quad \text{a.s.,}$$

and

$$E \int_0^t (t-s) g(s) \, d\beta(s) = 0$$

and

$$E \left\{ \int_0^t (t-s) g(s) \, d\beta(s) \right\}^2 = \int_0^t (t-s)^2 E\{g^2(s)\} \, ds.$$

So, the iterated integral in (b) can be replaced by $\int_0^t (t-s) f(s, \mathbf{X}(s)) \, d\beta(s)$ and hence it suffices to establish (a).

Recall from the theory of the Itô integral, (cf. Gihman and Skorokhod [18]), there exists a sequence $\{g_n\}$ of simple processes such that

$$E\left\{ \left| \int_0^t g(s) \, d\beta(s) - \int_0^t g_n(s) \, d\beta(s) \right|^2 \right\} = \int_0^t E|g(s) - g_n(s)|^2 \, ds \to 0,$$

as $n \to \infty$. This sequence can be chosen such that (1) $\int_0^t g_n(s) \, d\beta(s) \to \int_0^t g(s) \, d\beta(s)$, in $L_2$ and a.s.; (2) $\int_0^t g_n^2(s) \, ds \to \int_0^t g^2(s) \, ds$, a.s.; (3)

$$\int_0^t g_n(s) \, d\beta(s) = \sum_{k=0}^{n_m-1} g_{n,k}(\omega) [\beta(t_{n,k+1}, \omega) - \beta(t_{n,k}, \omega)],$$

$$\int_0^t g_n^2(s) \, ds = \sum_{k=0}^{n_m-1} g_{n,k}^2(\omega) [t_{n,k+1} - t_{n,k}],$$

where $0 = t_{n,0} < t_{n,1} < \cdots < t_{n,n_m-1} < t_{n,n_m} = t$ is a partition $[0, t]$; (4) if $g$ is majorized by a constant $M$ on $[0, T] \times R^2$, then $|g_n| \le M$, for all $n$; and (5) for each fixed $n$ and $k = 0, 1, \ldots, n_m$, $g_{n,k}$ is $\mathscr{A}_{t_{n,k}}$-measurable. For the convenience of writing, denote the integrals in (3) by

$$J_m = \sum_{k=0}^{m-1} g_k \Delta_{k+1} \beta \quad \text{and} \quad L_m = \sum_{k=0}^{m-1} g_k^2 \Delta_{k+1} s,$$

respectively. Let $\alpha \ge 1$ and $Q_m = \lambda J_m - 2^{-1} \lambda^2 L_m$. Then,

$$\alpha Q_m = \alpha Q_{m-1} + \alpha \rho_m \quad \text{with} \quad \rho_m = \lambda g_{m-1} \Delta_m \beta - 2^{-1} \lambda^2 g_{m-1}^2 \Delta_m s,$$

where $Q_{m-1}$ is $\mathscr{A}_{m-1}$-measurable. Now

$$E\{\exp[\alpha Q_m]\} = E\{E_{\mathscr{A}_{m-1}}[\exp \alpha Q_m]\} = E\{\exp[\alpha Q_{m-1}] E_{\mathscr{A}_{m-1}}\{\exp[\alpha \rho_m]\}\}.$$

But,

$$E_{\mathscr{A}_{m-1}}\{\exp[\alpha\rho_m]\} = \exp[2^{-1}\alpha(\alpha-1)\lambda^2 g_{m-1}^2 \Delta_m s]$$
$$= 1, \quad \text{if} \quad \alpha = 1,$$
$$\leqslant \exp[2^{-1}\alpha(\alpha-1)\lambda^2 M^2 \Delta_m s], \quad \text{if} \quad \alpha > 1.$$

So, for $\alpha \geqslant 1$,

$$1 \leqslant E\{\exp[\alpha Q_m]\} \leqslant \exp\{2^{-1}\alpha(\alpha-1)\lambda^2 M^2 \Delta_m s\} E\{\exp[\alpha Q_{m-1}]\}$$
$$\leqslant \cdots \leqslant \exp[2^{-1}\alpha(\alpha-1)\lambda^2 M^2 t].$$

Taking $\alpha = 1$, we obtain (a) for approximating sums. From this it follows that

$$\left\{\exp\left\{\alpha\left[\lambda \int_0^t g_n(s)\, d\beta(s) - 2^{-1}\lambda^2 \int_0^t g_n^2(s)\, ds\right]\right\}\right\}$$

forms a uniformly integrable sequence. From (1) and (2),

$$\lambda \int_0^t g_n(s)\, d\beta(s) - 2^{-1}\lambda^2 \int_0^t g_n^2(s)\, ds,$$
$$\xrightarrow{\text{a.s.}} \lambda \int_0^t g(s)\, d\beta(s) - 2^{-1}\lambda^2 \int_0^t g^2(d)\, ds,$$

for each $\lambda$. If $Q$ denotes the limit of $Q_n$, then from the uniform integrability

$$1 = \lim_{n\to\infty} E\{e^{Q_n}\} = E\{\lim_{n\to\infty} e^{Q_n}\} = E\{e^Q\}.$$

This completes the proof.

Instead of a bounded $f$ let us take, in Theorem 3.12, an $h$ such that $P\{\int_0^T h^2(t)\, ds < \infty\} = 1$. After establishing (a) for approximating sum, let $n \to \infty$ and use Fatou's lemma to obtain

$$E\left\{\exp\left[\lambda \int_0^t h(s, \mathbf{X}(s))\, d\beta(s) - 2^{-1}\lambda^2 \int_0^t h^2(s, X(s))\, ds\right]\right\} \leqslant 1.$$

The following Corollary to Theorem 3.12 gives some bounds for the exponential moments.

**Corollary 3.13** *Let the coefficients $m(\cdot,\cdot,\cdot)$ and $\sigma(\cdot,\cdot,\cdot)$ be bounded by a positive constant $M = M(T)$. Then,*

(a) $E\{\exp[\lambda x(t)]\} \leqslant 2^{-1} L \exp[\lambda(\xi_0 + \eta_0 t)], \quad \text{for all} \quad \lambda,$
(b) $E\{\exp[\lambda |x(t)|]\} \leqslant L \exp\{\lambda |\xi_0 + \eta_0 t|\}, \quad \lambda \geqslant 0,$

*where* $L = 2\exp\{2^{-1}|\lambda| Mt^2 + 6^{-1}\lambda^2 M^2 t^3\},$

(c) $E\{\exp[\lambda \dot{x}(t)]\} \leq 2^{-1} L^* \exp[\lambda \eta_0]$,   for all $\lambda$,
(d) $E\{\exp[\lambda |\dot{x}(t)|]\} \leq L^* \exp[\lambda |\eta_0|]$,   $\lambda \geq 0$,

where $L^* = 2 \exp\{2^{-1} |\lambda| Mt + 2^{-1} \lambda^2 M^2 t\}$.

*Proof* To simplify the writing put

$$\mu = \int_0^t \int_0^s m(u, X(u)) \, du \, ds,$$

$$S = \int_0^t \int_0^s \sigma(u, X(u)) \, d\beta(u) \, ds.$$

Then,

$$E\{\exp[\lambda x(t)]\} = E\{(\exp[\lambda(\xi_0 + \eta_0 t)])(\exp \lambda \mu)(\exp \lambda S)\}.$$

$$E\{e^{\lambda S}\} \leq \exp\left[2^{-1} \lambda^2 M^2 \int_0^t (t-s)^2 \, ds\right] \leq \exp[6^{-1} \lambda^2 M^2 t^3],$$

$$|\lambda \mu| \leq M \int_0^t \int_0^s du \, ds = 2^{-1} |\lambda| Mt^2,$$

and hence

$$E\{e^{\lambda x(t)}\} \leq 2^{-1}\{2 \exp[2^{-1} |\lambda| Mt^2 + 6^{-1} \lambda^2 M^2 t^3]\} \exp[\lambda(\xi_0 + \eta_0 t)].$$

If $\lambda > 0$,

$$E\{\exp[\lambda x(t)]\} \leq E\{\exp[\lambda x(t)]\} + E\{\exp[-\lambda x(t)]\}$$
$$\leq L \exp \lambda |\xi_0 + \eta_0 t|.$$

The rest of the corollary follows similarly.

In Section III.A we presented conditions under which the variance $\mathscr{D}^2[x(t)]$ of the solution of a random integrodifferential equation tends to infinity. In the present Itô integrodifferential case, if $m \equiv 0$ and $\eta_0(\omega) \neq 0$, for example, then

$$\mathscr{D}^2[x(t)] \geq \mathscr{D}^2[x(t) - \xi_0] \geq \mathscr{D}^2[t\eta_0] = t^2 \mathscr{D}^2(\eta_0) \uparrow \infty,$$

as $t \uparrow \infty$. But one can show that $\mathscr{D}^2[x(t)] \uparrow \infty$ and $\mathscr{D}^2[\dot{x}(t)] \uparrow \infty$, as $t \uparrow \infty$, if the diffusion coefficient $\sigma$ is bounded away from zero. So, let $\sigma \geq \alpha > 0$. Let $\mu_0(\cdot) = m(\cdot, \cdot, \cdot)$, $\sigma_0(\cdot) = \sigma(\cdot, \cdot, \cdot)$. Now,

$$\mathscr{D}^2[\dot{x}(t) - \eta_0] = E\left\{\int_0^t \mu_0(s) \, ds\right\}^2 + \int_0^t E\{\sigma_0^2(s)\} \, ds - E^2 \int_0^t \mu_0(s) \, ds$$

$$= \mathscr{D}^2\left[\int_0^t \mu_0(s) \, ds\right] + \int_0^t E\{\sigma_0^2(s)\} \, ds,$$

$$x(t) - \xi_0 = \eta_0 t + \int_0^t (t-s)\mu_0\,ds + \int_0^t (t-s)\sigma_0(s)\,d\beta(s),$$

$$E[x(t) - \xi_0] = E\left\{\eta_0 t + \int_0^t (t-s)\mu_0(s)\,ds\right\}.$$

Put $A = \eta_0 t + \int_0^t (t-s)\mu_0(s)\,ds$, and $B = \int_0^t (t-s)\sigma_0(s)\,d\beta(s)$. But $E[B^2] = \int_0^t (t-s)^2 E[\sigma_0^2(s)]\,ds$ and $E[AB] = 0$.

$$\mathscr{D}^2[\dot{x}(t)] \geqslant \mathscr{D}^2[\dot{x}(t) - \eta_0] \geqslant \int_0^t E\{\sigma_0^2(s)\}\,ds \geqslant \alpha^2 t \to \infty, \qquad \text{as } t \uparrow \infty;$$

$$\mathscr{D}^2[x(t)] \geqslant \mathscr{D}^2[x(t) - \xi_0] = E[A^2] + E[B^2] - E^2[A]$$

$$= \mathscr{D}^2[A] + \int_0^t (t-s)^2 E\{\sigma_0^2(s)\}\,ds \geqslant \alpha^2 (t^3/3) \to \infty, \qquad \text{as } t \uparrow \infty.$$

The next theorem gives the asymptotic behavior of the position $x(t)$ and velocity $\dot{x}(t)$ as $t \to \infty$.

**Theorem 3.14** *Let $\sigma(t, \mathbf{X})$ be independent of $\mathbf{X} \in R^2$ such that $\sigma(t, \mathbf{X}) \equiv \sigma(t) > 0$ and $\int_0^\infty \sigma^2(t)\,dt = \infty$.*

(a) *If $m \equiv 0$, then the velocity $\{\dot{x}(t), t \geqslant 0\}$ is a Brownian motion after the change of the time scale.*

(b) *Let $m(t, \mathbf{X}) \geqslant \alpha$ (or $m(t, \mathbf{X}) \leqslant -\alpha$) for all sufficiently large $t$ and all $\mathbf{X} \in R^2$, for some $\alpha > 0$. If there is a $\gamma > 0$ such that*

$$\log\log \int_0^t \sigma^2(s)\,ds \leqslant 2^{-1}(\alpha - \gamma)^2 t^2, \tag{3.35}$$

*for all large $t$, then almost all sample paths satisfy*

$$\lim_{t \to \infty} \dot{x}(t) = \infty \quad \text{or} \quad \lim_{t \to \infty} \dot{x}(t) = -\infty,$$

$$\lim_{t \to \infty} x(t) = \infty \quad \text{or} \quad \lim_{t \to \infty} x(t) = -\infty.$$

*Proof* (a) This is a standard result on stochastic integrals (see McKean [42]). In fact, the stochastic integral $\int_0^t \sigma(s)\,d\beta(s, \omega)$ is a continuous square integrable martingale and by changing to the intrinsic time we get our claim.

(b) Define $S(t) = \int_0^t \sigma(s)\,d\beta(s)$. Since $\int_0^\infty \sigma^2(t)\,dt = \infty$, it follows from the law of iterated logarithms that

$$P\{\overline{\lim_{t \to \infty}} S(t) = \infty\} = 1_A = P\{\underline{\lim_{t \to \infty}} S(t) = -\infty\}.$$

Define $M(t) = \int_0^t m(s, \mathbf{X}(s))\,ds$. By the hypothesis on $m$, for any $\gamma > 0$, there is a $T_0 = T_0(\gamma, \omega) > 0$ such that $M(t) \geqslant (\alpha - \gamma)t$ [or $M(t) \leqslant (\alpha - \gamma)t$], for all $t \geqslant T_0$ with probability 1. Since

$$\dot{x}(t) = \eta_0 + M(t) + S(t),$$

it follows from (3.35) and the law of iterated logarithms that

$$P\{\lim_{t\to\infty} \dot{x}(t) = \infty\} = 1 \quad \text{or} \quad P\{\lim_{t\to\infty} \dot{x}(t) = -\infty\} = 1.$$

Integration yields the remainder of the conclusion.

Under the hypothesis of (a) of the above theorem, we saw that $\dot{x}(t)$ is a Brownian motion after a time change. Consequently, we see that almost all sample paths of $\dot{x}(t)$ are of unbounded variation in any $[0, T]$. Actually, we can do away with the said hypothesis and still establish the claim on the path variation.

**Theorem 3.15** *If $\sigma(t, \mathbf{X}) \neq 0$ for all $(t, \mathbf{X}) \in R_+ \times R^2$, then almost all sample paths of the Markov process $\mathbf{X}(t, \omega) = (x(t, \omega), \dot{x}(t, \omega))$ are of unbounded variation in any interval $[a, b] \subset R_+$. Same holds for $\dot{x}(t, \omega)$.*

*Proof* Let $\pi$ be collection of all partitions $\pi$ of an arbitrarily fixed interval $[a, b]$. Define

$$y(\omega) = \int_a^b \sigma^2(t, \mathbf{X}(t, \omega)) \, dt.$$

Then, by a result of Wong and Zakai, we can find a sequence $\{n(k) \uparrow \infty, k \uparrow \infty\}$ of positive integers such that

$$P\left\{\lim_{k\to\infty} \sum_{m=0}^{u(k)} |\Delta x(m, k)|^2 = y(\omega)\right\} = 1, \tag{3.36}$$

where the upper limit $u(k) = 2^{n(k)} - 1$ and

$$\Delta x(m, k) = x(a + 2^{-n(k)}(b-a)(m+1)) - x(a + 2^{-n(k)}(b-a)m).$$

From the uniform continuity of $x(t)$, for every $\varepsilon > 0$ there exists a $\delta = \delta(\varepsilon, \omega)$ such that

$$|x(t, \omega) - x(s, \omega)| < \varepsilon y(\omega) \tag{3.37}$$

if $|t - s| < \delta$, $s, t \in [a, b]$. So, there is a $K_0 > 0$ such that if $k \geq K_0$, $2^{-n(k)} < \delta$, and for $m = 1, 2, \ldots, 2^{n(k)}$,

$$|\Delta x(m, k)|^2 \leq \varepsilon y(\omega) |\Delta x(m, k)|. \tag{3.38}$$

From the limit in (3.36), there is a $K_1$ such that for $k \geq K_1$

$$\sum_{m=0}^{u(k)} |\Delta x(m, k)|^2 > y(\omega)/2.$$

Let $K = \max[K_0, K_1]$. Then, for each $k \geq K$,

$$\sum_{m=0}^{u(k)} |\Delta x(m, k)| \geq \sum_{m=0}^{u(k)} [|\Delta x(m, k)|^2 / \varepsilon y(\omega)] \geq (y(\omega)/2)(\varepsilon y(\omega))^{-1} = \tfrac{1}{2}\varepsilon.$$

Since $\varepsilon$ is arbitrary the theorem follows from this.

In recalling the basic facts about the Itô integrodifferential equation, we note that $x(t,\omega)$ is an absolutely continuous process such that $\mathbf{X}(t,\omega) = (x(t,\omega), \dot{x}(t,\omega))$ is a continuous Markov process in $R^2$. Conversely, we pose the problem that under what conditions on a given continuous two-dimensional Markov process there exist functions $M(t, \mathbf{X})$ and $\sigma(t, \mathbf{X})$ and some Brownian motion such that $\mathbf{X}$ solves system (3.31). The following theorem answers this converse problem. We omit the proof of this theorem and refer the reader to Doob [9, p. 286].

**Theorem 3.16** *Let* $\mathbf{X}(t,\omega) \{(x(t,\omega), x^*(t,\omega)) \mathcal{A}_t, t \in [0,T]\}$ *be an adapted family such that*

(a) $x(t,\omega)$ *is an absolutely continuous process and* $x^*(t,\omega)$ *is a continuous process;*

(b) $E\{|x(t)|^2 + |x^*(t)|^2\} < \infty$, *for all* $t \in [0,T]$;

(c) *there is an integrable process* $y(s)$ *adapted to* $\mathcal{A}_s$ *such that, for* $0 \leqslant s < t \leqslant T$, $E\{(|x(t)|^2 + |x^*(t)|^2 \mid \mathcal{A}_s\} \leqslant y(s)$, a.s.;

(d) *there are two Borel functions* $m(\cdot,\cdot,\cdot)$ *and* $\sigma(\cdot,\cdot,\cdot)$ *on* $[0,T] \times R^2$ *that are continuous in the pair* $\mathbf{X} = (x, x^*)$ *such that*

$$|m(t,\mathbf{X})| \leqslant K(1+|\mathbf{X}|^2)^{\frac{1}{2}}, \quad 0 \leqslant \sigma(t,\mathbf{X}) \leqslant K(1+|\mathbf{X}|^2)^{\frac{1}{2}}$$

*for some constant* $K$; *and*

(e) *there exists a monotone nondecreasing function* $f(\cdot)$ *with* $\lim_{h \to 0} f(h) = 0$ *such that*

(i) $E\{|x(s+h) - x(s) - hx^*(s)| \mid \mathcal{A}_s\} \leqslant hf(h)[1+|\mathbf{X}(s)|^2]$,

(ii) $\left| E\{x^*(s+h) - x^*(s) \mid \mathcal{A}_s\} - \int_s^{s+h} m(u, \mathbf{X}(u))\, du \right| \leqslant hf(h)[1+|\mathbf{X}(s)|^2]$,

(iii) $\left| E\{[x^*(s+h) - x^*(s)]^2 \mid \mathcal{A}_s\} - \int_s^{s+h} \sigma^2(u, X(u))\, du \right|$

$\leqslant hf(h)[1+|\mathbf{X}(s)|^2]$.

*Then* (1) $\mathbf{X}(t) - \mathbf{X}(0)$ *is a Markov process;* (2) *the* $L_2$-*derivative* $\dot{x}(t)$ *of the* $x(t)$-*process exists for each* $t \in [0,T]$ *and* $P\{\dot{x}(t) = x^*(t)\} = 1$, $0 \leqslant t \leqslant a$; *and* (3) *if* $\sigma(t, \mathbf{X})$ *for all* $(t, X) \in [0,T] \times R^2$, *there is a Brownian motion* $\{\beta(t), 0 \leqslant t \leqslant T\}$ *such that the components of the process* $\mathbf{X}(t,\omega) = (x(t,\omega), \dot{x}(t,\omega))$ *satisfy system* (3.31) *of Itô integrodifferential equations.*

## IV. Small Perturbations

In understanding and solving some problems associated with certain physical or biological systems, it often becomes necessary to study suitable perturbations of the given system. The perturbation techniques are plentiful in the study of the deterministic equations. In several applications, the dynamical systems are described by a deterministic process together with a "small" random spread about it. The solution of the corresponding stochastic equations will depend, possibly, on a small parameter $\varepsilon > 0$, where $\varepsilon$ characterizes the size of the fluctuation. An important problem is to study the behavior of the solution as $\varepsilon \to 0$. This section illustrates certain methods associated with the problems of this type.

Consider the following perturbed random integrodifferential equation

$$\dot{x}(t,\omega) = \xi(t,\omega) + k(t)[x(t,\omega) + S_1 x(t,\omega)]$$
$$+ \int_0^t K(t,s)[x(s,\omega) + S_2 x(s,\omega)]\, ds + S_3 x(t,\omega),$$
$$x(0,\omega) = x_0(\omega), \qquad (4.1)$$

where $S_1$, $S_2$, and $S_3$ are nonlinear operators mapping $\mathscr{L}_1[R_+, L_2(\Omega)]$ into itself. Equations of the form (4.1) arise, for example, in the dynamics of nuclear reactors (cf. Gyftopoulos [21]). The mild mean-square solution of Eq. (4.1) is given by

$$x(t,\omega) = \rho(t,0)x_0(\omega) + \int_0^t \rho(t,s)[\xi(s,\omega)$$
$$+ k(s)\{S_1 x(s,\omega) - S_2 x(s,\omega)\} + S_3 x(s,\omega)]\, ds$$
$$- \int_0^t \rho_s(t,s) S_2 x(s,\omega)\, ds. \qquad (4.2)$$

For $\xi \in \mathscr{L}_1[R_+, L_2(\Omega)]$ and $0 \leqslant t < \infty$, define

$$(T\xi)(t) = \int_0^t \rho(t,s)\xi(s)\, ds$$

and

$$(T^*\xi)(t) = -\int_s^t \rho_s(t,s)\xi(s)\, ds.$$

The Banach space $\mathbf{C} = \mathbf{C}[R_+, L_2(\Omega)]$ of bounded continuous functions is a subspace of $\mathscr{L}_1$ with a topology stronger than the topology of the Frechet space $\mathscr{L}_1$. Therefore, if $T$ and $T^*$ map $\mathbf{C}$ into itself, then, due to the admissibility theory (cf. Massera and Schaffer [39]), $T$ and $T^*$ are closed operators. Now, the closed graph theorem yields $T$ and $T^*$ continuous. In Eq. (4.1) we consider only "small" operators $S_i$, $i = 1, 2, 3$. A mapping $S: \mathbf{C} \to \mathbf{C}$ is called *small* relative to $\mathbf{C}$ if (i) $S(0) = 0$, and (ii) for each $\varepsilon > 0$, there exists a $\delta > 0$ such that $\|Sx - Sy\|_\mathbf{C} \leqslant \varepsilon \|x - y\|_\mathbf{C}$ whenever $\|x\|_\mathbf{C} \leqslant \delta$ and $\|y\|_\mathbf{C} \leqslant \delta$.

**Theorem 4.1** *Assume that*

(i) $S_1$, $S_2$, and $S_3$ are small relative to $\mathbf{C}[R_+, L_2(\Omega)]$,
(ii) $T$ and $T^*$ map $\mathbf{C}$ into itself, and
(iii) $\xi \in \mathbf{C}[R_+, L_2(\Omega)]$ and $\rho(t,0) \in \mathbf{C}[R_+, R]$.

*Then, for each $\varepsilon > 0$, there is $\delta > 0$ such that the perturbed equation* (4.1) *has a unique solution $x \in \mathbf{C}[R_+, L_2(\Omega)]$ with $\|x\|_\mathbf{C} \leq \varepsilon$ whenever $\|x_0\|_2 \leq \delta$ and $\|\xi\|_\mathbf{C} \leq \delta$.*

Banach's contraction principle will yield the theorem once one observes that the operation $U$ defined on $\mathbf{C}$ by

$$Uy(t,\omega) = \rho(t,0)x_0(\omega) + \int_0^t \rho(t,s)[\xi(s,\omega)]$$
$$+ k(s)\{S_1 y(s,\omega) - S_2 y(s,\omega)\} + S_3 y(s,\omega)] \, ds$$
$$- \int_0^t \rho_s(t,s) S_2 x(s,\omega) \, ds$$

maps and $\varepsilon_0$-ball of $\mathbf{C}$ into itself and is a contraction. (We omit the details.)

Next we treat, using the tool of asymptotic expansion, the following perturbed equation

$$\dot{x}(t,\omega) = \xi(t,\omega) + k(t)x(t,\omega) + \int_0^t K(t,s)x(s,\omega) \, ds$$
$$+ \varepsilon f(t, x(t,\omega)), \qquad x(0,\omega) = x_0(\omega), \qquad (4.3)$$

where $\varepsilon > 0$ is a dimensionless parameter describing the size of the fluctuation in the system and $f(t,x)$ is a bounded function such that $f$ and all its partial derivatives are bounded and Lebesgue measurable in $t$, and are Lipschitzian in $x$. The mild solution of the Eq. (4.3) is

$$x(t,\omega) = \rho(t,0)x_0(\omega) + \int_0^t \rho(t,s)\xi(s,\omega) \, ds$$
$$+ \varepsilon \int_0^t \rho(t,s) f(s, x(s,\omega)) \, ds. \qquad (4.4)$$

Fix an $\varepsilon_0 > 0$, and for each $\varepsilon \in [0, \varepsilon_0]$, let $x_\varepsilon(t,\omega)$ denote the mild solution of Eq. (4.3). Then, for each $\varepsilon \in [0, \varepsilon_0]$, $x_\varepsilon(t,\omega)$ is given by (4.4). We are interested in the asymptotic behavior of the process $x_\varepsilon(t,\omega)$ as $\varepsilon \to 0$.

Let us assume that there is an expansion of $x_\varepsilon(t,\omega)$ given as follows:

$$x_\varepsilon(t) = y_0(t) + \varepsilon y_1(t) + \varepsilon^2 y_2(t) + \cdots + \varepsilon^n y_n(t) + \cdots. \qquad (4.5)$$

Also, assume that, for fixed $y_0, y_1, \ldots$, the function $f(t, y) = f(t, y_0 + \varepsilon y_1 + \cdots)$

RANDOM INTEGRODIFFERENTIAL EQUATIONS 151

can be expanded into a power series in $\varepsilon$ as follows:

$$f(t, y) = f(t, y_0) + \sum_{n=1}^{\infty} \varepsilon^n \sum_{\substack{n_1 + \cdots + n_k = n \\ 1 \leq k \leq n}} \frac{y_{n_1} y_{n_2} \cdots y_{n_k}}{k!} \left[ \frac{\partial^k f(t, \theta)}{\partial \theta^k} \right]_{\theta = y_0} \quad (4.6)$$

To simplify the writing, introduce

$$f_0(t; y_0) = f(t, y_0),$$

$$f_n(t; y_0, \ldots, y_n) = \sum_{\substack{n_1 + \cdots + n_k = n \\ 1 \leq k \leq n}} \frac{y_{n_1} y_{n_2} \cdots y_{n_k}}{k!} \left[ \frac{\partial^k f(t, \theta)}{\partial \theta^k} \right]_{\theta = y_0}.$$

Then, (4.6) becomes, $f(t, y) = \sum_{n=0}^{\infty} f_n(t; y_0, \ldots, y_n) \varepsilon^n$. Using these expressions in the Eq. (4.4) we get

$$x_\varepsilon(t) = \sum_{n=0}^{\infty} y_n(t) \varepsilon^n$$

$$= \int_0^t \rho(t, s) \xi(s) \, ds + \sum_{n=1}^{\infty} \int_0^t \varepsilon^n \rho(t, s) f_{n-1}(s; y_0, \ldots, y_{n-1}) \, ds \quad (4.7)$$

and, consequently,

$$y_0(t, \omega) = \rho(t, 0) x_0(\omega) + \int_0^t \rho(t, s) \xi(s, \omega) \, ds,$$

$$y_n(t, \omega) = \int_0^t \rho(t, s) f_{n-1}(s; y_0(s), \ldots, y_{n-1}(s)) \, ds, \quad n \geq 1. \quad (4.8)$$

Let $\xi(t, \omega)$ be a Gaussian process and $x_0(\omega)$ be a Gaussian random variable independent of $\xi$. Then, the following lemma is clear and says, under the above conditions and the asymptotic expansion, that the mild solution $x_\varepsilon(t)$ of Eq. (4.3) is, to a first-order approximation, a Gaussian process.

**Lemma 4.2** *Let $\xi(t, \omega)$ be a Gaussian process, and let $x_0(\omega)$ be a Gaussian random variable independent of $\xi(t, \omega)$. Then, the process $y_0(t, \omega)$ is a Gaussian process.*

Since $y_0(t)$ is Gaussian, for any $p > 0$, we have $E\{\sup_{0 \leq t \leq T} |y_0(t)|^p\} < \infty$. Being true for $n = 0$, let $E\{\sup_{0 \leq t \leq T} |y_n(t)|^p\} < \infty$ for $n = m$. From (4.8), the boundedness assumption on $f$ and the joint continuity of $\rho(t, s)$, it follows that $E\{\sup_{0 \leq t \leq T} |y_{m+1}(t)|^p\} < \infty$. So, by the induction principle it follows that, if $z_n = (y_0, y_1, \ldots, y_n)$, then

$$E\{\sup_{0 \leq t \leq T} |z_n|^p\} < \infty, \quad n \geq 0, \quad p > 0. \quad (4.9)$$

Our interest is in the asymptotic behavior of the process $x_\varepsilon(t, \omega)$ as $\varepsilon \downarrow 0$.

Towards this end define $\varepsilon^{n+1}\psi_{n+1}(t,\omega;\varepsilon) = x_\varepsilon(t,\omega) - \sum_{k=0}^n \varepsilon^k y_k(t,\omega)$. We claim that $E\{\sup_{0\leq t\leq T}|\psi_{n+1}(t,\omega;\varepsilon)|^2\} \leq M < \infty$. From relations (4.4), (4.7), and (4.8),

$$\varepsilon \int_0^t \rho(t,s) f(s, x_\varepsilon(s))\, ds$$
$$= \sum_{k=1}^n \varepsilon^k \int_0^t \rho(t,s) f_{k+1}(s; y_0(s), \ldots, y_{k+1}(s))\, ds + \varepsilon^{n+1}\psi_{n+1}(t,\omega;\varepsilon);$$

and

$$\varepsilon^{n+1}\psi_{n+1}(t,\omega;\varepsilon)$$
$$= \varepsilon \int_0^t \rho(t-s) \bigg\{ f(s; y_0(s) + \varepsilon y_1(s) + \cdots + \varepsilon^n y_n(s))$$
$$\quad - \sum_{k=1}^n \varepsilon^{k-1} f_{k-1}(s; y_0(s), \ldots, y_{k-1}(s)) \bigg\} ds$$
$$= \varepsilon \int_0^t \rho(t,s) \varepsilon^n f_n(s, y_0(s) + \theta y_1(s) + \cdots + \theta^n y_n(s), y_1(s), \ldots, y_n(s))\, ds,$$

$(0 \leq \theta \leq \varepsilon)$, by Taylor's formula. Hence, $\psi_{n+1}(t;\varepsilon) = \int_0^t \rho(t,s) f_n(s, \theta)\, ds$, where $f_n(t, \theta) = f_n(t; y_0(s) + \theta y_1(s) + \cdots + \theta^n y_n(s), y_1(s), \ldots, y_n(s))$. Recalling the hypotheses on $f(t, x)$ and noting that

$$\big||f_n(t,\theta)| - |f_n(t)|\big| \leq |f_n(t,\theta) - f_n(t)| \leq \theta \{\sup_k |y_k|^{n+1}\},$$

we get

$$\psi_{n+1}(t;\varepsilon) \leq \int_0^t \rho(t,s)[f_n(s) - \theta\{\sup_k |y_k|^{n+1}\}]\, ds.$$

From the joint continuity of $\rho(t,s)$, this yields our claim. Hence, we have the following theorem.

**Theorem 4.3** *Let $f(t,x)$ be a bounded function such that $f$ and all its partial derivatives (in $x$) up to order $(n+1)$ are bounded Lebesgue measurable in $t$ on $[0,T]$, $T > 0$, for each $x$ and Lipschitzian in $x$. Let $x_\varepsilon(t,\omega)$ be a mild solution of the perturbed integrodifferential Eq. (4.3) and the $y_n$ be given by relations (4.8). Then,*

$$x_\varepsilon(t,\omega) = y_0(t,\omega) + \varepsilon y_1(t,\omega) + \cdots + \varepsilon^n y_n(t,\omega) + \varepsilon^{n+1}\psi_{n+1}(t,\omega;\varepsilon)$$

*with probability* 1. *Moreover,*

$$P\{\sup_{\substack{0\leq t\leq T \\ 0\leq \varepsilon \leq \varepsilon_0}} \psi_{n+1}(t,\omega;\varepsilon) < \infty\} = 1,$$

*and*

$$E\{\sup_{0\leq t\leq T} |\psi_{n+1}(t,\omega;\varepsilon)|^2\} \leq M < \infty.$$

Next we consider a dynamical system perturbed by white noise:

$$\dot{x}(t) = \dot{x}_0 + \int_0^t m(x(s), \dot{x}(s))\, ds + \varepsilon \beta(t)$$

$$x(t) = x_0 + \int_0^t \dot{x}(s)\, ds. \tag{4.10}$$

(Just to stress, once again, that the applications of these systems are plenteous, we mention that such systems arise, among others, in problems associated with the penetration of cosmic particles in matter and in the study of electrical circuits with a nonlinear relaxation mechanism and weak thermal noise.) We assume that $m(\cdot, \cdot)$ is Lipschitzian so that a unique solution $\mathbf{x}_\varepsilon(t) = (x_\varepsilon(t), \dot{x}_\varepsilon(t))$ exists for system (4.10). Following Dubrovskii [10] we treat this system in order to study the limit behavior of the normalized invariant measure of the process $\mathbf{x}_\varepsilon(t)$; the first-order differential equation case has been extensively treated by Ventcel' and Freidlin [56], (see also Varadhan [55]). First some notations are in order. For $0 < T < \infty$, $C^T = C[0, T]$ is the Banach space of continuous functions with *sup* norm, $C_0^T = \{f \in C^T : f(0) = 0\}$, for $\mathbf{x}_0 \in R^2$ and $T > 0$, define an operator $A = A(\mathbf{x}_0, T)$ on $C_0^T$ by $Af = (g, \dot{g})$, where $(g, \dot{g})$ satisfies (4.10) with the standard Brownian motion $\beta$ replaced by $f \in C_0^T$, and $W = W(\mathbf{x}_0, T) = A[C_0^T]$ denotes the set of all admissible paths initiating from $\mathbf{x}_0$. Define $\|\mathbf{x}\| = |x| + |\dot{x}|$, (for $\mathbf{x} \in R^2$), and $\|\mathbf{y}\|_t = \sup_{0 \leq s \leq t} \|y(s)\|$, for $\mathbf{y} \in W$. It is a simple exercise to show that $A$ is continuous from $(C_0^T, \|\cdot\|_t)$ to $(W, \|\cdot\|_t)$. Note that $A$ is invertible and $(A^{-1}\mathbf{y})(t) = y(t) - y_0 - \int_0^t m(y(s), \dot{y}(s))\, ds$, for any $\mathbf{y} \in W(\mathbf{x}_0, T)$. Define, next, the action functional, on $C_0^T$, for the Wiener process: $F(f; 0, T) = \int_0^T [df(s)/ds]^2\, ds$ if $f$ is absolutely continuous and the integral is finite, and $F(f; 0, T) = \infty$ otherwise, and define the action along $\mathbf{y} \in W$ by $\alpha(\mathbf{y}) = F([A(\mathbf{x}_0, T)]^{-1}\mathbf{y}; 0, T)$. Note that

$$\alpha(\mathbf{y}) = \int_0^T [(d^2 y(s)/ds^2) - m(y(s), \dot{y}(s))]^2\, ds,$$

whenever $\mathbf{y}$ is absolutely continuous and the integral exists. For two sets $B_1$ and $B_2$ in $R^2$, define $V(B_1, B_2) = \inf \alpha(\mathbf{y})$, where the *inf* is taken over all admissible paths $\mathbf{y}$ whose initial points are in $B_1$ and terminal points $\mathbf{y}(T) \in B_2$, $0 < T < \infty$. Now, two points $\xi, \eta$ of the phase plane are said to be *equivalent* if $V(\xi, \eta) = 0 = V(\xi, \eta)$. It is shown [10] that the system (4.10) satisfies the following property: *For any compact (equivalence class) $K$ containing at least two points and for any $h > 0$, there is a $\delta > 0$ such that if the distance $d(\mathbf{x}, K) < \delta$, then $V(\mathbf{x}, K) < h$ and $V(K, \mathbf{x}) < h$.*

We shall need the following assumptions made on the random integrodifferential system (4.10). $(A_\omega)$: There are a finite number of compact equivalence classes $K_1, \ldots, K_\lambda$ such that each $\omega$-limit set of the system is contained in one of these $K_i$. (Recall that the $\omega$-*limit set* or *positive* limit set of the system is the set of limit points of any solution $\mathbf{x}(t)$ of (4.10) as $t \to \infty$. The $\alpha$-*limit set* or *negative limit set* is similarly defined as $t \to -\infty$.) $(A_\partial)$: There exists an increasing sequence of regions $\{D_n, n \geq 0\}$ such that

(i) $\bigcup_n D_n = R^2$,
(ii) if the system $\mathbf{x}(t)$ initiates from some point $x \in \partial D_n$, then $\mathbf{x}(t) \in D_n$ for sufficiently small $t > 0$,
(iii) $d(D_n, D_{n+1}) \geq r > 0$, and
(iv) $\sup_{x \in \partial D_n} \inf \{t: \mathbf{x}(t) \in \partial D_n\} \leq t_0 < \infty$, for all systems initiating from $\mathbf{x} \in \partial D_n$.

Next we state several lemmas. These lemmas and their proofs are essentially given in Ventcel' and Freidlin [56], also see Dubrovskii [10].

**Lemma 4.4** (i) *For any $h$, $\delta > 0$ and $C > 0$, there exists an $\varepsilon_0 > 0$ such that, for all $\varepsilon$, $0 < \varepsilon < \varepsilon_0$, and for all $\mathbf{x} \in R^2$ we have*

$$P_\mathbf{x}^\varepsilon \{\|\mathbf{x}_\varepsilon - \mathbf{g}\| < \delta\} \geq \exp\{-(\alpha(\mathbf{g}) + h)/2\varepsilon^2\},$$

*where $\mathbf{g} \in W(\mathbf{x}, T)$, $\int_0^T [1 + d^2 g(t)/dt^2]^2 \, dt < C$, and $P_\mathbf{x}^\varepsilon$ is the probability conditioned on the paths originating from $\mathbf{x}$.*
(ii) *For any $h$, $\delta > 0$, small $\varepsilon > 0$ and all $\mathbf{x} \in R^2$, we have*

$$P_\mathbf{x}^\varepsilon \{d_T(\mathbf{x}^\varepsilon, G_a(\mathbf{x}, T)) \geq \delta\} \leq 2 \exp\{[-a + h(a + T)]/2\varepsilon^2\}$$

*for any $a$ and $T$, where $G_a(\mathbf{x}, T) = \{\mathbf{g} \in W(\mathbf{x}, T): \alpha(\mathbf{g}) \leq a\}$ and $d_T(\mathbf{f}, G_a(\mathbf{x}, T)) = \inf_{\mathbf{G} \in G_a} \|\mathbf{f} - \mathbf{g}\|_T$.*

**Lemma 4.5** (i) *If $K$ is any compact subset of the phase plane not entirely containing any positive limit set of the system (4.10), then there exist a $c_0 > 0$ and a $T_0 > 0$ such that, for sufficiently small $\varepsilon > 0$, all $T > T_0$, and any $\mathbf{x} \in K$,*

$$P_\mathbf{x}^\varepsilon \{\tau_{\partial K} > T\} \leq \exp\{-c_0(T - T_0)/2\varepsilon^2\},$$

*where $\tau_{\partial K}$ is the hitting time of the boundary $\partial K$.*
(ii) *Under the condition $(A_\partial)$, there exist a $c_1$ and a $T_1$ such that, for all sufficiently small $\varepsilon > 0$, all $\mathbf{x} \in \partial D_1$ and all $T > T_1$,*

$$P_\mathbf{x}^\varepsilon \{\tau_{\partial D_0} > T\} \leq \exp\{-c_1(T^{\frac{1}{2}} - T_1^{\frac{1}{2}})/2\varepsilon^2\}.$$

From here on until the end of the proof of Theorem 4.7 we assume that the condition $(A_\omega)$ holds, and that $\mathbf{x}_\varepsilon(t)$ is a strong Feller process and is recurrent relative to some compact set. To obtain the asymptotic behavior of the invariant measure $\mu_\varepsilon$ of $\mathbf{x}_\varepsilon(t)$ one constructs a Markov chain and applies a result of Khaśminskii [34]. (Recall that a finite measure $\mu_\varepsilon$ defined on the Borel sets of $R^2$ is said to be *invariant* of the strong Feller process $\mathbf{x}_\varepsilon(t)$ if for any bounded continuous function $f(\mathbf{x})$ and $t > 0$, we have $\int T_t^{(\varepsilon)} f(\mathbf{x}) \, d\mu_\varepsilon(\mathbf{x}) = \int f(\mathbf{x}) \, d\mu_\varepsilon(\mathbf{x})$, where $\{T_t^{(\varepsilon)}\}$ is the transition semigroup of $\mathbf{x}_\varepsilon$.) Enclose each of the compactum $K_i$ by a pair $\mathbf{n}_i$ and $\mathbf{N}_i$ of neighborhoods such that the boundaries $\mathbf{b}_i = \partial \mathbf{n}_i$ and $\mathbf{B}_i = \partial \mathbf{N}_i$ are smooth, $\mathbf{n}_i \cup \mathbf{b}_i \subset \mathbf{N}_i$, and $(\mathbf{N}_i \cup \mathbf{B}_i) \cap (\mathbf{N}_j \cup \mathbf{B}_j) = \emptyset$, $i \neq j$. Let $\mathbf{b} = \bigcup_i \mathbf{b}_i$ and $\mathbf{B} = \bigcup_i \mathbf{B}_i$. Introduce the random times

$$\tau_0 = 0, \quad \sigma_n = \inf\{t \geq \tau_{n-1}: \mathbf{x}_\varepsilon(t) \in \mathbf{B}\}, \quad \tau_n = \inf\{t \geq \sigma_n: \mathbf{x}_\varepsilon(t) \in \mathbf{b}\},$$

and consider the Markov chain $\{\xi_n = \mathbf{x}_\varepsilon(\tau_n), n \geq 0\}$. It is well known (cf. Khaśminskii [34]) that the invariant measure $\mu_\varepsilon$ of $\mathbf{x}_\varepsilon$ can be expressed (up to a constant factor) in terms of the invariant measure $v_\varepsilon$ of the Markov chain $\{\xi_n\}$ as follows:

$$\mu_\varepsilon(B) = \int_\mathbf{b} v_\varepsilon(d\mathbf{y}) E_\mathbf{y}^\varepsilon \int_0^{\tau_1} I_B(\mathbf{x}_\varepsilon(t)) \, dt. \tag{4.11}$$

**Lemma 4.6** *Let $V_{ij} = V(K_i, K_j)$, (noting that $V(\mathbf{x}, \mathbf{y})$ has the same value $V_{ij}$ for all $\mathbf{x} \in K_i$ and $\mathbf{y} \in K_j$), and $K = \bigcup_{i=1}^\lambda K_i$. If for some region $D \supset K$ we have $V(K, \partial D) = \max_{1 \leq i,j \leq \lambda} V_{ij} = V > 0$, then, for any $h > 0$, there exists an $r > 0$, such that for any choice of neighborhoods $\mathbf{n}_i$ and $\mathbf{N}_i$ inside the $r$-neighborhoods of $K_i$ ($i = 1, \ldots, \lambda$), for sufficiently small $\varepsilon > 0$ and arbitrary $\mathbf{x} \in \mathbf{b}_i$ we get the inequalities*

$$\exp[(-V_{ij} - h)/2\varepsilon^2] \leq p^{(\lambda-1)}(\mathbf{x}, \mathbf{b}_j) \leq \exp[(-V_{ij} + h)/2\varepsilon^2] \tag{4.12}$$

*where $p^{(\lambda-1)}(\cdot, \cdot)$ is the $(\lambda - 1)$-step transition probability of the chain $\{\xi_n\}$.*

For a proof, see Lemma 6.1 in [56] and Lemma 3.5 in [10].

Let $\Lambda = \{1, 2, \ldots, \lambda\}$. For any $i \in \Lambda$, define a $\{i\}$-graph as follows: it is a graph consisting of arrows $m \to n$ such that

(i) precisely one arrow emanates from each point $m \in \Lambda \setminus \{i\}$, and
(ii) there are no closed cycles in the graph.

Let $\Gamma\{i\}$ denote the set of all $\{i\}$-graphs on $\Lambda$. Define

$$V\{i\} = \min_{\gamma \in \Gamma\{i\}} \sum_{(m \to n) \in \gamma} V_{mn},$$

$$V = \min_{1 \leq i \leq \lambda} V\{i\} \quad \text{and} \quad K_0 = \bigcup_{i: V\{i\} = V} K_i.$$

For the details of the following theorem we refer to Ventcel' and Freidlin [56] and Dubrovskii [10].

**Theorem 4.7** *Assume that the random integrodifferential system (4.10) satisfies the conditions $(A_\omega)$ and $(A_\partial)$ and that $\mathbf{x}_\varepsilon(t) = (x_\varepsilon(t), \dot{x}_\varepsilon(t))$ is a strong*

*Feller process. Then, for any closed set F, which does not intersect $K_0$, we have $\lim_{\varepsilon \to 0} \mu_\varepsilon(F) = 0$, where $\mu_\varepsilon$ is the normalized invariant measure of $\mathbf{x}_\varepsilon$.*

*Proof* From the condition $\{A_\partial\}$, and hence from Lemma 4.5(ii), it follows that there exist constants $c_1$ and $T_1$ such that for all $n \geq 0$, all $\mathbf{x} \in \partial D_{n+1}$, $T > T_1$ and sufficiently small $\varepsilon > 0$, $P_\mathbf{x}^\varepsilon\{\tau_{\partial D_n} > T\} \leq \exp\{-c_1(T^{\frac{1}{2}} - T_1^{\frac{1}{2}})/2\varepsilon^2\}$, and hence, $E_\mathbf{x}^\varepsilon \tau_{\partial D_n} \leq T_1 + (2\varepsilon^2/c_1) < T_1 + 1 = t_1$, say. This implies that the process $\mathbf{x}_\varepsilon$ is recurrent relative to the compactum $D_0 \cup D_0$. Consequently, we can construct the Markov chain $\{\xi_n, n \geq 0\}$ as soon as the boundaries $\mathbf{b}_i$ and $\mathbf{B}_i$ are obtained (for the details of how the neighborhoods $\mathbf{n}_i$ and $\mathbf{N}_i$ are chosen see Vent-tsel' and Freidlin [56, Section 8]).

If $\mathbf{g} \in W(\mathbf{x}, T)$ and $\alpha(\mathbf{g}) \leq h$, then it is easy to see that $\|\mathbf{g} - \mathbf{x}_\varepsilon(\cdot, \mathbf{x})\|_T \leq N^0 h^{\frac{1}{2}}$, for some constant $N^0 > 0$ depending only on $T$. This implies, on the account of condition $(A_\partial)$, that one can find a $v > 0$ such that, for all $n \geq 1$, $V(\partial D_n, \Delta_n) \geq v$, where $\Delta_n = \partial\{\mathbf{y}: d(\mathbf{y}, D_n) \leq r/2\}$. Hence, $V(\bigcup_i K_i, \partial D_m) > \max_{i,j} V_{ij}$, for some $m$, and consequently Lemma 4.6 is obtained.

Next we claim that

$$\sup_{\mathbf{x} \in \mathbf{B}} E_\mathbf{x}^\varepsilon \tau_1 < T < \infty, \tag{4.13}$$

where $T$ is independent of $\varepsilon$. Once (4.13) is obtained, Khaśminskii [34] provides us with the formula

$$\mu_\varepsilon(F) = \int_\mathbf{b} v_\varepsilon(d\mathbf{y}) E_\mathbf{y}^\varepsilon \int_0^{\tau_1} I_F(\mathbf{x}_\varepsilon(t)) \, dt,$$

and one can follow the proof of Theorem 8.1 in Ventcel' and Freidlin [56] to complete our proof. So, it remains only to establish (4.13). Noting that $D_0$ contains all positive limit sets of the random integrodifferential system (4.10) and, consequently, choosing $\mathbf{B}$ such that $\mathbf{B} \subset D_0$, we see that $E_\mathbf{x}^\varepsilon \tau_1 = E_\mathbf{x}^\varepsilon \tau_\mathbf{b}$, if $\mathbf{x} \in \mathbf{B}$. From Lemma 4.5(i) it follows that $\mathbf{x} \in \sup_{\mathbf{x} \in D_1 \setminus \mathbf{N}} E_\mathbf{x}^\varepsilon \tau_{D_1 \cup \mathbf{b}} \leq T_0 + 1 = t_0$, where $\mathbf{N} = \bigcup \mathbf{N}_i$. It is easy to see that $\sup_{\mathbf{x} \in D_0 \setminus \mathbf{N}} P_\mathbf{x}^\varepsilon\{\tau_\mathbf{b} > \tau_{\partial D_1}\} \leq \alpha < 1$. Let $\tau^N$ denote $\tau \wedge N$ for any stopping time $\tau$, $(N > 0)$. Then, for $\mathbf{x} \in \mathbf{B}$, we have

$$E_\mathbf{x}^\varepsilon \tau_\mathbf{b}^N \leq E_\mathbf{y}^\varepsilon \tau_{\mathbf{b} \cup \partial D_1} + E_\mathbf{x}^\varepsilon E_{\mathbf{x}_\varepsilon(\tau(\partial D_1))}^\varepsilon \tau_\mathbf{b}^N P_\mathbf{x}^\varepsilon\{\tau_\mathbf{b} > \tau_{\partial D_1}\}$$

$$\leq t_0 + \alpha E_\mathbf{x}^\varepsilon E_{\mathbf{x}_\varepsilon(\tau(\partial D_1))}^\varepsilon (\tau_{\partial D_0} + E_{\mathbf{x}_\varepsilon(\tau(\partial D_0))}^\varepsilon \tau_\mathbf{b}^N)$$

$$\leq t_0 + \alpha t_1 + \alpha \sup_{\mathbf{y} \in \partial D_0} E_\mathbf{y}^\varepsilon \tau_\mathbf{b}^N.$$

But,

$$E_\mathbf{y}^\varepsilon \tau_\mathbf{b}^N \leq E_\mathbf{y}^\varepsilon \tau_{\mathbf{b} \cup \partial D_1} + \alpha E_\mathbf{y}^\varepsilon E_{\mathbf{x}_\varepsilon(\tau(\partial D_1))}^\varepsilon \tau_\mathbf{b}^N$$

$$\leq t_0 + \alpha t_1 + \alpha \sup E_\mathbf{z}^\varepsilon \tau_\mathbf{b}^N,$$

so that

$$\sup_{x \in B} E_x^\varepsilon \tau_b^N \leq t_0 + \alpha t_1 + \frac{\alpha}{1-\alpha}(t_0 + \alpha t_1) = T < \infty.$$

Now, as $N \to \infty$, we get (4.13), and this completes the proof. (We omitted several details, and the interested reader should read Vent-tsel' and Freidlin [56]).

For an extension of the averaging method of Bogoliubov and Mitropolsky to the random integrodifferential equation

$$\dot{x}(t,\omega) = k(t)x(t,\omega) + \int_0^t K(t,s)x(s,\omega)\,ds + \varepsilon f(t,x(t,\omega)), \qquad x(0) = 0,$$

and for the diffusion type approximation of the process $x_\varepsilon(t,\omega)$ as $\varepsilon \to 0$, we refer to Kannan [30]. We also refer to a recent expository article of Papanicolaou [50] for a general account of diffusion-type approximations of a large class of transport processes.

## V. Vibrating String

This final section treats an example of a vibrating string forced by a white noise. The results presented here are taken from Cabaña [7] (also see, Feller [14], McKean [41], and Itô and McKean [24]). The classical equation of a damped vibrating string with an external force is given by

$$\frac{\partial^2 u(t,z)}{\partial t^2} + 2b\frac{\partial u(t,z)}{\partial t} = \frac{\partial^2 u(t,z)}{\partial z^2} + F(t,z), \qquad (5.1)$$

where $F$ is the external force. The usefulness of the classical theory suffers under some unnatural regularity conditions imposed on these equations. Feller derived (see [14]) a more general and intuitive form of the equation from a purely analytical hypothesis. We adopt Feller's generalized equation of a vibrating string. Consider the string as a one parameter family of pair of functions $(u(t,\cdot), \dot{u}(t,\cdot))$ defined on an interval $I(\ni 0)$, where $u(t,\cdot)$ is the position of the string at time $t$ and $\dot{u}(t,\cdot)$ is the velocity. Corresponding to the classical mass distribution and the effect of elastic force let us consider, respectively, a canonical measure $\mu$ on $I$ (see Section I.C) and a Borel measure $\eta$ on $I$, which is finite on closed intervals. Let $H_\mu$ denote the Hilbert space $L_2(I,\mu)$ with norm $\|\phi\|_\mu = [\int \phi^2 d\mu]^{1/2}$. Define an operator $L$ by

$$L\phi \cdot d\mu = d\phi' - \phi \cdot d\eta. \qquad (5.2)$$

The domain of $L$ is the set of all continuous functions $\phi$ such that $L\phi$ is continuous, the derivative $\phi' = d\phi/dx$ exists at each point where $\mu$ and $\eta$

are continuous (as distribution functions), the one-sided derivatives $\phi^+$ and $\phi^-$ exist at all points with the obvious continuity properties, and $\phi'$ is of bounded variation. The interpretation of (5.2) is that the both sides are equal as measures. In order to express $L$ in a more familiar form, we first note that there are two positive convex functions $\xi$ and $\zeta$ solving $L\phi = 0$ and that every solution of $L\phi = 0$ is a linear combination of $\xi$ and $\zeta$. Now $L$ can be put in the form

$$L\phi = \frac{1}{\psi}\frac{d}{d\mu}\left(\psi^2 \frac{d}{dx}\left(\frac{\phi}{\psi}\right)\right). \tag{5.3}$$

with $d\eta = d\psi'/\psi$, ($\psi = \xi$ or $\zeta$). Before rewriting the equation of the damped vibrating string as an integrodifferential equation, in terms of $L$, we present two examples of $L$, (see Feller [14]).

**Example 1** *Dirac function for the elastic force.* Let $I$ be an interval containing the origin. Set $d\mu = dx$, and let $\psi = 1 + (|x|/2)$. If $x$ represents the natural physical scale, then we have a string with uniform mass distribution and a constant horizontal component of tension. Here $\psi$ gives a possible position of equilibrium. That is, we have an elastic force concentrated at the origin, and the measure $\eta$ (define by $d\eta = d\psi'/\psi$) attaches unit weight to the origin and null weight to each interval not containing the origin. Then,

$$L\phi = \left(1 + \frac{|x|}{2}\right)^{-1} \frac{d}{dx}\left\{\left(1 + \frac{|x|}{2}\right)\phi' - \frac{1}{2}\phi \cdot \operatorname{sgn} x\right\}.$$

**Example 2** *Positive mass concentrated at the origin.* Let

$$\eta = 0 \quad \text{and} \quad \mu(x) = \begin{cases} x, & \text{if } x < 0 \\ 1 + x, & \text{if } x \geq 0. \end{cases}$$

We have a uniform mass distribution with the exception that the origin carries unit weight. Clearly, $L\phi = \phi''$ in every interval not containing the origin, and at the origin

$$\lim_{x \to 0} \phi''(x) = \phi^+(0) - \phi^-(0) = \lim_{x \downarrow 0} \phi'(x) - \lim_{x \downarrow 0} \phi'(x).$$

We rewrite the equation of the damped vibrating string, with an external force $F$ and initial conditions $u(0, \cdot)$, $\dot{u}(0, \cdot)$, as an integrodifferential system:

$$u(t, \cdot) = u(0, \cdot) + \int_0^t \dot{u}(s, \cdot)\, ds,$$

$$\dot{u}(t, \cdot) = \dot{u}(0, \cdot) + \int_0^t \{Lu(s, \cdot) - 2b\dot{u}(s, \cdot)\}\, ds + \int_0^t F(s, \cdot)\, ds. \tag{5.4}$$

The formal solution of system (5.4) is

$$\begin{bmatrix} u(t, \cdot) \\ \dot{u}(t, \cdot) \end{bmatrix} = e^{Bt} \begin{bmatrix} u(0, \cdot) \\ \dot{u}(0, \cdot) \end{bmatrix} + \int_0^t e^{B(t-s)} \begin{bmatrix} 0 \\ 1 \end{bmatrix} F(s, \cdot) \, ds, \qquad (5.5)$$

where $B = \begin{bmatrix} 0 & 1 \\ L & -2b \end{bmatrix}$. Instead of (5.5) we shall work with the Fourier transform of solution (5.5) (cf. Cabaña [7]). McKean has shown, (cf. McKean [41] and Itô and McKean [24]), that the operator $L$ induces a Fourier transform between $H_\mu$ and another Hilbert space $H_\nu$, defined below. From here on, the indices will run over the set $A = \{1, 2\}$, and if an expression involves a repeated index, then summation with respect to that index is understood.

Let $e_i(z, \lambda)$ be the solution of

$$Le_i(\cdot, \lambda) = \lambda e_i(\cdot, \lambda),$$

$$e_1(0, \lambda) = e_2^+(0, \lambda) = 1, \qquad e_2(0, \lambda) = e_1^+(0, \lambda) = 0.$$

There exists a Borel measure $\nu = [\nu_{ij}]$ from $(-\infty, 0]$ into $(2 \times 2)$-symmetric nonnegative definite matrices such that if $H_\nu$ denotes the Hilbert space $L_2((-\infty, 0], [d\nu_{ij}])$ with norm

$$\|\phi(\cdot)\| = \|(\phi_1(\cdot), \phi_2(\cdot))\| = \left( \int_{-\infty}^0 \phi_i(\lambda) \phi_j(\lambda) \, d\nu_{ij}(\lambda) \right)^{\frac{1}{2}},$$

then the Fourier transform from $H_\mu$ to $H_\nu$ is defined by

$$u \to (u_i^*(\cdot)) = \left( \int_I u(z) e_i(z, \cdot) \, d\mu(z) \right) \in H_\nu \qquad (5.6)$$

for $u \in H_\mu$. Here, we have the inverse of (5.6) given by

$$u^* = (u_i^*) \to u(\cdot) = \int_{-\infty}^0 u_i^*(\lambda) e_j(\cdot, \lambda) \, d\nu_{ij}(\lambda) \in H_\mu, \qquad (5.7)$$

and the Phancherel theorem $\|u\|_\mu = \|u^*\|$ holds. We will be working with a Hilbert space whose norm can be described in terms of energy. Let $H$ be the Hilbert space of functions $\phi(\cdot) = (\phi_1(\cdot), \phi_2(\cdot))$ on $(-\infty, 0]$ with norm

$$\|\|\phi(\cdot)\|\| = \left\{ \int_{-\infty}^0 (-\lambda) \phi_i(\lambda) \phi_j(\lambda) \, d\nu_{ij}(\lambda) \right\}^{\frac{1}{2}}$$

and inner product denoted by $\langle \cdot, \cdot \rangle$. Now define the Hilbert space $H_\mathscr{E}$ of pairs $(\phi; \psi) = ((\phi_i), (\psi_i))$, $\phi \in H$, $\psi \in H_\nu$, with norm

$$\|(\phi; \psi)\|_\mathscr{E} = \{\|\|\phi\|\|^2 + \|\psi\|^2\}^{\frac{1}{2}}.$$

Here $\frac{1}{2} \|(\phi; \psi)\|_\mathscr{E}^2$ is said to be the *energy* of $(\phi; \psi) \in H_\mathscr{E}$, and the energy can be decomposed into the *potential energy* $\frac{1}{2} \|\|\phi\|\|^2$ and *kinetic energy* $\frac{1}{2} \|\psi\|^2$.

Now let

$$B_\lambda = \begin{bmatrix} 0 & 1 \\ \lambda & -2b \end{bmatrix} \quad \text{and} \quad \begin{bmatrix} \alpha_{11}(t,\lambda) & \alpha_{12}(t,\lambda) \\ \alpha_{21}(t,\lambda) & \alpha_{22}(t,\lambda) \end{bmatrix} = \exp(B_\lambda t).$$

Then, the formal solution (5.5) of system (5.4) can be given in the transformed version as follows:

$$u_i^*(t,\lambda) = \alpha_{11}(t,\lambda) u_i^*(0,\lambda) + \alpha_{12}(t,\lambda) \dot u_i^*(0,\lambda)$$

$$+ \int_0^t \int \alpha_{1j}(t,\lambda) \alpha_{j2}(-s,\lambda) e_i(z,\lambda) F(s,z) \, ds \, d\mu(z),$$

$$\dot u_i^*(t,\lambda) = \alpha_{21}(t,\lambda) u_i^*(0,\lambda) + \alpha_{22}(t,\lambda) \dot u_i^*(0,\lambda)$$

$$+ \int_0^t \int \alpha_{2j}(t,\lambda) \alpha_{j2}(-s,\lambda) e_i(z,\lambda) F(s,z) \, ds \, d\mu(z). \quad (5.8)$$

The following proposition shows that the energy is conserved for a free string, and it describes how the damping reduces energy (cf. Feller [14] and Cabaña [7]).

**Proposition 5.1** *Define a function* $G^* = (U^*, \dot U^*)$ *by*

$$U^*(t) = U^*(t,\cdot) = (U_i^*(t,\cdot)) = (\alpha_{11}(t,\cdot) u_i^*(0,\cdot) + \alpha_{12}(t,\cdot) \dot u_i^*(0,\cdot)),$$

$$\dot U^*(t) = \dot U^*(t,\cdot) = (\dot U_i^*(t,\cdot)) = (\alpha_{21}(t,\cdot) u_i^*(0,\cdot) + \alpha_{22}(t,\cdot) \dot u_i^*(0,\cdot)). \quad (5.9)$$

*Then*, (a) (*conservation of energy*): *if* $b = 0$, *the energy of* $G^*$ *is constant* (b) (*decay of energy due to damping*): *for any* $b \geqslant 0$,

$$\tfrac{1}{2}\|G^*(t,\cdot)\|_\mathscr{E}^2 + 2b \int_0^t \|\dot u^*(s,\cdot)\|^2 \, ds$$

*is constant.*

*Proof* It is enough to prove (b), because (a) is a special case of (b). We begin by rewriting the matrix $[\alpha_{ij}(t,\lambda)]$ as follows:

$$\alpha_{ij}(t,\lambda) = \begin{bmatrix} e^{-bt}(\cos \rho t + b\rho^{-1} \sin \rho t) & e^{-bt}\rho^{-1} \sin \rho t \\ e^{-bt}\lambda \rho^{-1} \sin \rho t & e^{-bt}(\cos \rho t - b\rho^{-1} \sin \rho t) \end{bmatrix}$$

for $-\infty < \lambda < -b^2$ and $\rho = (-\lambda - b^2)^{\frac{1}{2}}$;

$$\alpha_{ij}(t,\lambda) = \begin{bmatrix} e^{-bt}(\cosh rt + br^{-1} \sinh rt) & e^{-bt} r^{-1} \sinh rt \\ e^{-bt}\lambda r \sinh rt & e^{-bt}(\cosh rt - br^{-1} \sinh rt) \end{bmatrix}$$

for $-b < \lambda \leqslant 0$ and $r = (\lambda + b^2)^{\frac{1}{2}}$; and finally

$$\alpha_{ij}(t,-b^2) = \begin{bmatrix} e^{-bt}(1+bt) & te^{-bt} \\ \lambda t e^{-bt} & e^{-bt}(1-bt) \end{bmatrix}.$$

Noting that
$$(\alpha_{ii}(t,\lambda))^2 \leq 4e^{4t^-}, \quad i = 1, 2,$$
$$-\lambda\alpha_{11}^2(t,\lambda) \leq e^{4t^-}, \quad (-\lambda)^{-1}\alpha_{21}^2(t,\lambda) \leq e^{4t^-}, \quad (5.10)$$

where $t^- = \max\{0, -t\}$, it follows that

$$\alpha_{11}(t,\cdot)u^*(0,\cdot) \in H, \quad \alpha_{12}(t,\cdot)\dot{u}^*(0,\cdot) \in H, \quad \alpha_{21}(t,\cdot)u^*(0,\cdot) \in H_v$$

and
$$\alpha_{22}(t,\cdot)\dot{u}^*(0,\cdot) \in H_v.$$

Consequently $U^* \in H$, $\dot{U}^* \in H_v$ and hence $G^* \in H_{\mathscr{E}}$. Since

$$\begin{bmatrix} \dot{\alpha}_{11} & \dot{\alpha}_{12} \\ \dot{\alpha}_{21} & \dot{\alpha}_{22} \end{bmatrix} = \frac{\partial}{\partial t}\exp(B_\lambda t)$$

$$= B_\lambda \exp(B_\lambda t) = \begin{bmatrix} \alpha_{21} & \alpha_{22} \\ \lambda\alpha_{11} - 2b\alpha_{21} & \lambda\alpha_{12} - 2b\alpha_{22} \end{bmatrix},$$

we have that

$$(\partial/\partial t) U^*(t,\lambda) = \dot{U}^*(t,\lambda) \quad \text{and} \quad (\partial/\partial t)\dot{U}^*(t,\lambda) = \lambda U^*(t,\lambda) - 2b\dot{U}^*(t,\lambda).$$

Therefore

$$\frac{\partial}{\partial t}\{-\lambda U_i^*(t,\lambda) U_j^*(t,\lambda) + \dot{U}_i^*(t,\lambda)\dot{U}_j^*(t,\lambda)\} = -4b\dot{U}_i^*(t,\lambda)\dot{U}_j^*(t,\lambda)$$

and hence

$$\frac{\partial}{\partial t}\left\{\|\|U^*(t)\|\|^2 + \|\dot{U}^*(t)\|^2 + 4b\int_0^t \|\dot{U}^*(s)\|^2\,ds\right\} = 0,$$

which gives (b). This completes the proof.

Next, we study some of the properties of the (formal) solution of the equation

$$\frac{\partial^2 u}{\partial t^2} + 2b\frac{\partial u}{\partial t} = Lu + \frac{\partial^2 \beta}{\partial t\,\partial\mu} \quad (5.11)$$

of a vibrating string driven by an external force $(\partial^2\beta/\partial t\,\partial\mu)$ of plane white noise type, where $\beta$ is a $\mu$-Brownian motion (see Section I.C). Define the stochastic integrals $\eta_{ji}$ as follows:

$$\eta_{ji}(t) = \int_0^t \int \alpha_{j2}(-s,\cdot)\gamma\, e_i(z,\cdot)\, d\beta(s,z), \quad (i,j = 1,2), \quad (5.12)$$

where the factor $\gamma(\lambda)$ is subject to the constraint that the integral

$$k = \int_{R^1} \int_{-\infty}^{0\mathscr{E}} \gamma^2(\lambda) e_i(z,\lambda) e_j(z,\lambda)\, dv_{ij}(\lambda)\, d\mu(z) < \infty. \quad (5.13)$$

Recalling the definition of $G^* = (U^*, \dot{U}^*)$, the transformed version of the formal solution of Eq. (5.11) is given by

$$u^*(t) = U^*(t) + \alpha_{1j}(t, \cdot)\eta_j(t),$$
$$\dot{u}^*(t) = \dot{U}^*(t) + \alpha_{2j}(t, \cdot)\eta_j(t). \tag{5.14}$$

Let $f^*(t) = (u^*(t), \dot{u}^*(t))$ denote the process given by (5.14) where $f^*(0)$ is, by assumption, an $\mathscr{A}_0$-measurable function such that $P\{\|f^*(0)\|_\mathscr{E} < \infty\} = 1$. From the estimates in (5.10) it follows that

$$\eta_{1i}(t) \in H, \qquad \eta_{2i}(t) \in H_v, \qquad \text{a.s.,} \qquad \text{for all } t,$$
$$u^*(t) \in H, \qquad \dot{u}^*(t) \in H_v, \qquad f^*(t) \in H_\mathscr{E}, \qquad \text{for all } t, \quad \text{a.s.} \tag{5.15}$$

**Theorem 5.2** *Define a process $z(t)$ by*

$$z(t) = \|f^*(t)\|_\mathscr{E}^2 + 4b \int_0^t \|\dot{u}^*(s)\|^2 \, ds - kt. \tag{5.16}$$

*Then,* (a) *$z(t)$ has a stochastic integral representation*

$$z(t) = z(0) + 2 \int_0^t \int \langle \dot{u}^*(s), \gamma e(z, \cdot)\rangle \, d\beta(t, z).$$

*where $e = (e_i)$,* (b) *the process $\{z(t), \mathscr{A}_t, t \geq 0\}$ is a martingale,* (c) *for $|2k\alpha t| < 1$, we have*

(i) $\quad E\{[z(t)+kt]^n \mid \mathscr{A}_0\} = \sum_{j=0}^n \frac{(2n)!}{(n-j)! \, 2^{n-j} (2j)!} [z(0)]^j (kt)^{n-j},$

(ii) $\quad E\{e^{\alpha z(t)} \mid \mathscr{A}_0\} \leq (1 - 2\alpha kt)^{-\frac{1}{2}} \exp\left\{-\alpha kt + \frac{\alpha z(0)}{1 - 2\alpha kt}\right\},$

(d) *for each positive $c$ and $0 < \alpha < (2kt)^{-1}$, we have*

$$P\{\|f^*(s)\|_\mathscr{E}^2 \leq c + ks, \, s \in (0, t)\} > 1 - \frac{e^{-\alpha(c+kt)}}{(1-2k\alpha t)^{\frac{1}{2}}} E\left\{\exp \frac{\alpha \|f^*(0)\|_\mathscr{E}^2}{1 - 2\alpha kt}\right\}.$$

*Proof* (a) Define a function $\phi(t, \eta_1, \eta_2)$ on $R_+^1 \times H_\mathscr{E}$ by

$$\phi(t, \eta_1, \eta_2) = \|\|U^*(t) + \alpha_{1j}(t, \cdot)\eta_j\|\|^2 + \|\dot{U}^*(t) + \alpha_{2j}(t, \cdot)\eta_j\|^2.$$

Noting that $D_0 \phi(t) = -4b\|\dot{u}^*(t)\|^2$, from Itô's lemma (see Theorem 1.4), applied to $\phi$, we get

$$\|f^*(t)\|_\mathscr{E}^2 = \|f^*(0)\|_\mathscr{E}^2 + \int_0^t (-4b) \|\dot{u}^*(s)\|^2 \, ds$$

$$= +2\int_0^t \int \langle u^*(s), \alpha_{1j}(s,\cdot)\alpha_{2j}(-s,\cdot)\gamma(\cdot)e(z,\cdot)\rangle \, d\beta(s,z)$$

$$+ 2\int_0^t \int \langle \dot{u}^*(s), \alpha_{2j}(s,\cdot)\alpha_{j2}(-s,\cdot)\gamma(\cdot)e(z,\cdot)\rangle \, d\beta(s,z)$$

$$+ \int_0^t \int \|\alpha_{2j}(s,\cdot)\alpha_{j2}(-s,\cdot)\gamma(\cdot)e(z,\cdot)\|^2 \, ds \, d\mu(z). \qquad (5.17)$$

Noting that

$$\alpha_{ij}(s,\cdot)\alpha_{jk}(-s,\cdot) = \begin{Bmatrix} 1 & \text{if } i=k \\ 0 & \text{if } i\neq k \end{Bmatrix}.$$

Itô's formula reduces to

$$\|f^*(t)\|_{\mathscr{E}}^2 = \|f^*(0)\|_{\mathscr{E}}^2 - 4b\int_0^t \|\dot{u}^*(s)\|^2 \, ds + kt$$

$$+ 2\int_0^t \int \langle \dot{u}^*(s), \gamma(\cdot)e(z,\cdot)\rangle \, d\beta(s,z),$$

from which (a) follows.

(b) That $\{z(t), \mathscr{A}_t, t \geqslant 0\}$ is a martingale follows from the properties of stochastic integrals.

(c) Since $\{z(t)\}$ is a martingale and

$$z(t) + kt = z(0) + \int_0^t k \, ds + 2\int_0^t \int \langle \dot{u}^*(s), \gamma e(z,\cdot)\rangle \, d\beta(s,z),$$

from Itô's formula, we get

$$(z(t)+kt)^n = [z(0)]^n + \int_0^t n(z(s)+ks)^{n-1} k \, ds$$

$$+ \int_0^t \int 2n[z(s)+ks]^{n-1} \langle \dot{u}^*(s), \gamma e(z,\cdot)\rangle \, d\beta(s,z)$$

$$+ \frac{n^2-n}{2} \int_0^t \int 4[z(s)+ks]^{n-2} \langle \dot{u}^*(s), \gamma e(z,\cdot)\rangle \, ds \, d\mu(z).$$

Hence, observing $\|\dot{u}^*(t)\|^2 \leqslant z(t)+kt$,

$$E\{[z(t)+kt]^n \mid \mathscr{A}_0\}$$

$$\leqslant [z(0)]^n + nk\int_0^t c_{n-1}(s) \, ds$$

$$+ 2(n^2-n)kE\left\{\int_0^t \int [z(s)+ks]^{n-2} \|\dot{u}^*(s)\|^2 \, ds \, d\mu(z) \,\Big|\, \mathscr{A}_0\right\}$$

$$\leqslant [z(0)]^n + nk \int_0^t c_{n-1}(s)\, ds + 2k(n^2-n) \int_0^t E\{[z(s)+ks]^{n-2}$$
$$\times \|\dot u(s)\|^2 \,|\, \mathscr{A}_0\}\, ds$$
$$\leqslant [z(0)]^n + kn(2n-1) \int_0^t c_{n-1}(s)\, ds,$$

where $c_n(t) = E\{[z(t)+kt]^n \,|\, \mathscr{A}_0\}$.

Now, noting that $c_0(t)=1$ and $c_n(t) = [z(0)]^n + kn(2n-1)\int_0^t c_{n-1}(s)\, ds$, we get (i). From (i),

$$E\{e^{\alpha z(t)} \,|\, \mathscr{A}_0\} = e^{-\alpha kt} \sum_{n=0}^{\infty} \frac{\alpha^n c_n(t)}{n!}$$

$$= e^{-\alpha kt} \sum_{n=0}^{\infty} \sum_{j=0}^{\infty} \frac{\alpha^n (2n)! [z(0)]^j [kt]^{n-j}}{n!(n-j)!\, 2^{n-j}(2j)!}$$

$$= e^{-\alpha kt} \sum_{m,j=0}^{\infty} \frac{\alpha^{j+m}[2(m+j)]!}{(m+j)!\, m!\, 2^m (2j)!} [z(0)]^j (kt)^m. \quad (5.18)$$

Also, for $|y| < \tfrac{1}{2}$,

$$(1-2y)^{-1} \exp[x/(1-2y)] = \sum_{j=0}^{\infty} x^j (j!)^{-1} (1-2y)^{-(j+(1/2))}$$

$$= \sum_{j,m=0}^{\infty} \frac{x^j (2j+2m)!\, y^m}{(j+m)!\, (2j)!\, m!\, 2^m}. \quad (5.19)$$

Now (ii) follows from (5.18) and (5.19).

(d) Applying Doob's inequality to the submartingale $e^{\alpha z(t)}$, $\alpha > 0$, recall that $z(t)$ is a martingale), we get $P\{\sup_{0 \leqslant s \leqslant t} z(s) > c\} \leqslant e^{-\alpha c} E\, e^{\alpha z(t)}$. Now, from the inequality (c)(ii), it follows that

$$P\left\{\sup_{0 \leqslant s \leqslant t} \|f^*(x)\|_{\mathscr{E}}^2 + 4b \int_0^s \|\dot u^*(\sigma)\|^2\, d\sigma - ks > c\right\}$$
$$\leqslant (1-2\alpha kt)^{-1} e^{-\alpha(c+kt)} E\{\exp[\alpha z(0)(1-2\alpha kt)^{-1}]\},$$

and consequently,

$$P\left\{\|f^*(s)\|_{\mathscr{E}}^2 + 4b \int_0^s \|\dot u^*(\sigma)\|^2\, d\sigma \leqslant c+ks,\ s \in (0,t)\right\}$$
$$> 1 - (1-2\alpha kt)^{-1} e^{-\alpha(c+kt)}$$
$$\times E\{\exp[\alpha z(0)(1-2\alpha kt)^{-1}]\},$$

which gives the desired bound for the energy. This completes the proof.

We refer to [7] where Cabaña has used such bounds for the energy to derive solutions for the double barrier problems for a class of stationary Gaussian processes which are represented as linear combinations of the positions and velocities of the string.

## REFERENCES

1. Bharucha-Reid, A. T., Random algebraic equations *in* "Probabilistic Methods in Applied Mathematics," Vol. 2., Academic Press, New York, 1970, pp. 1–52.
2. Bharucha-Reid, A. T., "Random Integral Equations." Academic Press, New York, 1972.
3. Bharucha-Reid, A. T., Random difference equations, Preprint, 1970.
4. Blumenthal, R. M., and Getoor, R. K., "Markov Processes and Potential Theory." Academic Press, New York, 1968.
5. Borcher, D. R., Second Order Stochastic Differential Equations and related Itô Process, Ph.D. thesis, Carnegie Inst. Tech., California, 1964.
6. Cabaña, E. M., On stochastic differentials in Hilbert spaces, *Proc. Amer. Math. Soc.* **20** (1969), 259–265.
7. Cabaña, E. M., The vibrating string forced by white noise, *Z. Wahrscheinlichkeitstheorie und Verw. Gebiete.* **15** (1970), 111–130.
8. Cramér, H., and Leadbetter, M. R., "Stationary and Related Stochastic Processes." Wiley, New York, 1967.
9. Doob, J. L., "Stochastic Processes." Wiley, New York, 1953.
10. Dubrovskii, V. N., On small random perturbations of a second order differential equation, *Theor. Probability Appl.* **18** (1973), 476–485.
11. Dynkin, E. B., "Markov Processes," Vols. I and II, Springer-Verlag, Berlin and New York, 1965.
12. Edwards, D. A., and Moyal, J. E., Stochastic differential equations, *Proc. Cambridge Phil. Soc.* **51** (1955), 663–677.
13. Ehrlich, P. R., and Birch, L. C., The "Balance of Nature" and "Population Control," *Amer. Nat.* **101** (1967), 97–107.
14. Feller, W., On the equation of the vibrating string, *J. Math. Mech.* **8** (1959), 339–348.
15. Ford, G. W., Kac, M., and Mazur, P., Statistical mechanics of assemblies of coupled oscillators, *J. Mathematical Phys.* **6** (1965), 504–515.
16. Friedman, A., "Stochastic Differential Equations and Applications," Vols. I and II. Academic Press, New York, 1975.
17. Gard, T. C., and Kannan, D., On a stochastic differential equation modelling of prey-predator evolution, *J. Appl. Probability* **13** (1976), 429–443.
18. Gihman, I. I., and Skorokhod, A. V., "Stochastic Differential Equations," Springer-Verlag, Berlin and New York, 1972.
19. Goldstein, J. A., Second order Itô process, *Nagoya Math. J.* **36** (1969), 27–63.
20. Grossman, S. I., Existence and Uniqueness of Solutions to Non-linear Volterra Integral and Integro Differential Equations, Ph.D. thesis, Brown Univ., Rhode Island, 1969.
21. Gyftopoulos, E. P., Theoretical and experimental criteria for nonlinear reactor stability, *Nucl. Sci. Eng.* **26** (1966), 26–33.
22. Hille, E., and Phillips, R. S., Functional Analysis and Semigroups, *Colloq. Publ., Amer. Math. Soc.*, Providence, Rhode Island, 1957.
23. Itô, K., On stochastic differential equations, *Mem. Amer. Math. Soc.* **4** (1961).
24. Itô, K., and McKean, H. P., "Diffusion Processes and their Sample Paths," Springer-Verlag, Berlin and New York, 1965.

25. Kannan, D., Operator valued stochastic intergal, II, *Ann. Inst. H. Poincaré* **8** (1972), 9–32.
26. Kannan, D., Wave propagation in one dimensional random media, *J. Math. Phys. Sci.* **8** (1974), 201–217.
27. Kannan, D., On some markov models of certain interacting populations, *Bull. Math. Biology* **38** (1976), 723–738.
28. Kannan, D., On generalized Langevin equation, *J. Mathematical and Physical Sci.* **11** (1977), in press.
29. Kannan, D., Volterra–Verhulst prey–predator systems with time dependent coefficients: Diffusion type approximation and periodic solution, *Bull. Math, Biology*, to appear.
30. Kannan, D., On a random integro-differential equation, (to appear).
31. Kannan, D., On second order Itô equations, (in preparation).
32. Kannan, D., and Bharucha-Reid, A. T., Operator valued stochastic integral, *Proc. Jap Acad.* **47** (1971), 472–476.
33. Kannan, D., and Bharucha-Reid, A. T., Random integral equation formulation of a generalized Langevin equation, *J. Statist. Phys.* **5** (1972), 209–233.
34. Khas'miniskii, R. Z., Erogodic properties of recurrent diffusion processes and stabilization of the solution to the Cauchy problem for parabolic equations, *Theor. Probability Appl.* **5** (1960), 179–196.
35. Kubo, R., The fluctuation-dissipation theorem, *Rep. Progr. Phys.* **29** (1966), 225–284.
36. Kunita, H., and Watanabe, S., On square integrable martingales, *Nagoya Math. J.* **30** (1967), 209–245.
37. Levin, J. J., and Nohel, J. A., On a system of integrodifferential equations occurring in reactor dynamics, *J. Math. Mech.* **9** (1960), 347–386.
38. Loéve, M., "Probability Theory," Van Nostrand–Reinhold, Princeton, New Jersey, 1963.
39. Massera, J. L., and Schäffer, J. J., "Linear Differential Equations and Function Spaces," Academic Press, New York, 1966.
40. May, R. M., "Stability and Complexity in Model Ecosystems," Princeton Univ. Press, New Jersey, 1973.
41. McKean, H. P., Elementary solutions for certain parabolic partial differential equation, *Trans. Amer. Math. Soc.* **82** (1956), 519–548.
42. McKean, H. P., "Stochastic Integrals." Academic Press, New York, 1969.
43. Meyer, P. A., "Probability and Potential." Ginn (Blaisdell), Boston, Massachusetts, 1966.
44. Miller, R. K., "Non-Linear Volterra Integral Equations." Benjamin, New York, 1971.
45. Mori, H., Transport, collective motion, and Brownian Motion, *Progr. Theoret. Phys.* **33** (1965), 423–455.
46. Moyal, J. E., Stochastic processes and statistical physics, *J. Roy. Statist. Soc., Ser. B*, **11** (1949), 150–210.
47. Nelson, E., "Dynamical Theories of Brownian Motion." Princeton Univ. Press, New Jersey, 1967.
48. Neveau, J., "Mathematical Foundations of Calculus of Probability." Holden-Day, San Francisco, 1965.
49. Padgett, W. J., and Tsokos, C. P., On a stochastic integro-differential equation of Volterra type, *SIAM J. Appl. Math.* **23** (1972), 496–512.
50. Papanicolaou, G. C., Asymptotic analysis of transport processes, *Bull. Amer. Math. Soc.* **81** (1975), 330–392.
51. Rao, A. N. V., and Tsokos, C. P., On the existence, uniqueness and stability behavior of a random solution to a nonlinear perturbed stochastic integro-differential equation, *Information and Control* **27** (1975), 61–74.
52. Soong, T. T., "Random Differential Equations in Sciences and Engineering." Academic Press, New York, 1973.

53. Srinivasan, S. K., and Vasudevan, R., "Introduction to Random Differential Equations and their Applications." Amer. Elsevier, New York, 1971.
54. Tsokos, C. P., and Padgett, W., "Random Integral Equations with Applications to Life Sciences and Engineering." Academic Press, New York, 1974.
55. Varadhan, S. R. S., Diffusion processes in a small time interval, *Comm. Pure Appl. Math.* **20** (1967), 659–685.
56. Ventcel', A. D., and Freidlin, M. I., On small random perturbations of a dynamical system, *Uspehi. Mat. Nauk.* **25** (1970), 3–55, (in Russian).
57. Volterra, V., "Leçons sur la Théorie Mathématique de la Lutte pour la Vie, Gauthier-Villars, Paris, 1939.
58. Wax, N., "Selected Papers on Noise and Stochastic Processes." Dover, New York, 1954
59. Wong, E., and Zakai, M., The oscillation of stochastic integrals, *Z. Wahrscheinlichkeitstheorie und Verw. Gebiete.* **4** (1965), 103–112.
60. Yosida, K., "Functional Analysis," Springer-Verlag, Berlin and New York, 1975.

# Equivalence and Singularity of Gaussian Measures and Applications

S. D. CHATTERJI

DÉPARTEMENT DE MATHÉMATIQUES
ECOLE POLYTECHNIQUE FEDERALE
DE LAUSANNE
1007 LAUSANNE, SWITZERLAND

V. MANDREKAR

DEPARTMENT OF STATISTICS AND PROBABILITY
MICHIGAN STATE UNIVERSITY
EAST LANSING, MICHIGAN

| | | |
|---|---|---|
| I. | Introduction | 169 |
| II. | General Problem of Equivalence and Singularity | 171 |
| | A. Introduction | 171 |
| | B. Applications | 172 |
| III. | Reproducing Kernel Hilbert Spaces and Gaussian Processes | 175 |
| | A. Introduction | 175 |
| | B. Gaussian Processes | 177 |
| IV. | Equivalence and Singularity of Gaussian Processes | 180 |
| V. | Conditions for Equivalence: Special Cases | 185 |
| | A. Introduction | 185 |
| | B. Gaussian Processes with Independent Increments | 186 |
| | C. Stationary Gaussian Processes | 187 |
| | D. Gaussian Measures on Banach Spaces | 188 |
| | E. Generalized Gaussian Processes Equivalent to Gaussian White Noise of Order $p$ | 189 |
| VI. | Applications | 189 |
| VII. | Concluding Remarks | 194 |
| | Appendix | 194 |
| | References | 195 |

## I. Introduction

Let $\mathscr{F}$ be a $\sigma$-field of subsets of a space $\Omega$ and $\{X_t, t \in T\}$ be a family of real-valued Borel-measurable functions on the probability space $(\Omega, \mathscr{F})$

indexed by an arbitrary set $T$. Let $I$ be a family of finite subsets of $T$ and $\mathscr{F}_i$ be the smallest $\sigma$-algebra with respect to which $\{X_t, t \in i\}$ are measurable for each $i \in I$ (henceforth, $\mathscr{F}_i = \sigma\{X_t, t \in i\}$). If $P_1$ and $P_2$ are two probability measures on $\mathscr{F}$ such that $P_2$ is absolutely continuous or mutually absolutely continuous with respect to $P_1$ on $\mathscr{F}_i$ for each $i \in I$, then our purpose is to determine conditions for the absolute continuity, mutual absolute continuity, or singularity of $P_1$ with respect to $P_2$ on $\sigma$-field $\mathscr{F}_T = \sigma\{X_t, t \in T\}$. A measure $\mu_1$ is said to be absolutely continuous with respect to $\mu_2$ on a sub-$\sigma$-field $\mathscr{A}$ of $\mathscr{F}$ if for all $A \in \mathscr{A}$ with $\mu_2(A) = 0$ we have $\mu_1(A) = 0$ and write $\mu_1 \ll \mu_2$. Also, we say that $\mu_1$ is singular with respect to $\mu_2$ ($\mu_1 \perp \mu_2$) on $\mathscr{A}$ if there exists a set $N \in \mathscr{A}$ such that $\mu_1(N) = 0$ and $\mu_2(\complement N) = 0$. In case

$$\Omega = \prod_{n=1}^{\infty} \Omega_n, \quad \mathscr{F} = \bigotimes_{n=1}^{\infty} \mathscr{F}_n,$$

i.e., $(\Omega, \mathscr{F})$ is a product-measurable space and $X_n(\omega) = \omega_n$, the $n$th coordinate, the problem of equivalence and singularity was first studied and completely solved by Kakutani [29] under the assumptions $P_1 = \bigotimes \mu_k$, $P_2 = \bigotimes \nu_k$ and $P_1 \equiv P_2$ (i.e., $P_1 \ll P_2$ and $P_2 \ll P_1$) on $\mathscr{F}_i$ ($i = \{1, 2, ..., n\}$). Recently, Nemetz [46] has obtained a generalization of [29]. In the case where $P_1$ and $P_2$ are two measures on $(\Omega, \mathscr{F})$ such that $\{X_t, t \in T\}$ is a Gaussian stochastic process under both $P_1$ and $P_2$, J. Hajék [20] and J. Feldman [12] showed that $P_1 \equiv P_2$ on $\mathscr{F}_T$ or $P_1 \perp P_2$ on $\mathscr{F}_T$. General conditions for equivalence of $P_1$ and $P_2$ were already available in I. Segal [63]. However, Hajék gave conditions in terms of $J$-divergence. Subsequently, Yu. A. Rozanov examined the relationship between Hajék–Feldman conditions and derived more verifiable conditions. Almost concurrently, Kallinapur–Oodaira [34] and Parzen [54] (see also Oodaira [50]) derived easily verifiable conditions purely in terms of the structure of the reproducing kernel Hilbert spaces associated with the covariance kernels under $P_1$ and $P_2$. However, their results depended on some extraneous assumptions, mainly because of the method used. This led to abstract generalizations of the problem [37, 45]. Also, all these results used in some form the Kakutani theorem described above. Finally, J. Neveu [48] gave the most general theorem on the equivalence and singularity for Gaussian processes using the concept of exponential map and direct sum of the tensor products of Hilbert spaces. Our purpose here is to present an elementary proof of the Nemetz theorem and by easy calculation derive both the Kakutani theorem and the theorem for Gaussian processes. This approach is due to S. D. Chatterji and V. Mandrekar [9] and brings out the natural role of reproducing kernels in these problems.

In Section III, we present the concept of Gaussian process and associated reproducing kernel Hilbert spaces. This is followed by the main results on equivalence and singularity in Sections IV and V, which also treat some special cases. The final section uses the main results to obtain several significant structural theorems on Gaussian measures.

We begin in the next section by considering general conditions for equivalence and singularity. As a general notation, $E_P$ will denote the integral with respect to $P$. We shall drop the subscript in cases where the measure in question is well understood.

## II. General Problem of Equivalence and Singularity

### A. Introduction

Let $\mathscr{F}$ be a $\sigma$-field of the subsets of a space $\Omega$ and $P_i$ ($i = 1, 2$) be two probability measures on $\mathscr{F}$, then the Lebesgue decomposition of $P_2$ with respect to $P_1$ can be written in the form ([58], p. 211)

$$P_2(A) = \int_A \frac{p_2}{p_1} dP_1 + P_2(A \cap (p_1 = 0)) \quad \text{for} \quad A \in \mathscr{F}, \tag{2.1}$$

where $p_i$ denote the *Radon–Nikodym* (R–N) *density* of $P_i$ with respect to $P_1 + P_2$.

As an immediate consequence of (2.1) we get

(i) $P_1 \perp P_2$    iff    $P_1[(p_2/p_1) = 0] = 1$,

(ii) $P_1 \ll P_2$    iff    $P_1[(p_2/p_1) > 0] = 1$.    (2.2)

To see (2.2)(ii): Under the condition $P_1[(p_2/p_1) > 0] = 1$ we get that $P_1(A) > 0$ implies $P_2(A) > 0$ giving $P_1 \ll P_2$ the reverse implication follows as

$$P_2(p_1 = 1) = 0 \Rightarrow P_1(p_1 = 1) = 0 \Rightarrow P_1(0 < p_1 < 1) = 1$$
$$\Rightarrow P_1[(p_2/p_1) > 0] = 1$$

since $p_1 + p_2 = 1$ a.e. $[P_1 + P_2]$.

Henceforth, we set $\rho = (p_2/p_1)$ and note that $\rho$ is the R–N density of the absolutely continuous part of $P_2$ with respect to $P_1$. Now $\rho \geq 0$ and as $\{\rho^\alpha, 0 < \alpha \leq \alpha_0\}$ satisfies $E_{P_1}(\rho^\alpha)^{1/\alpha_0}$ is finite for $\alpha_0 < 1$ we get [47] $\{\rho^\alpha, 0 < \alpha < \alpha_0\}$ is uniformly integrable. Since $\lim_{\alpha \to 0} \rho^\alpha = 1_{\{\rho > 0\}}$ a.e. $P_1$, we get, from (2.2)(ii),

(i) $P_1 \perp P_2$ on $\mathscr{F}$    iff    $E_{P_1} \rho^\alpha = 0$    for $0 < \alpha < 1$,    (2.3)

(ii) $P_1 \ll P_2$ on $\mathscr{F}$    iff    $\lim_{\alpha \to 0} E_{P_1} \rho^\alpha = 1$.

We remark that the validity of condition (2.3)(i) for one $\alpha$ implies it for all $\alpha$.

Let $I$ be a partially ordered set and $\{\mathscr{F}_i, i \in I\}$ be a nondecreasing family of sub-$\sigma$-fields of $\mathscr{F}$ such that $\mathscr{F} = \sigma(\bigcup_{i \in I} \mathscr{F}_i)$. Let $\rho_i$ denote the R–N derivative of the absolutely continuous part of $P_2$ with respect to $P_1$ on $\mathscr{F}_i$. Using the fact that $\rho_i$ is essential supremum of functions $g$ satisfying $\int_A g\, dP_1 \leq P_2(A)$ for $A \in \mathscr{F}_i$ [47] we get $\int_A \rho_i\, dP_1 \geq \int_A \rho_j\, dP$ for all $A \in \mathscr{F}_i$ for $j \geq i$. We therefore get that $\mathscr{E}^{\mathscr{F}_i}(\rho_j) \leq \rho_i$ for $j \geq i$, where $\mathscr{E}^{\mathscr{F}_i}$ denotes conditional expectation [47]. In view of the Jensen inequality we get $\mathscr{E}^{\mathscr{F}_i}(\rho_j^\alpha) \leq \rho_i^\alpha$ $(0 < \alpha < 1)$. Hence, we get that $E_{P_1}\rho_i^\alpha$ is a nonincreasing function of $i$ for each $\alpha$ $(0 < \alpha < 1)$. In order to express conditions (2.3) in terms of $E_{P_1}\rho_i^\alpha$ we need the following, where all expectations are with respect to $P_1$.

**Lemma 2.1** *Let $(\Omega, \mathscr{F})$ be a measurable space and $I$ a directed set. Let $\{\mathscr{F}_i, i \in I\}$ be a nondecreasing family of sub-$\sigma$-fields of $\mathscr{F}$ such that $\mathscr{F} = \sigma(\bigcup_{i \in I} \mathscr{F}_i)$. Let $\rho_i$ and $\rho$ be as above, then $\inf_i E\rho_i^\alpha = E\rho^\alpha$ for all $0 < \alpha < 1$.*

*Proof* Since $\rho$ is measurable with respect to $\mathscr{F}$ there exists a sequence $i1 < i2 < i3, \ldots$ of elements of $I$ such that $\rho$ is measurable $\sigma(\bigcup_j \mathscr{F}_{ij})$. It is known ([47], p. 41) that $\rho_{ij} \to \rho$ a.e. $[P_1]$. Since $\{\rho_{ij}^\alpha, j = 1, 2, \ldots\}$ is uniformly integrable we get $\lim_j E\rho_{ij}^\alpha = E\rho^\alpha$. This implies $E\rho^\alpha = \lim_j E\rho_{ij}^\alpha \geq \inf E\rho_i^\alpha \geq E\rho^\alpha$ since $E\rho_i^\alpha \geq E\rho^\alpha$ for all $i$.

*Remark* 2.1 We note that $\inf_I E\rho_i^\alpha = \lim_I E\rho_i^\alpha$.

From (2.3) and Lemma 2.1, we get the following simple extension of the result of T. Nemetz [46].

**Theorem 2.1** *Let $P_1$ and $P_2$ be two measures on a measurable space $(\Omega, \mathscr{F})$ and $I$ be a directed family of sets. If $\{\mathscr{F}_i, i \in I\}$ is a nondecreasing family of sub-$\sigma$-fields of $\mathscr{F}$ such that $\mathscr{F} = \sigma(\bigcup_I \mathscr{F}_i)$ and $\rho_i$ denote the R–N density of the absolutely continuous past of $P_2$ with respect to $P_1$ on $\mathscr{F}_i$ then*

(a) $P_1 \perp P_2$ on $\mathscr{F}$    iff $\lim_i \int \rho_i^\alpha\, dP_1 = 0$    for $0 < \alpha < 1$;

(b) $P_1 \ll P_2$ on $\mathscr{F}$    iff for every $\varepsilon > 0$, there exists an $\alpha(\varepsilon) \in (0, 1)$ such that $\int \rho_i^\alpha\, dP_1 > 1 - \varepsilon$ for all $\alpha \in [0, \alpha(\varepsilon)]$ and for all $i \in I$.

We end this section by giving some applications of the above results in case $I = \{1, 2, 3, \ldots\}$ $(= Z^+)$ and the usual ordering. In this case we note that $\rho = \lim_n \rho_n$ a.e. $[P_1]$.

### B. Applications

**1. Equivalence and Singularity Dichotomy and Zero–One Law [41], [42]** Let $\mathscr{F}_1 \subset \mathscr{F}_2 \subset \cdots$ be $\sigma$-algebras of subsets of $\Omega$ and $\mathscr{F} = \sigma(\bigcup_n \mathscr{F}_n)$. Let $P$ and $Q$ be two probability measures on $\mathscr{F}$ such that $Q \equiv P$ on $\mathscr{F}_1, \mathscr{F}_2, \ldots, \rho_n = (dQ|\mathscr{F}_n/dP|\mathscr{F}_n)$. Let $\mathscr{G} = \bigcap_n \sigma\{(\rho_{k+1}/\rho_k): k \geq n\}$. If $\mathscr{G}$ is $P$-trivial, i.e.,

$P(A) = 0$ or 1 for all $A \in \mathscr{G}$, then $P \perp Q$ or $P \ll Q$ on $\mathscr{F}$. Furthermore, we have $P \ll Q$ on $\mathscr{F}$ iff $E_P \rho_n^\alpha \nrightarrow 0$ for some $\alpha$ if $\mathscr{G}$ is $Q$-trivial.

*Proof* Since $\{\rho > 0\} \in \mathscr{G}$, from (2.2) we get the first conclusion. To obtain the second conclusion, use 2.3(i) and Lemma 2.1.

**2. Kraft's Theorem** [36]  Let $P$, $Q$, $\mathscr{F}_n$, $\mathscr{F}$, $\rho_n$, etc. be as in (1) above; then $P \perp Q$ iff $\lim_n E_P \rho_n^{\frac{1}{2}} \to 0$. This follows immediately from Lemma 2.1 and 2.3(i).

**3. Kakutani's Theorem** [29]  Let $\mu_k$ and $\nu_k$ be two probability measures on $(\Omega_k, \mathscr{A}_k)$ ($k = 1, 2, \ldots$). Let $\Omega = \prod_k \Omega_k$, $\mathscr{F} = \bigotimes_k \mathscr{A}_k$, $P_1 = \bigotimes_k \mu_k$, and $P_2 = \bigotimes_k \nu_k$. Let

$$\mathscr{B}_i = \bigotimes_{k=1}^{i} \mathscr{A}_k$$

and $\mathscr{F}_i = \tau_i^{-1}(\mathscr{B}_i)$, where $\tau_i$ is the projection map of $\Omega$ onto

$$\prod_{k=1}^{i} \Omega_k \qquad (i = 1, 2, \ldots).$$

Clearly, $\mathscr{F} = \sigma(\bigcup_i \mathscr{F}_i)$ and $\{\mathscr{F}_i, i = 1, 2, \ldots\}$ are nondecreasing. Let $\nu_k \equiv \mu_k$ on $\mathscr{A}_k$ for each $k$. Then $P_1 \equiv P_2$ on $\mathscr{F}_i$ and

$$\rho_i(\omega) = \prod_{k=1}^{i} \frac{d\mu_k}{d\nu_k}(\omega_k) \qquad \text{for } \omega = (\omega_1, \omega_2, \ldots) \in \Omega.$$

Since

$$\rho(\omega) = \prod_{k=1}^{\infty} \frac{d\mu_k}{d\nu_k}(\omega_k) \qquad \text{a.e. } [P_1]$$

and

$$\mathscr{G} = \bigcap_n \sigma\left\{\frac{d\nu_{k+1}}{d\mu_{k+1}}(\omega_{k+1}): k \geq n\right\}$$

is $P_1$-trivial by Kolmogorov's zero–one law [47] we get by Section II.B.1 that $P_1 \perp P_2$ or $P_1 \ll P_2$ on $\mathscr{F}$. By the symmetry of the problem and second conclusion of Section II.B.1 we get the Kakutani Theorem

$$P_1 \perp P_2 \quad \text{iff} \quad \prod_k \int \left(\frac{d\nu_k}{d\mu_k}\right)^{\frac{1}{2}} d\mu_k = 0$$

$$P_1 \equiv P_2 \quad \text{iff} \quad \prod_k \int \left(\frac{d\nu_k}{d\mu_k}\right)^{\frac{1}{2}} d\mu_k > 0. \qquad (2.4)$$

## 4. Equivalence and Singularity: Products of Discrete Measures

We note that in case 3, $\Omega_k = \{1, 2, 3, ..., N\}$ for all $k$ with $\mu_k$ and $\nu_k$ discrete probability measures given by positive numbers $p_{jk}$, $q_{jk}$ ($j = 1, 2, ..., N$). Then with

$$\lim_i E_{P_L} \rho_i^{\frac{1}{2}} \neq 0 \quad \text{iff} \quad \prod_i \frac{E_{P_1}(\rho_{i+1})^{\frac{1}{2}}}{E_{P_1}(\rho_i)^{\frac{1}{2}}} \quad \text{converges}$$

$$\text{iff} \quad \sum_i \left(1 - \frac{E_P \rho_{i+1}^{\frac{1}{2}}}{E_{P_i} \rho_i^{\frac{1}{2}}}\right) \quad \text{converges}. \tag{2.5}$$

Now

$$2\left(1 - \frac{E\rho_{i+1}^{\frac{1}{2}}}{E\rho_i}\right) = \sum_{j=1}^{N} (p_{j,i+1}^{\frac{1}{2}} - q_{j,i+1}^{\frac{1}{2}})^2.$$

Hence

$$P_1 \equiv P_2 \quad \text{iff} \quad \sum_k (p_{j,k}^{\frac{1}{2}} - q_{j,k}^{\frac{1}{2}})^2 < \infty \quad \text{for} \quad j = 1, 2, ..., N.$$

In particular, for all $k$, $p_{ik} = p_i$ and $q_{ik} = q_i$ then $P_1 \perp P_2$ iff $\exists\ i$ for which $p_i \neq q_i$.

These ideas lead to study of measures absolutely continuous or singular w.r.t. Lebesgue measure on the interval $[0, 1]$ [4, 5, 6] and equivalence and singularity of measures given by Markov chains [42]. The former set of problems [5, 6] are analytically sophisticated.

## 5. Product Gaussian Measures

In Section 3, let $\Omega_k = R$, $\mathscr{A}_k = \mathscr{B}(R)$, the Borel subsets of $R$, $\nu_k = G(m_{1k}, 1)$ and $\mu_k = G(m_{2k}, \sigma_k^2)$ for each $k$, where $G(m, \sigma^2)$ is the probability measure on $\mathscr{B}(R)$ with R–N density $(\sigma 2\pi^{\frac{1}{2}})^{-1} \exp\{-\frac{1}{2}[(\mu - m/\sigma)]^2\}$ with respect to the Lebesgue measure. Then by an easy calculation we get

$$E_{P_1} \rho_i^{\frac{1}{2}} = \prod_1^i \left(\frac{2\sigma_k}{1 + \sigma_k^2}\right)^{\frac{1}{2}} \exp\left[-\frac{1}{4} \sum_{k=1}^{i} \frac{(m_{1k} - m_{2k})^2}{1 + \sigma_k^2}\right].$$

Hence we get from Section 3 that

$$P_1 \equiv P_2 \quad \text{iff} \quad \text{(a)}\ 0 < c_1 \leq \sigma_k \leq c_2 < \infty, \forall k; \sum_k (1 - \sigma_k^2) < \infty;\ \text{and}$$

$$\text{(b)} \sum_{k=1}^{\infty} (m_{1k} - m_{2k})^2 \quad \text{is finite}. \tag{2.6}$$

This case was the motivation of the early work of I. Segal [63]. It plays an important role in the development of the free field theory [2, 64].

*Remark 2.2* In the case where $\sigma_k = 1$, $m_{1k} = 0$ and $m_{2k} = m_k$ we get $P_2(A) = P_1(A + m)$ ($= P_{1m}(A)$), $A \in \mathscr{B}(R)$, where $m = (m_1, m_2, ...) \in R^\infty$ and

$A+m = \{x+m; x \in A\}$. From (2.6) we get $P_1 \equiv P_2$ iff $\sum_{k=1}^{\infty} m_k^2 < \infty$. This fact is referred to as quasi-invariance of $P_1$ under translation by $l_2$, the space of square summable real sequences. As this property was basic in the development of Weyl–Von Neumann relations for dynamical systems with infinite degrees of freedom, the question of whether the Gaussian measure was the only measure on $R^\infty$, quasi-invariant under $l_2$ was raised. J. Feldman [14] showed that any product measure on $R^\infty$ constructed from identical factors $\mu$ with positive R–N density $p(x) = (d\mu/dx)$ with respect to Lebesgue measure and satisfying $\int [p'(x)^2/p(x)]\, dx$ finite is quasi-invariant under translation by $l_2$. In [65], Shepp characterized product measures $P$ on $R^\infty$ with identical factors $\mu$ for which $E = \{a \in R^\infty; P_a \equiv P\} = l_2$ as those for which $p(x) = (d\mu/dx)$ satisfies the above condition. The detailed structure of $E$ has been recently studied in [7, 8].

## III. Reproducing Kernel Hilbert Spaces and Gaussian Processes

### A. Introduction

This section is added in order to make the chapter self-contained. The readers familiar with this material may proceed to Section IV. We first start by defining the concept of a covariance, the form of which will allow us to construct various Gaussian processes.

**Definition 3.1** Let $T$ be any set and $C$ be a real-valued function on $T \times T$. Then $C$ is called a *covariance on T* if (a) $C(t,s) = C(s,t)$ for all $s, t \in T$ and (b) $\sum_{t,s \in i} a_t a_s C(t,s) \geq 0$ for all finite subsets $i$ of $T$ and $\{a_s, s \in i\}$ of $R$.

**Theorem 3.1** (*Aronszajn* [2]) *Let $T$ be any set and $C$ be a real-valued covariance on $T \times T$. Then there exists a unique Hilbert space $K(C)$ of functions on $T$, satisfying*

$$C(\cdot, t) \in K(C) \quad \text{for each} \quad t \in T;$$

$$[f, C(\cdot, t)] = f(t) \quad \text{for each} \quad f \in K(C) \quad \text{and} \quad t \in T. \tag{3.1}$$

*Here for each $t$, $C(\cdot, t)$ denotes the function of the first variable.*

*Proof* Let $R^T$ be the real linear space of all functions on $T$ to $R$ with coordinate-wise addition and scalar multiplication. Let $\mathcal{H}$ be the linear manifold in $R^T$ generated by $[C(\cdot, t), t \in T]$. On $\mathcal{H}$ define inner product

$$(f, g) = \sum_{s \in i, s' \in i'} a_s b_{s'} C(s, s') = \sum_{s \in i} a_s g(s) = \sum_{s' \in i'} b_{s'} f(s'), \tag{3.2}$$

where $f = \sum_{s \in i} a_s C(\cdot, s)$, and $g = \sum_{s' \in i'} b_{s'} C(\cdot, s')$ with $i, i'$ finite subsets of $T$. From the last equality in (3.2) we get that $(f, g)$ is independent of the representation of $f$ and $g$. From properties of $C$ we get $(f, f) \geq 0$ and $(f, g)$

is bilinear function on $\mathcal{H}$. Also, $f(t) = (f, C(\cdot, t))$ for each $t \in T$ and $f \in \mathcal{H}$ giving $|f(t)|^2 \leq (f,f) C(t,t)$. This implies $(f,f) = 0$ iff $(f(t) = 0$ for all $t$. Thus $(\mathcal{H}, (\ ))$ is a pre-Hilbert space. Let $\bar{\mathcal{H}}$ be completion of $\mathcal{H}$ under norm $(f,f)^{\frac{1}{2}}$. Define $K(C) = \{f \in R^T : f(t) = (C(\cdot, t), h_f)$ for $h_f \in \bar{\mathcal{H}}\}$. On $K(C)$, define $(f, g) = (h_f, h_g)$. Then $K(C)$ has all the properties and is uniquely determined by $C$.

**Definition 3.1** Let $T$ be any set. A class $K(C)$ of functions on $T$ forming a Hilbert space is called the *reproducing kernel Hilbert space* (for short, rkhs) of a covariance $C$ if it satisfies (3.1). The above theorem gives existence and uniqueness.

**Corollary 3.1** [3] *Let $T$ be any set; then for any real-valued function $C$ on $T \times T$, the following are equivalent*

(i)  $C$ is a covariance on $T \times T$.
(ii)  There exists a real Hilbert space $\mathcal{H}$ and a function $f$ on $T$ to $\mathcal{H}$ such that $C(t, t') = [f(t), f(t')]_{\mathcal{H}}$.
(iii)  There exists a family $(f_j)_{j \in J}$ ($J$ is some index set) such that $\sum_j f_j^2(t)$ is finite for each $t$, and $C(t, t') = \sum_j f_j(t) f_j(t')$ for $t, t' \in T$.

*Proof* Theorem 3.1 gives (i) $\Rightarrow$ (ii) and (ii) $\Rightarrow$ (i), which is obvious from properties of the inner product. Now (ii) $\Rightarrow$ (iii) follows from Parseval's identity [21] with $f_j(t)$, the $j$th Fourier coefficient of $f(t)$. For (iii) $\Rightarrow$ (ii) take any Hilbert space $H$ with cardinality of $J$ equal to cardinality of the orthonormal basis $\{e_j\}_{j \in J}$ in $H$. Take $f(t) = \sum_{j \in J} f_j(t) e_j$ and use Parseval's identity.

*Remark* 3.1 (a) Let $C_1$ and $C_2$ be two covariances on $T \times T$ then from Corollary 3.1(iii) we get that $C_1 \cdot C_2(t, s) = C_1(t, s) C_2(t, s)$ is a covariance. By Definition 3.1 the sum of covariances with nonnegative coefficients and limits are covariances. Hence we get the following.

If $C$ is a covariance then $\sum a_n C^n$ is a covariance for $a_n \geq 0$. (3.3)

(b) If $C_1$ is a covariance on $T_1$ and $C_2$ is a covariance on $T_2$ then $C_1 \otimes C_2\{[(t_1, t_2), (t_1', t_2')]\} = C_1(t_1, t_1') C_2(t_2, t_2')$ is a covariance on $T_1 \times T_2$ by Corollary 3.1(iii).

(c) Let $C_1$ and $C_2$ be as in (b) above and let $L$ be a bounded linear operator on the rkhs $K(C_1)$ into rkhs $K(C_2)$. Then $\Lambda(s, t) = L^* C_2(\cdot, t)(s)$ satisfies $(Lf)(t) = [f, \Lambda(\cdot \cdot t)]_{K(C_1)}$, $t \in T_2$. Conversely, a function $\Lambda$ on $T_1 \times T_2$ with $\Lambda(\cdot, t_2) \in K(C_1)$ for $t_2 \in T_2$ gives a bounded linear operator on $K(C_1)$ into $K(C_2)$ by the above equation.

We now compute some rkhs associated with several covariances.

**Example 3.1** Let $\mu$ be a $\sigma$-finite measure on a space $(\mathscr{X}, \mathscr{A})$ and $L_{2,c}(\mathscr{X}, \mathscr{A}, \mu)$ be the space of complex-valued $\mathscr{A}$-measurable functions, square integrable with respect to $\mu$. We consider $L_{2,c}(\mathscr{X}, \mathscr{A}, \mu)$ as a real Hilbert space with real inner product $\langle f, g \rangle = \operatorname{Re} \int f\bar{g}\, d\mu$. Let $T$ be an indexing set and $\{f(t, \cdot); t \in T\}$ be a subset of $L_{2,c}(\mathscr{X}, \mathscr{A}, \mu)$. If

$$C(t, t') = \operatorname{Re}\left(\int \overline{f(t,u)} f(t', u)\, d\mu\right),$$

then

$$K(C) = \left\{h\colon h(t) = \operatorname{Re}\int \overline{f(t,u)} g(u)\, d\mu \quad \text{for} \quad g \in M\right\},$$

where $M$ is the (closed) linear subspace of $L_{2,c}(\mathscr{X}, \mathscr{A}, \mu)$ generated by $\{f(t, \cdot); t \in T\}$, Re denotes the real part, and $(h_1, h_2)_{K(C)} = \langle g_1, g_2 \rangle$. This can be proved by checking properties (3.1).

**Example 3.2** Let $C_i$ be a covariance on $T_i$ ($i = 1, 2$) and $C_1 \otimes C_2$ be the covariance on $T_1 \times T_2$ as in Remark 3.1(b), then

$$K(C_1 \otimes C_2) = \left\{f\colon f(t_1, t_2) = \sum_{\alpha, \beta} a_{\alpha\beta} e_\alpha^{(1)}(t_1) e_\beta^{(2)}(t_2),\ \sum a_{\alpha\beta}^2 < \infty\right\}.$$

Here $\{e_\alpha^{(i)}, \alpha \in J_i\}$ is an orthonormal basis in $K(C_i)$ ($i = 1, 2$) and

$$(f, g)_{K(C_1 \otimes C_2)} = \sum_{\alpha, \beta} a_{\alpha\beta} a'_{\alpha\beta},$$

where

$$g = \sum a'_{\alpha\beta} e_\alpha^{(1)}(t_1) e_\beta^{(2)}(t_2).$$

**Example 3.3** Let $T = C_0^\infty(R^n)$, the space of infinitely differentiable functions with compact support on $R^n$. Let $C$ be any covariance on $T$ then $K(C)$ is a subset of the space of Schwarz distributions on $R^n$ if $C$ is a continuous bilinear form on $T$.

### B. Gaussian Processes

**Definition 3.2** Let $T$ be a set and $(\Omega, \mathscr{F}, P)$ be a probability space, then a family $\{X_t, t \in T\}$ of (real) random variables is called

(i) a *(centered) Gaussian process* if every finite real linear combination of elements of $\{X_t, t \in T\}$ is a Gaussian ramdon variable or

(ii) a Gaussian process with mean $m$ if $\exists$ an $m \in R^T$, such that $\{X_t - m(t), t \in T\}$ is a (centered) Gaussian process.

If $\{X_t, t \in T\}$ is a centered Gaussian process then $C_X(t, t')$ is a covariance on $T$ (Corollary 3.1) and

$$K(C_X) = \{f\colon f(t) = E_P X_t Y_f \quad \text{for a unique} \quad Y_f \in H(X)\}.$$

Here $H(X)$ is the linear subspace of $L_2(\Omega, \mathscr{F}, P)$ generated by $\{X_t, t \in T\}$. Conversely, we can associate a Gaussian process with a covariance.

**Lemma 3.1** *Let $C$ be a covariance on $T$ then there exists a Gaussian process $\{X_t, t \in T\}$ defined on a suitable probability space $(\Omega, \mathscr{F}, P)$ such that $C = C_X$.*

*Proof* Let $K(C)$ be the rkhs of $C$ and $\{e_j, j \in J\}$ be an orthonormal basis in $K(C)$. Define $\Omega = \prod_J \Omega_j$, $\mathscr{F} = \otimes_J F_j$ and $P = \otimes P_j$, where $\Omega_j = R$, $F_j = \mathscr{B}(R)$ and $P_j = G(0, 1)$ for $j \in J$. Also, let $\xi_j(\omega) = \omega_j$, $j \in J$. For $h \in K(C)$, $h = \sum_J (h, e_j) e_j$ define $\pi(h) = \sum_J (h, e_j) \xi_j$. Then by Parseval's identity we get $\pi(h)$ as a Gaussian random variable for each $h \in K(C)$. With $X_t = \pi(C(\cdot, t))$ we get $\{X_t, t \in T\}$ as a Gaussian process with covariance $C_X$.

We now use the above lemma to give some well-known Gaussian processes in terms of their covariances.

**1. Brownian Motion** Let $T = \{(t_1 \cdots t_n) = t \in R^n, t_i \geq 0\}$ and $C_1(t, t') = \prod_{i=1}^n \min(t_i, t_i')$. Take $X = R^n$, $A = \mathscr{B}(R^n)$, $\mu$ is a Lebesgue measure on $R^n$, and $f(t, u) = 1_{[0, t_1] \times \cdots \times [0, t_n]}(u_1 \cdots u_n)$ in Example 2.1. Then $C_1(t, t')$ is a covariance. For $n = 1$, the associated Gaussian process is called the *Wiener–Lévy Brownian motion* and for $n > 1$ the associated Gaussian process is called *Cameron–Yeh process*. Example 2.1 shows that the rkhs of $C_1$ is given by

$$K(C_1) = \left\{h: h(t) = \int_{-\infty}^{t_n} \cdots \int_{-\infty}^{t_1} g(u_1 \cdots u_n) \, d\mu, \, g \in L_2[R^n, \mathscr{B}(R^n), \mu]\right\},$$

with inner product $(h_1, h_2) = \int g_1(u) g_2(u) \, d\mu$.

**2. Lévy's Multiparameter Brownian Motion** [3] Let $T = R^n$, $C_2(t, t') = \frac{1}{2}\{|t| + |t'| - |t - t'|\}$. Let $f(t, u) = \|u\|^{-n+\frac{1}{2}}(\exp -i \sum_{k=1}^n t_k u_j - 1)$, $u \in R^n$, $t \in T - \{0\}$ and $f(o, u) \equiv 0$, then $C(t, t') = k_n \operatorname{Re} \int f(t, u) \overline{f(t', u)} \, d\mu$, where $k_n$ is a constant. Hence Example 2.1 shows

$$K(C_2) = \left\{h: h(t) = \operatorname{Re}\left[k_n \int_{R^n} \|u\|^{-n+\frac{1}{2}}\left[\exp\left(i \sum_k t_k u_k - 1\right)\right] g(u) \, d\mu\right],\right.$$
$$\left. g \in L_2 t(R^n)\right\}.$$

Here $(h_1, h_2) = \operatorname{Re} \langle k_n^{\frac{1}{2}} g_1, k_n^{\frac{1}{2}} g_2 \rangle_{L_2, t(R^n)}$.

**3. Gaussian Processes with Triangular Covariances** [70] Let $T = [a, b] \subseteq R$ and $C_3(s, t) = \phi(s) \psi(t)$, $s \leq t$, where $\phi, \psi$ are real-valued functions so that $(d/du)[\phi(u)/\psi(u)]$ exists and $(d/du)[\phi(u)/\psi(u)] > 0$, $\phi(a) = 0$. Let $f(t, u) = \psi(t) \{(d/du)[\phi(u)/\psi(u)]\}^{\frac{1}{2}} 1_{\{u \leq t\}}$. Then $C(t, t') = \int_a^b f(t, u) f(t', u) \, du$.

Hence

$$K(C_3) = \left\{h\colon h(t) = \int_a^b f(t,u)g(u)\,du,\ g \in L_2([a,b], \mathscr{B}[a,b], du)\right\},$$

with $(h_1, h_2) = \int_a^b g_1(u)g_2(u)\,du$ in view of Example 3.1.

**3. Gaussian Measure on a Banach Space** [40] Let $E$ be a real separable Banach space, $\mathscr{B}(E)$ the Borel subsets of $E$, and $E'$ the topological duel of $E$. A probability measure $P$ on $(E, \mathscr{B}(E))$ is called (*centered*) *Gaussian* if for each $e' \in E'$, $e'(\cdot)$ there is a (centered) Gaussian r.v. on $(E, \mathscr{B}(E), P)$. Let $T = E'$, then $\{t(\cdot),\ t \in T\}$ is a (centered) Gaussian process. Let $C(t, t') = \int t(x)t'(x)P(dx)$, then $C(t, t')$ is a covariance, from Corollary 3.1(ii). In Example 3.1 take $f(t,u) = t(u)$, $t \in T$, $u \in E$, $\mathscr{X} = E$, $\mathscr{A} = \mathscr{B}(E)$, $\mu = P$. Then

$$K(C_4) = \left\{h \mid h(t) = \int t(x)g(x)\,dP,\ g \in H(X)\right\},$$

where $(h_1, h_2) = \int g_1(x)g_2(x)\,dP$ and $H(X)$ is the (closed) linear subspace of $L_2(E, \mathscr{B}(E), P)$ generated by $\{(t, \cdot),\ t \in T\}$. It is well known [5] that $\int \|x\|_E^2 P(dx)$ is finite and hence each element $h \in K(C)$ can be represented by $h = \int xg(x)\,dP \in E$. Here the integral is in the sense of Bochner [24].

**5. Stationary Gaussian Processes** Let $T$ be a locally compact abelian group and $C_5(t, s) = R(t - s)$, where $R$ is a continuous nonnegative definite function on $T$. Then $R(t) = \operatorname{Re}\{\int_{\hat{T}} t(u)\mu(du)\}$, where $\mu$ is a nonnegative finite measure on the dual $\hat{T}$ of $T$ [61]. Hence $R(t-s) = \operatorname{Re}[\int t(u)s(u)\mu(du)]$. We get, by Example 3.1, $K(C_5) - \{h\colon h(t) = \operatorname{Re}[\int t(u)g(u)\mu(du)]$, where $g \in L_{2,\blacksquare}(\hat{T}, \mu)\}$ and $(h_1, h_2) = \int g_1(u)g_2(u)\mu(du)$.

**6. Generalized Gaussian Processes** Let $T = C_0^\infty(G)$, where $G$ is a bounded domain with smooth boundary in $R^n$. We denote for $\phi \in C_0^\infty(G)$, $(D^\alpha \phi)(x) = [\partial^{|\alpha|}\phi(x)]/(\partial x_1^{\alpha_1} \cdots \partial x_n^{\alpha_n})$ for $|\alpha| = \sum_{j=1}^n \alpha_j$, $\alpha_i$ nonnegative integers. We showed in Example 2.3 that every bilinear function $C$ is a covariance. The associated Gaussian process is called a *generalized Gaussian process* [16].

In the case where $C_W(t, t') = \int_G t(u)t'(u)\,du$ we call the associated process *Gaussian white noise* [22] and note that $K(C_W) = L_2(G, du)$.

For $C_{W_m}(t, t') = \sum_{|\alpha| \le m} \int_G (D^\alpha t)(u)(D^\alpha t')(u)\,du$, the associate process is called *Gaussian white noise of order n* and $K(C_{W_n})$ is the Sobolev space of $H_0^m(G)$ [1].

We note that in the case where $m \ge [n/2]+1$, we get $H_0^m(G)$ representable by continuous functions on $G$, hence the generalized process $X_t = \int_G t(u)Y_u\,du$, where $\{Y_u,\ u \in G\}$ is a Gaussian process and the integral is in the sense of Bochner.

## IV. Equivalence and Singularity of Gaussian Processes

The following notation will be used in this section.

Let $(\Omega, \mathscr{A})$ be a probability space $T$, an index set, and $I = \{i: i \text{ finite subset of } T\}$ ordered under set inclusion. Let $\{X_t, t \in T\}$ be a family of real random variables on $(\Omega, \mathscr{A})$ and denote by $\mathscr{F}_i = \sigma\{X_t, t \in i\}$ and $\mathscr{F} = \sigma\{X_t, t \in T\}$.

Suppose $P_1$ and $P_2$ are two measures on $\mathscr{F}$ such that $\{X_t, t \in T\}$ is a Gaussian process on $(\Omega, \mathscr{F}, P_l)$ $(l = 1, 2)$. Denote by $C_1(t, s) = E_{P_1} X_t X_s$ and $C_2(t, s) = E_{P_2}(X_t - m(t))(X_s - m(s))$ for $s, t \in T$, where $E_{P_1} X_t = 0$ and $E_{P_2} X_t = m(t)$.

Let $\mathscr{M}$ be the linear submanifold of the vector space of $\mathscr{F}$-measurable functions generated by $\{X_t, t \in T\}$ and $\mathscr{M}_i$ be the submanifold of $\mathscr{M}$ generated by $\{X_t, t \in i\}$ for each $i \in I$. We extend the functions $C_l$ ($l = 1, 2$) to $\mathscr{M}$ by defining $C_l(u, v) = \sum_{i,j} a_s b_t C_l(s, t)$, where $u = \sum_{s \in i} a_s X_s$ and $v = \sum_{t \in j} b_t X_t$ $(i, j \in I)$ for $l = 1, 2$. Hence $C_1, C_2$ so extended are non-negative bilinear forms on $\mathscr{M}$ and, in particular, on $\mathscr{M}_i$ for each $i \in I$. For a fixed $i$, $\mathscr{M}_i$ is finite dimensional; this allows us to choose ([38]) $\{u_{i1} \cdots u_{in_i}\} \subset \mathscr{M}_i$ such that $C_1(u_{ij}, u_{ik}) = \delta_{jk}$ and $C_2(u_{ij}, u_{ik}) = \delta_{jk} \lambda_{ik}$ with $\lambda_{ik} > 0$ $(k = 1, 2, \ldots, n_i)$ since the rank of $C_1$ equals the rank of $C_2$ on $\mathscr{M}_i$, as $P_1 \perp P_2$ in the contrary case. Let $m_{ik} = E_{P_2} u_{ik}$.

It is known that $P_1 \equiv P_2$ or $P_1 \perp P_2$ on $\mathscr{F}_i$. If $P_1 \equiv P_2$ on $\mathscr{F}_i$ we denote $\rho_i$ the R–N density of $P_2$ with respect to $P_1$ on $\mathscr{F}_i$ following Section II. With the above notation we then have the following lemma.

**Lemma 4.1** *The following are equivalent*:

(a) $P_1$ *is not singular with respect to* $P_2$ *on* $\mathscr{F}$;

(b) $\sup\limits_i \sum\limits_{k=1}^{n_i} \dfrac{(1 - \lambda_{ik}^{\frac{1}{2}})^2}{2\lambda_{ik}}$ and $\sup\limits_i \sum\limits_{k=1}^{n_i} m_{ik}^2 < \infty$.

(c) *There exists numbers* $r_1, r_2$ $(0 < r_1 \leq r_2 < \infty)$ *such that*
  (i) $0 < r_1 \leq \lambda_{ik} \leq r_2$ *for all* $k = 1, 2, \ldots, n_i$ *and* $i \in I$,
  (ii) $\sup\limits_i \sum\limits_{k=1}^{n_i} (1 - \lambda_{ik})^2$ *is finite*, (4.1)
  (iii) $\sup\limits_i \sum\limits_{k=1}^{n_i} m_{ik}^2$ *is finite*.

*Proof* From (2.3) $P_1 \perp P_2$ on $\mathscr{F}$ iff $E_{P_1} \rho^{\frac{1}{2}} = 0$. Hence, by Lemma 2.1, (a) is equivalent to $\inf_i E_{P_1} \rho_i^{\frac{1}{2}} > 0$. But

$$E_{P_1} \rho_i^{\frac{1}{2}} = \prod_{k=1}^{n_i} \left(\frac{2\lambda_{ik}^{\frac{1}{2}}}{1 + \lambda_{ik}}\right)^{\frac{1}{2}} \exp\left[-\frac{1}{4} \sum_{1}^{n_i} \frac{m_{ik}^2}{(1 + \lambda_{ik})^2}\right].$$

Hence (a) is equivalent to

$$\sup_i \prod_{j=1}^{n_i} \left(\frac{1+\lambda_{ik}}{2\lambda_{ik}^{\frac{1}{2}}}\right) \quad \text{and} \quad \sup_i \sum_{j=1}^{n_i} \frac{m_{ik}^2}{(1+\lambda_{ik})^2} \quad \text{are finite.} \quad (4.2)$$

Since $[(1+\lambda_{ik})/2\lambda_{ik}^{\frac{1}{2}}] = 1 + [(1-\lambda_{ik}^2)^2/2\lambda_{ik}^{\frac{1}{2}}]$ we get that the first condition in (4.2) is equivalent to the first condition in (b), giving (a) equivalent to (b).

Now $[(1+\lambda_{ik})/2\lambda_{ik}^{\frac{1}{2}}] > 1$ and hence we get that the first condition in (4.2) implies that $\exists\, r_1, r_2$ real numbers ($0 < r_1 \leqslant r_2 < \infty$) such that (c)(i) is satisfied. Thus we get that (b) and (c)(i) are equivalent to (c). But (b) being equivalent to (4.2), (b) implies (c)(i) thus completing the proof.

*Remark* 4.1  We note that Lemma 4.1(b) is equivalent to

$$\sup_i \sum_{k=1}^{n_i} \frac{(1-\lambda_{ik})^2}{\lambda_{ik}} \quad \text{and} \quad \sup_i \sum_{k=1}^{n_i} m_{ik}^2 \quad \text{are finite.}$$

**Lemma 4.2**  *Conditions* (4.1) *imply that for every* $\varepsilon > 0\ \exists\ \alpha(\varepsilon) \in (0,1)$ *such that for all* $i \in I$, $E_{P_1}\rho_i^\alpha > 1-\varepsilon$ *for* $\alpha \in (0, \alpha(\varepsilon))$.

*Proof*  Let us note

$$\rho_i = \left[\left(\prod_{k=1}^{n_i} \lambda_{ik}\right)^{-1}\right]^{\frac{1}{2}} \exp\left\{-\frac{1}{2}\sum_{k=1}^{n_i}\left[\frac{(u_{ik}-m_{ik})^2}{\lambda_{ik}} - u_{ik}^2\right]\right\}.$$

Hence

$$E_{P_1}\rho_i^\alpha = \left[\prod_{1}^{n_i} \frac{\lambda_{ik}^{1-\alpha}}{(\alpha+(1-\alpha)\lambda_{ik})}\right]^{\frac{1}{2}} \exp\left\{-\frac{\alpha(1-\alpha)}{2}\sum_{k=1}^{n_i}\frac{m_{ik}^2}{(\alpha+(1-\alpha)\lambda_{ik})}\right\}.$$

Taking the log of both sides,

$$-\log E_{P_1}\rho_i^\alpha = \frac{1}{2}\left[\sum_{k=1}^{n_i} \log(1-\alpha(1-\lambda_{ik}^{-1})) - \alpha\log(1-(1-\lambda_{ik}^{-1}))\right]$$
$$+ \frac{\alpha(1-\alpha)}{2}\sum_{k=1}^{n_i}\frac{m_{ik}^2}{(\alpha+(1-\alpha)\lambda_{ik})}.$$

Since for $-1 < a_1 < y < a_2 < \infty$ ($a_1 < 1 < a_2$) we have ([9])

$$y - dy^2 \leqslant \log(1-y) \leqslant y - cy^2 \quad \text{with} \quad 0 < c < d \quad (4.3)$$

and we get, using (4.1)(i), for a constant $d_1$,

$$-\log E_{P_1}\rho_i^\alpha \leqslant \alpha(d-c\alpha)\sum_{k=1}^{n_i}(1-\lambda_{ik}^{-1})^2 + \alpha d_1 \sum_1^{n_i} m_{ik}^2. \quad (4.4)$$

Choose $\alpha_0$ such that $(d-c\alpha) > 0$ for $\alpha \leqslant \alpha_0$, then (4.4) and (4.1) imply $\lim_{\alpha \to 0} \inf_i E_{P_1}\rho_i^\alpha = 1$.

Hence by Theorem 2.1 we get (4.1) implies $P_1 \ll P_2$ on $\mathscr{F}$. By the symmetry of the problem (4.1) implies $P_1 \equiv P_2$ on $\mathscr{F}$.

**Theorem 4.1** *Dichotomy* Either $P_1 \perp P_2$ on $\mathscr{F}$ or $P_1 \equiv P_2$ on $\mathscr{F}$. Further $P_1 \equiv P_2$ on $\mathscr{F}$ iff (4.1) holds.

Let us define for each $i \in I$,

$$J_i = E_{P_1}(-\log \rho_i) + E_{P_2}(\log \rho_i).$$

**Corollary 4.1** (*Hajek* [20]) (a) $P_1 \perp P_2$ on $\mathscr{F}$ iff $\sup_i J_i = \infty$ and (b) $P_1 \equiv P_2$ on $\mathscr{F}$ iff $\sup_i J_i$ is finite.

*Proof* From the form of $\rho_i$ given in the proof of Lemma 4.2 we get, using the fact that the distribution of $u_{ik}$ is $G(0,1)$ under $P_1$,

$$E_{P_1} \log \rho_i = \frac{1}{2} \prod_1^{n_i} \left( \log \frac{1}{\lambda_{ik}} - \frac{1}{\lambda_{ik}} + 1 - \frac{m_{ik}^2}{\lambda_{ik}} \right).$$

Similarly, we get

$$E_{P_2} \log \rho_i = \frac{1}{2} \sum_{k=1}^{n_i} (-\log \lambda_{ik} + \lambda_{ik} - 1 + m_{ki}^2).$$

Hence,

$$J_i = \frac{1}{2} \sum_1^{n_i} \left[ \frac{(1-\lambda_{ik})^2}{\lambda_{ik}} + \frac{m_{ik}^2}{1+\lambda_{ik}} \right]$$

Hence $\sup_i J_i < \infty$ iff (4.1) holds in view of Lemma 4.1 and Remark 4.1. Now Theorem 4.1 implies the result.

We know that since $\{u_{i1}, u_{i2}, \ldots, u_{in_i}\}$ generate $\mathscr{M}_i$ for each $i \in I$, $X_t = \sum_{k=1}^{n_i} a_{ik}(t) u_{ik}$, for each $t \in i$. This implies that for $k = 1, \ldots, n_i$, $a_{ik}(t) = C_1(x_t, u_{ik})$, and $\lambda_{ik} a_{ik}(t) = C_2(x_t, u_{ik})$. By the choice of $\{u_{i1}, \ldots, u_{in_i}\}$ we get that $\{u_{i1}, \ldots, u_{in_i}\}$ is an orthonormal basis in $\mathscr{M}_i$ as a subspace of $L_2(\Omega, \mathscr{F}_i, P_1)$ and

$$\left\{ \frac{u_{ik} - m_{ik}}{\lambda_{ik}^{\frac{1}{2}}}, k = 1, 2, \ldots, n_i \right\}$$

is an orthonormal basis in $\mathscr{M}_i$ as a subspace of $L_2(\Omega, \mathscr{F}_i, P_2)$. Let $K(C_l^i)$ ($i \in I$) be the rkhs of $C_l^i$ = restriction of $C_l$ to $i \times i$ ($l = 1, 2$). By the above remarks we have that $\{a_{ik}, k = 1, 2, \ldots, n_i\}$ is an orthonormal basis in $K(C_1^i)$ and $\{\lambda_{ik}^{\frac{1}{2}} a_{ik}, k = 1, 2, \ldots, n_i\}$ is an orthonormal basis in $K(C_2^i)$. Therefore (see Section III) $\{\lambda_{ik}^{\frac{1}{2}} a_{ik} a_{ij}, i, j = 1, 2, \ldots, n_i\}$ is an orthonormal basis in $K(C_2^i \otimes C_1^i)$ and $\{a_{ik} a_{ij}, i, j = 1, 2, \ldots, n_i\}$ is an orthonormal basis in $K(C_1^i \otimes C_1^i)$. Now, for $t, s \in i$

$$C_1^i(t,s) - C_2^i(t,s) = \sum_{i=1}^{n_i} a_{ik}(t)a_{ik}(s) - \sum_{i=1}^{n_i} \lambda_{ik} a_{ik}(t)a_{ik}(s)$$

$$= \sum_{k,j} [(1/\lambda_{ik}^{\frac{1}{2}}) - \lambda_{ik}^{\frac{1}{2}}] \delta_{kj} \lambda_{ik}^{\frac{1}{2}} a_{ik}(t) a_{ij}(s).$$

Hence,

$$C_1^i - C_2^i \in K(C_2^i \otimes C_1^i),$$

$$\|C_1^i - C_2^i\|_{K(C_2^i \otimes C_1^i)} = \sum_{k=1}^{n_i} \frac{(1-\lambda_{ik})^2}{\lambda_{ik}} \quad \text{for } i \in I. \tag{4.5}$$

Similarly,

$$C_1^i - C_2^i \in K(C_1^i \otimes C_1^i),$$

$$\|C_1^i - C_2^i\|_{K(C_1^i \otimes C_1^i)} = \sum_{k=1}^{n_i} (1-\lambda_{ik})^2. \tag{4.6}$$

Furthermore, condition (4.1) on $\lambda_{ik}$ implies that there exist $\gamma_1, \gamma_2$ with $0 < \gamma_1 \leq \gamma_2 < \infty$ such that for $u \in \mathcal{M}_i$ ($i \in I$)

$$\gamma_1 C_1(u,u) \leq C_2(u,u) \leq \gamma_2 C_1(u,u), \tag{4.7}$$

giving

$$\gamma_1 C_1 \ll C_2 \ll \gamma_2 C_2; \tag{4.7'}$$

conversely, (4.7') $\Rightarrow$ (4.7). Here $r_1 C_1 \ll C_2$ means that $C_2 - r_1 C_1$ is a covariance. We now state the main theorem.

**Theorem 4.2** *The following conditions are equivalent.*

(a) $P_1 \equiv P_2$ on $\mathscr{F}$.
(b) $C_1 - C_2 \in K(C_2 \otimes C_1)$ and $m \in K(C_1)$.
(c) (i) There exists $\gamma_1, \gamma_2$, $(0 < \gamma_1 \leq \gamma_2 < \infty)$
    such that $\gamma_1 C_1 \ll C_2 \ll \gamma_2 C_1$;
  (ii) $C_1 - C_2 \in K(C_1 \otimes C_1)$; and (iii) $m \in K(C_1)$.
(d) $C_2 - C_1 \in K(C_1 \otimes C_2)$ and $m \in K(C_2)$.
(e) (i) There exists $\delta_1, \delta_2$, $(0 < \delta_1 \leq \delta_2 < \infty)$
    such that $\delta_1 C_2 \ll C_1 \ll \delta_2 C_2$;
  (ii) $C_2 - C_1 \in K(C_2 \otimes C_2)$ and (iii) $m \in K(C_2)$.

*Violation of any condition implies $P_1 \perp P_2$ on $\mathscr{F}$.*

*Proof* In view of the symmetry it suffices to prove the equivalence of (a), (b), and (c). To prove the equivalence of (a) and (c) we observe that from (4.6), Theorem 4.1, (4.1), and (4.7) we get (a) is equivalent to (4.7') and

$$\sup_i \|C_1^i - C_2^i\|_{K(C_1^i C_2^i)}$$

and

$$\sup_i \|\text{Restriction}_i m\|_{K(C_1^i)} = \sup_i \left\|\sum_{k=1}^{n_i} m_{ik}a_{ik}\right\|_{K(C_1)}$$

is finite. Now (4.7′) is equivalent to (4.1)(i) and the remaining two conditions are equivalent to (c)(ii) and (iii), giving (a) equivalent to (c) by proposition 3 (Appendix). By (4.5) and the first part of the proof, (b) is equivalent to

$$\sup_i \sum_{k=1}^{n_i} \frac{(1-\lambda_{ik})^2}{\lambda_{ik}} \quad \text{and} \quad \sup_i \sum_{k=1}^{n_i} m_{ik}^2$$

being finite. By Lemma 4.1 and Remark 4.1 we get (b) ⇔ (a).

Let $(\Omega, \mathcal{A})$ be a probability space and $\{X_t, t \in T\}$ be the family of random variables on $(\Omega, \mathcal{A})$. Let $I, \mathcal{F}_i$ ($i \in I$) and $\mathcal{F}$ be as defined before and $P_1$ and $P_2$ be two measures on $\mathcal{F}$ such that $\{X_t, t \in T\}$ is a Gaussian process on $(\Omega, \mathcal{F}, P_l)$ ($l = 1, 2$) with $E_{P_l} X_t = m_l(t)$ and $C_l(t,s) = E_{P_l}(X_t - m_l(t))(X_s - m_l(s))$, $l = 1, 2$. Then $P_1 \equiv P_2$ iff $P_1' \equiv P_2'$ where $P_l'$ is the measure induced on $\mathcal{F}$ by $\{X_t - m_1(t), t \in T\}$ under $P_l$ ($l = 1, 2$). Since $E_{P_1'} X_t = 0$, $E_{P_2'} X_t = m_2(t) - m_1(t)$ and $C_l(t,s) = E_{P_l'}(X_t - m_l(t))(X_s - m_l(s))$ we get, by the above theorem.

**Theorem 4.3** *The following are equivalent.*

(a) $P_1 \equiv P_2$ on $\mathcal{F}$.
(b) $C_1 - C_2 \in K(C_2 \otimes C_1)$ and $m_2 - m_1 \in K(C_1)$.
(c) (i) There exists $\gamma_1, \gamma_2$ ($0 < \gamma_1 \leq \gamma_2 < \infty$)
such that $\gamma_1 C_1 \ll C_2 \ll \gamma_2 C_1$
(ii) $C_1 - C_2 \in K(C_1 \otimes C_1)$; and (iii) $m_2 - m_1 \in K(C_1)$.
(d) $C_2 - C_1 \in K(C_1 \otimes C_2)$ and $m_1 - m_2 \in K(C_2)$
(e) (i) There exists $\delta_1, \delta_2$ ($0 < \delta_1 \leq \delta_2 < \infty$)
such that $\delta_1 C_2 \ll C_1 \ll \delta_2 C_2$;
(ii) $C_2 - C_1 \in K(C_2 \otimes C_2)$ and (iii) $m_1 - m_2 \in K(C_2)$.

Parts (a) ⇔ (b) of the above theorem is due to Parzen [54] (see also Neveu [48]).

Using Theorem 4.3(c)(i) and Proposition 1 (Appendix) we get that there exists a nonnegative definite linear bounded operator $L$ on $K(C_1)$ into $K(C_1)$ with $(Lf)(t) = (f, C_2(\cdot, t))_{K(C_1)}$. Hence

$$((1-L)f)(t) = (f, C_1(\cdot, t) - C_2(\cdot, t))_{K(C_1)}.$$

Here 1 denotes the identity operator on $K(C_1)$. Using Theorem 4.3(c) and Proposition 2 (Appendix) we get with notation as above.

**Theorem 4.3** $P_1 \equiv P_2$ on $\mathscr{F}$ iff

(a) (i) There exist $\gamma_1, \gamma_2$ $(0 < \gamma_1 \leq \gamma_2 < \infty)$ such that $\gamma_1 C_1 \ll C_2 \ll \gamma_2 C_1$, (ii) $m_2 - m_1 \in K(C_1)$, and (iii) $(1-L)$ is Hilbert–Schmidt on $K(C_1)$ into $K(C_1)$ or iff.

(b) (i) The operator $L$ on $K(C_1)$ into $K(C_1)$ given by $(Lf)(t) = (f, C_2(\cdot, t))_{K(C_1)}$ with $f \in K(C_1)$ such that $L$ is nonnegative and $(1-L)$ is Hilbert–Schmidt (ii) 1 is not an eigenvalue of $(1-L)$ and (iii) $m_2 - m_1 \in K(C_1)$.

The equivalence of (a) and (b) follows from the fact that in the case $(1-L)$ Hilbert–Schmidt, one is not an eigenvalue of $(1-L)$ is equivalent to $L$ being invertible.

*Remark 4.2* The condition (a)(iii) can be restated by saying $L$ has pure point spectrum and $\sum_n (1-\lambda_n)^2 < \infty$ for nonzero eigenvalues $\{\lambda_n\}$ of $L$. In this form the theorem was proved by Kallianpur and Oodaira [34] (see also [50]).

Let us denote by $H_l(X)$ the completion of $\bigcup_{i \in I} \mathscr{M}_i$ in $L_2(\Omega, \mathscr{F}, P_l)$ ($l = 1, 2$) and let $\Lambda$ be the bounded linear operator on $H_1(X)$ into $H_2(X)$ defined as $\Lambda u = u$ for $u \in \bigcup_i \mathscr{M}_i$. This is possible in view of (4.7). Also, by (4.7) we get that there exists $\gamma_1, \gamma_2$ $(0 < \gamma_1 \leq \gamma_2 < \infty)$ such that

$$\gamma_1 \|u\|^2_{H_1(X)} \leq \|\Lambda u\|^2_{H_2(X)} \leq \gamma_2 \|u\|_{H_1(X)} \quad \text{for all} \quad u \in H_1(X). \quad (4.8)$$

Hence $\Lambda$ is a bounded operator with bounded inverse. We note that $\Lambda$ is a linear extension of $\Lambda X_t = X_t$ on $H_1(X)$ into $H_2(X)$. Denote by $\Pi_l$ the isometry between $K(C_l)$ into $H_l(X)$ ($l = 1, 2$) and set $\tilde{\Lambda} = \Pi_2^* \Lambda \Pi_1^*$. Then $\tilde{\Lambda}^* \tilde{\Lambda} = \Pi_1^* \Lambda^* \Lambda \Pi_1$. Also, for $t, s \in T$, $(\Pi_1^* \Lambda^* \Lambda \Pi_1 C_1(\cdot, t), C_1(\cdot, s))_{K(C_1)} = (\Lambda X_t, \Lambda X_s)_{H_2(X)} = C_2(t, s)$. Now if (4.7') (or (4.8)) is satisfied we get $\Pi_1^* \Lambda^* \Lambda \Pi_1(C_1(\cdot, t)) = L(C_1(\cdot, t))$ for $t \in T$, where $I_.$ is as before since $L(C_1(\cdot, t)(s)) = (C_1(\cdot, t), C_2(\cdot, s))_{K(C_1)} = C_2(t, s)$. Since (4.7) and (4.8) are equivalent and $(1-L) = \Pi_1^*(1_1 - \Lambda^* \Lambda)\Pi_1$ [$1_1$ is identity on $H_1(X)$], we get by Theorem 4.3

**Theorem 4.4** [12, 59] $P_1 \equiv P_2$ on $\mathscr{F}$ iff

(i) $\Lambda$ is one–one bounded, with bounded inverse on $H_1(X)$ onto $H_2(X)$ and

(ii) $(1_1 - \Lambda^* \Lambda)$ is Hilbert–Schmidt.

(iii) $m_2 - m_1 \in K(C_1)$.

## V. Conditions for Equivalence: Special Cases

### A. Introduction

In this section, $P = G(m, C)$ will mean that $\{X_t, t \in T\}$ is a Gaussian process under $P$ with $m(t) = E_P X_t$ and $C(s, t) = E_P(X_t - m(t))(X_s - m(s))$.

## 1. Gaussian Processes Equivalent to Wiener-Lévy Brownian Motion [66, 28]

Let $P_1$ be $G(0, C_1)$ with $C_1(t,s) = \min(t,s) = \int 1_{(0,t]}(u) 1_{(0,s]}(u) \, du$ $(s,t \in R_+)$. Then $P_2 = G(m,C)$ is equivalent to $P_1$ iff

(a) $m(t) = \int_0^t f(u) \, du$ for some $f \in L_2(R_+)$
(b) $C(t,s) = \min(t,s) - \int_0^t \int_0^s g(u,v) \, du \, dv$ for some symmetric $g \in L_2(R_+^2)$
(c) 1 is not an eigenvalue of the integral operator

$$I(h)(u) = \int g(u,v) h(v) \, dv.$$

The *proof* is immediate from Theorem 4.3 and Example 3.1 as

$$(C_1 \otimes C_1)((t_1, t_2), (s_1, s_2))$$

is equal to $\int\int 1_{(0,t_1] \times (0,s_1]}(u,v) 1_{(0,t_2] \times (0,s_2]}(u,v) \, du \, dv$.

## 2. Gaussian processes equivalent to Cameron-Yeh process [52]

Let $T = R_+^n$ and $P_1 = G(0, C_1)$ with $C_1(t,s) = \prod_{k=1}^n \min(t_i, s_i)$ for $t = (t_1, \ldots, t_n)$, $s = (s_1, \ldots, s_n)$. Then $P_2 = G(m,C)$ is equivalent to $P_1$ iff

(a) $m(t) = \int_T R_t(u) f(u) \, du$ for some $f \in L_2(T)$,
  where $R_t(u) = 1_{(0,t_1] \times \cdots \times (0,t_n]}(u_1 \cdots u_n)$.
(b) $C(t,s) = C_1(t,s) - \int_T \int_T R_t(u) R_s(v) g(u,v) \, du \, dv$
  with some $g \in L_2(T \times T)$ and symmetric.
(c) 1 is not an eigenvalue of the integral operator

$$L(h)(u) = \int g(u,v) h(v) \, du$$

on $L_2(T)$ into $L_2(T)$.

*Proof* follows again from Theorem 4.2, Theorem 4.3 and Example 3.1, Sect. III.A using form of $C_1 \otimes C_1$.

### B. Gaussian Processes with Independent Increments [67]

Let $(\mathscr{X}, \mathscr{A})$ be a measurable space. Let $T = \mathscr{A}$. A Gaussian process $\{X_t, t \in T\}$ on $(\Omega, \mathscr{F}, P)$ is said to have independent increments if $\exists$ a $\sigma$-finite measure $\mu$ on $\mathscr{A}$ such that $E_P X_t = m(t)$ and $EX_t X_s = \mu(t \cap s)$ for $t, s \in \{A \in \mathscr{A}, \mu(A) < \infty\}$. Let $P_1 = G(0, \mu_1(t \cap s))$ and $P_2 = G(m, \mu_2(t \cap s))$. Then $P_1 \equiv P_2$ iff

(a) $\mu_1^{(c)} = \mu_2^{(c)}$, where $\mu_i^{(c)}$ denotes the nonatomic part of $\mu_i$
(b) $\mu_1$ and $\mu_2$ have the same atoms $\{a_n, n = 1, 1, \ldots\}$ $a_n \in \mathscr{X}$ and

$$\sum_n \left[ 1 - \frac{\mu_2(\{a_n\})}{\mu_1(\{a_n\})} \right]^2 < \infty,$$

(c)  $m \ll \mu_1$ with $(dm/d\mu_1) \in L_2(\mathcal{X}, \mathcal{A}, \mu_1)$.

*Proof* We first note that since $C_1(t,s) = \int 1_t(u) 1_s(u) d\mu_1$ we get by Example 2.1 and Theorem 4.2 that $P_1 \equiv P_2$ iff

(a')  $\mu_2(t \cap s) = \mu_1(t \cap s) - \iint 1_t(u) 1_s(v) g(u,v) d\mu_1 d\mu_2$
  for some $g \in L_2(\mu_1 \otimes \mu_2)$
(c')  $m(t) = \int 1_t(u) f(u) du$ for some $f \in L_2(\mu_1)$.

Here we use the fact that

$$(C_1 \otimes C_2)[(t_1, t_2)(s_1, s_2)] = \iint 1_{t_1 \times s_1}(u,v) 1_{t_2 \times s_2}(u,v) d\mu_1 d\mu_2$$

and Example 3.1. Now (c) is equivalent to (c') above. In view of (a') we get the signed $\int_t \int_s g(u,v) d\mu_1 d\mu_2$ is zero off diagonal as it is a function of $t \cap s$. If $\mu_1$ is nonatomic then we get from (a') $\mu_1(t \cap s) = \mu_2(t \cap s)$, i.e., $\mu_1 = \mu_2$. In the case where $\exists a \in \mathcal{X} \ni \mu_1\{a\} \neq 0$ and $\mu_2\{a\} = 0$ again we get from (a') $\mu_1 = \mu_2$ as the integral term vanishes, giving contradiction. Hence, $\mu_1$ and $\mu_2$ have positive measures at the same $\{a_n\} \subset \mathcal{X}$. Since $\mu_1$ and $\mu_2$ are $\sigma$-finite there exist at most countable $\{a_n, n = 1, 2, \ldots\} \subseteq \mathcal{X}$ with $\mu_1\{a_n\} = \mu_2\{a_n\}$. In this case we get that (a') is equivalent to (a) $\mu_1^{(c)} = \mu_{(c)}^2$ and (b') $\mu_2(\{a_n\}) = \mu_1(\{a_n\}) - g(a_n, a_n) \mu_1\{a_n\} \mu_2\{a_n\}$. Now (b') is equivalent to (b) since $\sum_n g^2(a_n, a_n) \mu_1\{a_n\} \mu_2\{a_n\}$ converges by the assumption $g \in L_2(\mu_1 \otimes \mu_2)$.

## C. *Stationary Gaussian Processes* [13, 67]

Let $T$ be a locally compact abelian group with separable dual and operation $+$. Let $P_1 = G(0, C_1)$ and $P_2 = G(0, C_2)$, where $C_i(t,s) = R_i(t-s)$, where $R_i(t)$ is a continuous nonnegative definite function on $T$ and $\mu_1$ and $\mu_2$, be associated spectral measures, then

$P_1 \equiv P_2 \quad \Leftrightarrow \quad$ (a) $\mu_1^{(c)} = \mu_2^{(c)}$ where $\mu_i^{(c)}$ denotes the nonatomic part of $\mu_i$ ($i = 1, 2$).
(b) $\mu_1$ and $\mu_2$ have the same atoms $\{a_n, n = 1, 2, \ldots\}$ in

$$\sum_n \left(1 - \frac{\mu_2\{a_n\}}{\mu_1\{a_n\}}\right)^2 < \infty.$$

*Proof* From Section III.B.5 and Theorem 4.2, we get $P_1 \equiv P_2$ iff

$$R_2(t-s) = R_1(t-s) - \int_{\hat{T}} \int_{\hat{T}} \langle t, u \rangle \langle s, v \rangle g(u,v) d\mu_1 d\mu_2$$

with $g \in L_2(\hat{T} \times \hat{T}, \mu_1 \otimes \mu_2)$. Since the last term depends on $t-s$ we get that the signed measure $\int_A \int_B g(u,v)\, d\mu_1\, d\mu_2$ for $A \times B \subset \hat{T} \times \hat{T}$ is zero off diagonal by Bochner's theorem and the uniqueness of Fourier transform. An argument as in the proof in (b) above gives that $\mu_1$ and $\mu_2$ have the same atoms $\{a_n, n = 1, 2, \ldots\} \subseteq \hat{T}$. Hence we get

$$R_2(t) = R_1(t) - \sum_n e^{ita_n} g(a_n, a_n) \mu_1\{a_n\} \mu_2\{a_n\}.$$

Using uniqueness of the Fourier transform we get

$$\mu_2(A) = \mu_1(A) - \sum_{a_n \in \xi} g(a_n, a_n) \mu_1\{a_n\} \mu_2\{a_n\}.$$

Now the proof is completed as for Section IV.B.

D. *Gaussian Measures on Banach Spaces*

Let $E$ be a real separable Banach space and $\mathscr{B}(E)$ the Borel subsets of $E$. For a (centered) Gaussian measure $P$ on $E$, let $H(P)$ denote the subspace of $L_2(E, \mathscr{B}(E), P)$ generated by $\{x'(\cdot), x' \in E'\}$. We denote by $C_P$ the operator on $E' \to E$ defined by $C_P x' = \int x'(x) x\, dP$, where the integrals are in the sense of Bochner (see Section III.B.4). We note that $y'(C_P x') = C_P(x', y')$ is the covariance operator of the Gaussian process $\{x'(\cdot), x' \in E'\}$. We denote $P$ by $G(0, C_P)$. Let $Q$ be $G(m, C_Q)$ with mean $m = \int x\, dQ \in B$ and $C_Q(x') = \int x'(x-m)(x-m)\, dQ$. Then we get the following.

(A)  $P \equiv Q$  iff  (a) $m = \int x g_0(x)\, dP$ for $g_0 \in H(P)$
  (b) $C_P - C_Q = G_0$, where $G_0$ is an operator on $E'$ into $E$ given by $G_0(x') = \iint f_0(x, y) x'(x) y\, dP\, dQ$, where $f_0$ lies in the linear subspace of

$$L_2(E \times E, \mathscr{B}(E) \otimes \mathscr{B}(E), P \otimes Q)$$

generated by $\{x'(x) y'(y); x', y' \in E'\}$ and is symmetric.

*Proof*  From Theorem 4.2 and Example 3.1 we get

$P \equiv Q$  iff  (a') $x'(m) = \int x'(x) g(x)\, dP$
  (b') $C_P(x', y') = C_Q(x', y') = \iint f_0(x, y) x'(x) y'(y)\, dP\, dQ$.

Now conditions (a) and (b) are equivalent to conditions (a') and (b').

(B) The above conditions can be written, using Theorem 4.3.

$P \equiv Q$  iff  (a) $m = \int x g_0(x)\, dP$ for $g_0 \in H(P)$
  (b) $C_P - C_Q = \tilde{G}$, where $\tilde{G}(x') = \iint \tilde{g}(x, y) x'(x) y\, dP\, dP$ with $\tilde{g}$ symmetric and as in (b) above with $P \otimes P$ for $P \otimes Q$; and
  (c) The operator taking $f \to \int \tilde{g}(x, y) f(y)\, dP$ on $H(P)$ into $H(P)$ does not have eigenvalue unity.

With $E = C[0,1]$ and restricting to $x' = \varepsilon_t$, the unit mass at $t$, this condition in (B) gives the result of Shepp [66].

(C) $E = l_p$ $(p \geq 1)$ [37]. Let $C_P = (s_{ij}^P)_{i,j=1}^{\infty}$, $C_Q = (s_{ij}^Q)$ be matrices of covariance operators of measure $P = G(0, C_P)$, $Q = G(m, C_Q)$ and assume that $C_P$ is diagonal. Let $e_i = (0, 0, ..., 1, ...)$. We note that

$$H(P) = \{\sum a_i e_i(x) : \sum a_i^2 s_{ii}^P < \infty\}$$

and the subspace of $L_2(E \times E, \mathcal{B}(E) \otimes \mathcal{B}(E), P \otimes P)$ is

$$= \left\{\sum_{i,j} b_{ij} e_i(\cdot) e_j(\cdot); \sum_{i,j} b_{ij}^2 s_{ii}^P s_{jj}^P < \infty\right\}.$$

Hence $P \equiv Q$ iff (a) $e_i(m) = a_i s_{ii}^P$ for all $i$ with $\sum a_{ii} s_{ii}^P$ finite
(b) $s_{ij}^Q = s_{ij}^P - b_{ij} s_{ii}^P s_{jj}^P$ for some $b_{ij}$ with

$$\sum_{i,j} b_{ij}^2 s_{ii}^P s_{jj}^P < \infty$$

(c) the operator $B$ on $l_2$ given by matrix $b_{ij}$ has no eigenvalue equal to 1.

E. *Generalized Gaussian Processes Equivalent to Gaussian White Noise of Order $p$* [26]

Let $T = C_0^{\infty}(V)$ for $V$ an open subset of $R^n$ and $C_P(t,s) = \sum_{|\alpha| \leq rp} \int D^\alpha t D^\alpha s \, dx$. Let $P = G(0, C)$ and $Q = G(m, C_0)$, where $m$ is a continuous function on $T$ and $C_Q(\cdot, \cdot)$ is a continuous bilinear form on $T \times T$ under Schwartz's topology. The proof of the following is immediate from Theorem 4.3 and Section III.B.6.

$P \equiv Q$ iff (a) $m(t) = \int_V f(u) t(u) \, du$ for some $f \in H_0^p$
(b) $C_P(t,s) - C_Q(t,s) = \int_V \int_V g_0(u,v) t(u) s(v) \, du \, dv$ for some $g_0 \in H_0^p(V \times V)$ symmetric,
(c) the operator $G_0$ on the Sobolev space $H_0^m(0)$ into $H_0^m(0)$ given by $(G_0 f)(v) = \int g_0(u,v) f(v) \, dv$ does not have eigenvalue unity.

## VI. Applications

Let us consider the case when $P_0$ and $P_m$ are measures on $(\Omega, \mathcal{A})$ such that the family $\{X_t, t \in T\}$ of random variables is Gaussian under $P_0$ and $P_m$. Let

$$E_{P_0} X_t = 0, \quad E_{P_0} X_t X_s = C(t,s), \quad E_{P_m} X_t = m(t)$$

and

$$E_{P_m}(X_t - m(t))(X_s - m(s)) = C(t,s).$$

Then by our theorem $P_0 \equiv P_m$ iff $m \in K(C)$. Denote by $\pi$ the map on $K(C)$ to $H_{P_0}(X)$ and by $d_m = \exp[\pi(m) - \frac{1}{2}\|m\|^2]$. Since $\pi(m)$ is $G(0, \|m\|^2)$ we get $E_{P_0} d_m = E_{P_m} 1 = 1$ and $E_{P_0} d_m d_{m'} = \exp(m, m')$, $m, m' \in K(C)$. We now show that $d_m = dP_m/dP_0$. It includes results of [19, 45, 55, and 71] in view of last section and Section III.B.3.

**Theorem 6.1** $P_m(A) = \int_A d_m \, dP_0$ for $A \in \sigma\{\pi(m), m \in K(C)\} = \mathscr{F}_T$.

*Proof* Let $\{e_k, k \in J\}$ be an orthonormal basis in $K(C)$. Then

$$\sigma\{\pi(m), m \in K(C)\} = \sigma\left\{\bigcup_{i \in I} \sigma\{\pi(e_k); k \in i\}\right\},$$

where $I$ is the family of finite subsets of $J$ directed by inclusion. Now for $A \in \mathscr{F}_i = \sigma\{\pi(e_k); k \in i\}$ ($i$ fixed),

$$Q(A) = E_{P_0}[1_A d_m]$$
$$= E_{P_0}\left(1_A \exp\left\{\sum_{k \in i}[(m, e_k)\pi(e_k) - \frac{1}{2}(m, e_k)^2]\right\}\right) \exp\left[-\frac{1}{2}\sum_{k \notin i}(m, e_k)^2\right].$$

However by direct calculation

$$Q(A) = P_m(A) \cdot \exp\left[-\frac{1}{2}\sum_{k \notin i}(m, e_k)^2\right];$$

i.e.,

$$\frac{dP_m}{dP_0} \quad \text{on} \quad \mathscr{F}_i = E_{P_0}^{\mathscr{F}_i}\left(\frac{dQ}{dP_0}\right) \exp\left[\frac{1}{2}\sum_{k \notin i}(m, e_k)^2\right]. \tag{6.1}$$

Now $E(E_{P_0}^{\mathscr{F}_i}(dQ/dP_0))^2 \leq \exp(m, m)$, giving uniform integrability of $E_{P_0}^{\mathscr{F}_i}(dQ/dP_0)$. Hence $dP_m/dP_0$ on $\mathscr{F}_i$ converges to $dQ/dP_0$ by (6.1) in $L_1(\Omega, \mathscr{F}, P_0)$.

**Lemma 6.1** [33, 39] *The set* $\{d_m, m \in K(C)\}$ *generates* $L_2(\Omega, \mathscr{F}, P_0)$.

*Proof* Suppose not, then there exists $f \in L_2(\Omega, \mathscr{F}, P_0)$ such that $E_{P_0} d_m f = 0$ for all $m \in K(C)$. Using $\mathscr{F}_i$ as in Theorem 6.1 we get, using $E_{P_0}(d_m f) = E_{P_0}[E_{P_0}^{\mathscr{F}_i}(d_m f)]$, that

$$0 = \exp\left[-\frac{1}{2}\sum_{k \notin i}(m, e_k)^2\right]$$
$$\times \int_\Omega \exp\left\{\sum_{k \in i}(m, e_k)\pi(e_k) - \frac{1}{2}\sum_{k \in i}(m, e_k)^2\right\} E_{P_0}^{\mathscr{F}_i}(f) \, dP_0.$$

Hence there exists a Borel measurable function $h$ on $R^{\text{Card}(i)}$ and square integrable with respect to Gaussian product measure $\mu$ on $R^{\text{Card}(i)}$ with component $G(0, 1)$ satisfying

$$\int h(x_k, k \in i) \exp\left[\sum_{k \in i} m_k x_k - \frac{1}{2}\sum_{k \in i}(m, e_k)^2\right] d\mu = 0$$

for all $m_k$, $k \in i$). Hence $h = 0$ a.e. $P_0$ i.e. $E_{P_0}^{\mathscr{F}_i}(f) = 0$ for all $i$. Since $f$ is square integrable, $E^{\mathscr{F}_i}(f)$ converges to $f$ in $L_2(\Omega, \mathscr{F}, P_0)$ giving $f = 0$.

**Corollary 6.1** *Let $\Gamma(m, m') = \exp(m, m')_{K(C)}$ for $m, m' \in K(C)$. Then $\Gamma$ is a covariance on $K(C)$ and*

$$H(\Gamma) = \{g \colon g(m) = E_{P_0} \exp(\pi(m) - \tfrac{1}{2}\|m\|^2) f, \; f \in L_2(\Omega, \mathscr{F}, P_0)\}.$$

*Proof* The first part follows as $(m, m')_{K(C)}$ is a covariance and (3.3). Second part is a consequence of Example (a), Sect. III.A.

Let us recall that $\pi(K(C)) = H_{P_0}(X)$. We denote this by $H(X)$. From Corollary 6.1 we get $H(\Gamma)$ is isometric to $L_2(\Omega, \mathscr{F}, P)$ under the map $\Gamma(\cdot, m) \to \exp[\pi(m) - \tfrac{1}{2}\|m\|^2]$. Since $m \to \Gamma(\cdot, m)$ is a continuous map we get $\pi(m) \to e^{\pi(m)}$ is a continuous map of $H(X)$ onto $L_2(\Omega, \sigma(H(x)), P)$.

Let $\{\xi_i = \pi(e_i), i \in J\}$, where $\{e_i, i \in J\}$ is an orthonormal basis in $K(C)$. We now extend the original proof of N. Wiener [72], to give homogeneous chaos expansion.

We note that $\exp(ux - \tfrac{1}{2}u^2 x^2) = \sum_{n=0}^{\infty} (u^n/n!) h_n(x)$, where $h_n$ is $n$th Hermite polynomial. Let $i$ be a finite subset of $J$. Then with $\{\xi_k, k \in J\}$ as in the above corollary,

$$\prod_{k \in i} \{\exp a_k \xi_k - \tfrac{1}{2} a_k^2 \xi_k^2\} = \prod_{k \in i} \sum_{n_k = 0}^{\infty} \frac{a_k^n}{n!} h_n(\xi_k).$$

Since $\pi(m) \to \exp(\pi(m))$ is continuous we get by Lemma 5.3 that

$$\left( \prod_{k \in J} (1/(n_k!)^{\frac{1}{2}}) h_{n_k}(\xi_k); \; \sum_k n_k < \infty \right)$$

generates $L_2(\Omega, \mathscr{F}, P)$.

*Triviality* (a) The system

$$\left\{ \prod_{k \in J} (n_k!)^{-\frac{1}{2}} h_{n_k}(\xi_k) \colon \sum_k n_i = n \right\}$$

is orthonormal in $L_2(\Omega, \mathscr{F}, P)$.

(b) The systems

$$\left\{ \prod_{k \in J} (n_k!)^{-\frac{1}{2}} h_{n_k}(\xi_k) \colon \sum n_k = n \right\} \quad \text{and} \quad \left\{ \prod_{k \in J} (l_k!)^{-\frac{1}{2}} h_{l_k}(\xi_k) \colon \sum l_k = l \right\}$$

are orthogonal for $n \neq l$.

*Proof* (b) Let $Z_n$, $Z_l$ be any two elements of the first and second systems, respectively. Since $n \neq l \;\exists\; k \in J \ni l_k \neq n_k$. This implies

$$E_P Z_l Z_n = \text{Const } E_P l_k^{-\frac{1}{2}} h_{l_k}(\xi_k) \cdot n_k^{-\frac{1}{2}} h_{n_k}(\xi_k) = 0.$$

The proof of (a), being trivial, is omitted. Let $H_n$ be the subspace of $L_2(\Omega, \mathscr{F}, P)$ generated by

$$\left\{ \prod_{k \in J} (n_k!)^{-\frac{1}{2}} h_{n_k}(\xi_k) : \sum n_k = n \right\}.$$

**Theorem 6.2** [30] $L_2(\Omega, \mathscr{F}, P) = \sum_{n=0}^{\infty} \otimes H_n$. *In particular,* $f \in L_2(\Omega, \mathscr{F}, P)$ *has expansion,*

$$f = Ef + \sum_{n=1}^{\infty} \sum_{\sum n_{k_j} = n} \sum_{(k_1 \cdots k_r)} a_{n_{k_1} \cdots n_{k_r}}^{k_1 \cdots k_r} \prod_{j=1}^{r} h_{n_{k_j}}(\xi_{k_j}),$$

where $(k_1 \cdots k_r)$ is a finite subset of $J$ such that $\sum_1^r n_{k_j} = n$ and $n_{k_j} > 0$. Here convergence is in $L_2(\Omega, \mathscr{F}, P)$.

*Remark 6.1* (a) If $P$ is a Gaussian measure on a real separable Banach space $(E, \mathscr{B}(E))$ with covariance $C(x', y')$. Then $\mathscr{B}(E) = \sigma\{x'(\cdot), x' \in E'\}$ and hence we get in this case the known result in abstract Wiener space.

(b) If $X$ is a generalized white noise process [22] then with $T = C_0^{\infty}(R^n)$ we get the expansion in [23] and [22].

We get the following by Corollary 6.1.

**Theorem 6.3** [33, 39, 48] *Each* $f \in H(\Gamma)$ *can be uniquely represented by*

$$f(m) = C_m + \sum_{n=1}^{\infty} \sum_{n_k = u} \sum_{(k_1 \cdots k_r)} a_{n_{k_1} \cdots n_{k_r}}^{k_1 \cdots k_r} \prod_{j=1}^{r} \frac{(m, e_{n_{kj}})^{n_{kj}}}{n_{kj}!^{\frac{1}{2}}},$$

where $C_m$ is a constant $\sum \sum \sum (a_{n_{k_1} \cdots n_{k_r}}^{k_1 \cdots k_r})^2$ and $\sum \sum \sum \prod [(m, e_{n_j})^{2n_{kj}}/n_{kj}!]$ converge where the sum is over all indices.

**Zero–One Law** [31] Let $P_0, P_m$ and $C$ be as at the beginning of this section. A function $f$ in $L_2(\Omega, \mathscr{F}, P_0)$ is constant iff $E_{P_0} f = E_{P_m} f$ for all $m \in K(C)$.

*Proof* ($\Rightarrow$) Since $m \in K(C)$, we get $P_m \equiv P_0$ on $\mathscr{F}$. Hence, $d_m$ is $\mathscr{F}$ measurable. Since $E_{P_0} d_m = 1$, we get $E_{P_0} f = E_{P_m} f$ for $f$ constant. For the converse we observe from Remark 6.1 that $\text{Proj}_{H_0} d_m = 1$. Hence the condition implies that $E_{P_0}(1 - \text{Proj}_{H_0}) d_m f = 0$, giving $(1 - \text{Proj}_{H_0}) f = 0$ by Lemma 6.1, i.e., $f = \text{Proj}_{H_0} f = Ef$, a constant.

*Remark 6.2* Examining the proof of Lemma 6.1 we get that $f \in L_2(\Omega, \mathscr{F}, P_0)$ is constant iff $E_{P_0} f = E_{P_m} f$ for all $m \in K(C)$, which are finite linear combinations of the elements $\{e_j, j \in J\}$.

Let $(E, \mathscr{B}(E))$ be a separable Banach space and $P$ be a Gaussian measure on $\mathscr{B}(E)$ with covariance $C$ on $E'$. Let $K(C)$ be the reproducing kernel

Hilbert space as in Section III.B.5. Let $\bar{K}$ be the closure of $K(C)$ in $E$. Then, clearly, $\bar{K} \in \mathscr{B}(E)$ being a closed subset of $E$ and

$$A = \bigcap_m \bigcup_n \left\{ x \in E : \left\| x - \sum_1^n \pi(e_i) e_i \right\|_E < (1/m) \right\} \subseteq \bar{K},$$

where $\{e_i, i = 1, 2, \ldots\}$ is an orthonormal system in $K$. Here we use the fact that $K(C)$ is separable, using the representation in Section III.B.5, since $E$ is separable. Now $A \in \mathscr{B}(E)$ and $P(A) = 1$ [40]. Hence $P(\bar{K}) = 1$.

**Definition** [53] Let $\mathscr{X}$ be a separable metric space and $\mu$ be a probability measure on $(\mathscr{X}, \mathscr{B}(\mathscr{X}))$ a closed set $B$ is called the *support* of $\mu$ if $\mu(B) = 1$ and for each open set $0$ containing $x$ belonging to $B$ $\mu(0) > 0$ and write $B = \operatorname{supp} \mu$.

**Theorem 6.4** [32] *Let $P$ be a (centered) Gaussian probability measure on $[E, \mathscr{B}(E)]$. Then $\operatorname{supp} P = \bar{K}$.*

*Proof* As shown above $P(\bar{K}) = 1$. For $x \in \bar{K}$ suppose there exists an open set $0$ such that $P(0) = 0$. Then for each element $m$ in $M$ the submodule of $K$ generated by $\{e_i, i = 1, 2, \ldots\}$ with rational coefficients we get $P_m(0) = 0$ giving $P[\bigcup_m (0-m)] = 0$. But $\bar{K} \subset \bigcup_m (0-m)$ giving contradiction.

**Corollary 6.2** ([32, 69]) *Let $P$ be a (centered) Gaussian measure on $(E, \mathscr{B}(E))$ with the covariance operator $C$ defined on $E'$ by $C(x') = \int \langle x', x \rangle x \, dP$. Then the following are equivalent.*

(i) *Every nonzero linear functional on $E'$ has a nondegenerate Gaussian distribution.*
(ii) $C(E')$ *is dense in $E$.*
(iii) $C$ *is strictly positive, i.e., $x'(Cx') = 0$ iff $x' = 0$.*
(iv) *The absolute value of the characteristic functional of $P$ given by $\phi_P(x') = E_P \exp(i(x'(x)))$ is unity only for $x' = 0$.*
(v) *The $P$ measure of every nonvoid open subset of $E$ is positive.*

*Proof* Since $x'(\cdot)$ is $G[0, x'(Cx')]$ we get (i) $\Leftrightarrow$ (iii) $\Leftrightarrow$ (iv). Now $C(E') = K(C)$ giving (ii) $\Leftrightarrow$ (v) iff $\sup P = K(C) = E$ by Theorem 6.4. Further (iii) $\Leftrightarrow$ (ii) is a consequence of the Hahn–Banach theorem. Finally, suppose (ii) holds and (iii) fails then there exists $y' \neq 0$ in $E'$ such that $y'(Cy') = 0$. But positive definiteness of $C(x', y')$ implies that $|y'(Cx')|^2 \leq x(Cx') y'(Cy')$ $\forall x' \in E$, i.e., $\exists y' \neq 0$ in $E'$ such that $y'$ annihilates $C(E')$, contradicting (ii). This completes the proof.

## VII. Concluding Remarks

We have not made an attempt here to compute the densities in the general case. This problem is closely related to the methods developed in the study of Meyer–Doob–Fisk decomposition of super-martingales [44]. An excellent expository paper with extensive bibliography (S. Orey [51]) already exists in the literature. Orey also shows that such methods allow one to study the equivalence and singularity problem for the diffusion processes. Similar stochastic integral methods for the general Markov processes have been developed by Kunita [38]. All these methods stemmed from the results of Girsanov [17] and T. Hitsuda [25] who attempted to get nonanticipating representations of equivalent Gaussian processes. The latter problem has extensive literature [51]. An attack on the representation problem through purely Hilbert space methods is available (Kailath [27], Kallianpur–Oodaira [34]). We have avoided the presentation of his approach as it will entail heavy dependence on a difficult book [18] and would cause the paper to be unreadable (or at least difficult to read!). Also, in view of the availability of Newman [49], we have not presented the case of general processes with independent increments. The analogues of Theorem 5.12 are known for measures on more abstract linear spaces [10]. In view of H. Sato [62], the problem in most of the cases reduces to the one studied for separable Banach spaces.

## Appendix

The following propositions [2] were used in Section IV. In Section III, it was shown that every real-valued function $\Lambda$ on $T \times T$ such that $\Lambda(\cdot, t) \in K(C)$ corresponds to a bounded linear $L$ operator on $K(C)$. Further relationships between $L$ and $\Lambda$ are given by the following.

**Proposition 1** Let $L$ be a bounded linear operator on $K(C)$ into $K(C)$ and $\Lambda(t', t) = L^*(C(\cdot, t))(t')$. Then

(a) $L$ is self-adjoint off $\Lambda$ is symmetric;

(b) $L$ is a nonnegative definite operator iff $\Lambda$ is a covariance on $T$ and $\Lambda \ll kC$ ($k \geqslant 0$); and

(c) $L$ is a nonnegative definite operator with bounded inverse iff there exist $k_1, k_2$ ($0 < k_1 < k_2 < \infty$) such that $k_1 C \ll \Lambda \ll k_2 C$ and $\Lambda$ is a covariance on $T$.

**Proposition 2** Let $C_1$ and $C_2$ be two covariance kernels on $T_1$ and $T_2$, respectively, and $(C_1 \otimes C_2)$ be as in Section III, then $f \in K(C_1 \otimes C_2)$ satisfies $f(\cdot, t_2) \in K(C_1)$. If further $(L_f g)(t_2) = (g, f(\cdot, t_2))_{K(C_1)}$ for $g \in K(C_1)$, we get that $L_f$ is a Hilbert–Schmidt operator on $K(C_1)$ into $K(C_2)$. Conversely, every Hilbert–Schmidt operator on $K(C_1)$ into $K(C_2)$ is given by $f \in K(C_1 \otimes C_2)$. Finally, $\|f\|_{K(C_1 \otimes C_2)}$ equals the Hilbert–Schmidt norm of $L_f$.

**Proposition 3** Let $T$ be any set and $\{T_\alpha, \alpha \in I\}$ be a family of subsets of $T$ such that $T = \bigcup_\alpha T_\alpha$. Let $C$ be a covariance on $T$ then $f \in K(C)$ iff $f_\alpha =$ Restriction of $f$ to $T_\alpha \in K(C^{T_\alpha})$ and $\sup_\alpha \|f_\alpha\|_{K(C^{T'})}$ is finite where $C^{T_\alpha}$ is the restriction of $C$ to $T_\alpha \times T_\alpha$.

## ACKNOWLEDGMENTS

This work was started in the Fall of 1974, while Mandrekar lectured on the subject at the Département de Mathématiques, EPF-Lausanne [43]. Also, during the last several years he had the benefit of being associated with Professor R. D. LePage who also contributed to this work with several discussions. This work owes a lot to the authors of various papers on the subject whose ideas we attempted to assemble here coherently. In the end, readable representation of this manuscript was made possible (at least initially) by Mrs. Noralee Burkhardt through typing under extreme pressure.

## REFERENCES

1. Agmon, S., "Lectures on elliptic boundary value problems." Van Nostrand-Reinhold, Princeton, New Jersey 1965.
2. Aronszajn, N., Theory of reproducing kernels, *Trans. Amer. Math. Soc.* **68** (1950), 337–404.
3. Cartier, P., Introduction à l'étude des mouvement Browniens a plusieurs parametres, *Sémin. Prob. V, Lecture Notes,* **191** (1971), 58–75.
4. Chatterji, S. D., Certain induced measures and fractional dimentions of their "support," *Z. Wahrscheinlichkeitstheorie und Verw. Gebiete.* **3** (1964), 184–192.
5. Chatterji, S. D., Densities of certain measures, *Indag. Math.* **27** (1965), 754–759.
6. Chatterji, S. D., Continuous functions representable as sums of independent random variables, *Z. Wahrscheinlichkeitstheorie und Verw. Gebiete.* **13** (1969), 338–341.
7. Chatterji, S. D., and Mandrekar, V., Sur la quasi-invariance des measure sous les translations, *C.R. Acad. Sci. Paris Ser. A* **281** (1975), 581–583.
8. Chatterji, S. D., and Mandrekar, V., Quasi-invariance of measures under translations *Math. Zeit.* **154** (1977), 19–29.
9. Chatterji, S. D., and Mandrekar, V., Singularity and absolute continuity of measures (to appear).
10. Dudley, R. M., Feldman, J., and LeCam, L., On seminorms probabilities, and abstract Wiener spaces, *Ann. of Math.* **93** (1971), 390–408.
11. Dunford, N., and Schwartz, J. T., "Linear Operators II." Wiley (Interscience) New York, 1962.
12. Feldman, J., Equivalence and perpendicularity of Gaussian processes, *Pacific J. Math.* **9** (1958), 699–708 [*Correction, Pacific J. Math.* **10** (1959), 1295–1296].
13. Feldman, J., Some classes of equivalent Gaussian processes on an interval, *Pacific J. Math.* **10** (1960), 1211–1220.
14. Feldman, J., Examples of non-Gaussian quasi-invariant distributions in Hilbert space, *Trans. Amer. Math. Soc.* **99** (1961), 342–340.
15. Fernique, X., Integrabilité des vecteurs Gaussiens, *C.R. Acad. Sci. Paris* **270** (1970), 1698–1699.
16. Gelfand, I. M., and Vilenkin, N. Ja., "Generalized Functions." Vol. 4. Some Application of Harmonic Analysis, Academic Press, New York, 1964.
17. Girsanov, I. V., On transforming a certain class of stochastic processes by absolutely continuous substitution of measures, *Theory Probability Appl.* **5** (1960), 285–301.
18. Gohberg, I. C., and Krein, M. G., Theory and applications of Volterra operators in Hilbert space, *Amer. Math. Soc.* Providence, Rhode Island, 1970.

19. Golosov Fu. I., and Templeman A. A., Liklihood ratio for hypothesis on the trend of certain Gaussian processes, *Soviet Math. Dokl.* **4** (1963), 1796–1799.
20. Hajek, J., On a property of normal distribution of any stochastic processes, *Math. Statist. Prob.* **1** (1958–1961), 245–252.
21. Halmos, P. R., "Introduction to Hilbert space and spectral multiplicity." Chelsea, New York, 1951.
22. Hida, T., "Stationary Stochastic Processes." Princeton Univ. Press, Princeton, New Jersey, 1970.
23. Hida, T., and Ikeda, N., Analysis on Hilbert space with reproducing kernel arising form multiple Wiener integral, *Proc. 5th Berkeley Symp. Vol. II*, **1** (1976), 117–144.
24. Hille, E., and Phillips, R. S., "Functional Analysis and Semigroups." *Amer. Math. Soc. Coll. Publ.* **16**, Providence, Rhode Island, 1957.
25. Hitsuda, M., Representation of Gaussian processes equivalent to Weiner process, *Osaka J. Math.* **5** (1968), 299–312.
26. Inoue, K., Equivalence of measures for some class of Gaussian random fields, *J. Multivariate Analysis* **6** (1976), 295–308.
27. Kailath. T., Likelihood ratios for Gaussian processes, *IEEE Trans.* IT-**16** (1970), 276–288.
28. Kailath. T., On measures equivalent to Wiener measure, *Ann. Math. Statist.* **38** (1967), 261–263.
29. Kakutani, S., On equivalence of infinite product measures, *Ann. of Math.* **49** (1948), 214–224.
30. Kakutani, S., Spectral analysis of stationary Gaussian processes, *Proc. 4th Berkeley Symp.* **2** (1961), 239–247.
31. Kallianpur, G., Zero–one law for Gaussian processes, *Trans. Amer. Math. Soc.* **149** (1970), 199–211.
32. Kallianpur, G., Abstract Wiener processes and their reproducing kernel Hilbert space, *Z. Wahrscheinlichkeitstheorie und Verw. Gebiete.* **17** (1971), 113–123.
33. Kallianpur, G., The role of reproducing kernel Hilbert spaces in the study of Gaussian processes, *in* "Advances in Probability, Vol. II" (P. Ney, ed.), Dekker, New York, 1970, pp. 49–83.
34. Kallianpur, G., and Oodaira, H., The equivalence and singularity of Gaussian processes, "Proc. Symp. on Time Series Analysis," Wiley, New York, 1963, pp. 279–291.
35. Kallianpur, G., and Oodaira, H., Non-anticipative representations of equivalent Gaussian processes, *Ann. Prob.* **1** (1973), 104–122.
36. Kraft, C., Some conditions for consistance and uniform consistance of statistical procedures, *Univ. of California Publ. Stat.* **2** (1955), 125–141.
37. Kuelbs, J., Gaussian measures on a Banach space, *J. Functional Analysis* **5** (1970), 354–367.
38. Kunita, H., Absolute continuity of Markov processes, "Séminaire de Probabilites X. Lecture Notes 511," Springer-Verlag, Berlin and New York, pp. 44–77.
39. LePage, R. D., An Isometry Related to Unbiased Estimation, Ch.V, thesis, University of Minnesota, 1967.
40. LePage, R. D., Note relating Bochner integrals and reproducing kernels to series expansion on a Gaussian Banach space, *Proc. Amer. Math. Soc.* **32** (1972), 285–288.
41. LePage, R. D., and Mandrekar, V., Equivalence and singularity dichotomies from zero-one laws, *Proc. Amer. Math. Soc.* **31** (1972), 251–254.
42. LePage, R. D., and Mandrekar, V., On likelihood ratios of measures given by Markov chains, *Proc. Amer. Math. Soc.* **52** (1975), 377–380.
43. Mandrekar, V., Multiparameter Gaussian processes and their Markov property, Lecture Notes, EPF-Lausanne, 1975.
44. Meyer, P. A., "Probability and Potentials." Ginn (Blaisdell), Boston, Massachusetts, 1966.

45. Mirskaja, T. I., Pabadinskaite, A. S., and Templ'man, A. A., Hilbert spaces of certain reproducing kernels and the equivalence of Gaussian measures, *Math. Stat. Prob.* **11** (1973), 121–131.
46. Nemetz, T., Equivalence and singularity dichotomies of probability measures, *in* "Limit Theorems of Probability Theory," Keszhely, Hungary, 1974, pp. 183–191.
47. Neveu, J., "Discrete Parameter Martingales." North Holland, Amsterdam, 1974.
48. Neveu, J., "Processus aleatoires gaussiens," Univ. of Montreal Press, Canada, 1968.
49. Newman, C. M., On the orthogonality of independent increment processes, *in* "Topics in Probability Theory" (D. W. Stroock and S. R. S. Varadhan, eds.), Courant Institute of Mathematical Science. New York Univ., New York, 1973.
50. Oodaira, H., The equivalence of Gaussian stochastic processes, Research Memorandum No. 103, Department of Statistics and Probability, Michigan State Univ., 1963.
51. Orey, S., Radon-Nikodym derivatives of probability measures: Martingale methods, *J. Tokyo Univ.* (1974), 1–38.
52. Park, W. J., A Multiparameter Gaussian process, *Ann Math. Stat.* **41** (1970), 1582–1595.
53. Parthasarathy, K. R., "Probability Measures on Metric Spaces," Academic Press, New York, 1967.
54. Parzen, E., Probability density functionals and reproducing kernel Hilbert spaces, *Proc. Symp. Time Series Analysis*, Wiley, New York, 1963, pp. 155–169.
55. Pitcher, T. S., Likelihood ratios of Gaussain processes, *Ark. Mat.* **4** (1960), 35–44.
56. Rao, C. R., and Maitra, S. K., "Generalized Inverse of Matrices and Its Applications." Wiley, New York, 1971.
57. Rao, C. R., and Varadarajan, V. S., Discrimination of Gaussian processes, *Ser A* **25** (1963), 303–330.
58. Royden, H. L., "Real Analysis." MacMillan, New York, 1963.
59. Rozanov, Yu. A., On density of one Gaussian measure with respect to another, *Theory Prob. Appl.* **7** (1962), 82–87.
60. Rozanov, Yu. A., Infinite-dimensional Gaussian distributions, *Proc. Steklov Inst. Math.*, No. 108, Amer. Math. Soc., Providence, Rhode Island, 1971.
61. Rudin, W., "Fourier Analysis on Groups." Wiley (Interscience), New York, 1962.
62. Sato, H., Banach support of a probability measure in a locally convex space, *In* "Probability in Banach Spaces" (Oberwolfach No. 528) Springer-Verlag, Berlin and New York, 1976.
63. Segal, I., Tensor algebras over Hilbert spaces, *Trans. Amer. Math.Soc.* **81** (1956), 106–134.
64. Segal, I., Distributions in Hilbert space and canonical systems of operators. *Trans. Amer. Math. Soc.* **88** (1958), 12–41.
65. Shepp, L. A., Distinguishing a sequence of random variables from a translate of itself, *Ann. Math. Stat.* **36** (1965), 1107–1112.
66. Shepp, L. A., Radon-Nykodym derivatives of Gaussian measures, *Ann. Math. Stat.* **37** (1966), 321–354.
67. Skorohod, A. V., On differentiability of measures which correspond to stochastic processes I. Processes with independent increments, *Theory Prob. Appl.* **2** (1957), 417–443.
68. Skorohod, A. V., On the densities of probability measures in functional spaces, *Proc. 5th Berkeley Sym. II.* **1** (1967), 163–182.
69. Vahaniya, N. N., The topological support of Gaussian measure in Banach space, *Nagoya Math. J.* **57** (1975), 59–63.
70. Varberg, D. E., On equivalence of Gaussian measures, *Pacific J. Math.* **11** (1961), 751–762.
71. Varberg, D. E., Gaussian measures and a theorem of T. S.Pitcher, *Proc. Amer. Math. Soc.* **13** (1962), 799–807.
72. Wiener, N., "Non-Linear Problems in Random Theory," MIT Univ. Press, Cambridge, Massachusetts, 1958.

# Stochastic Riemannian Geometry

*MARK A. PINSKY*

DEPARTMENT OF MATHEMATICS
NORTHWESTERN UNIVERSITY
EVANSTON, ILLINOIS

|      |                                                    |     |
|------|----------------------------------------------------|-----|
| I.   | Introduction                                       | 199 |
| II.  | Brownian Motion                                    | 201 |
|      | A. Laplacian of a Riemannian Manifold              | 201 |
|      | B. Brownian motion on $M$                          | 202 |
|      | C. Isotropic Transport Process on $M$              | 205 |
|      | D. Convergence to Brownian Motion                  | 209 |
| III. | Semilocal Properties                               | 210 |
|      | A. Geodesic Polar Coordinates                      | 210 |
|      | B. Monotone Dependence on Curvature                | 211 |
|      | C. Small Perturbations of the Metric               | 213 |
| IV.  | Asymptotic Properties, $t \to 0$                   | 216 |
|      | A. Near Points                                     | 216 |
|      | B. Distant Points                                  | 218 |
| V.   | Asymptotic Properties, $t \to \infty$              | 221 |
|      | A. Compact Manifolds                               | 221 |
|      | B. Open Manifolds of Negative Curvature—Strong Laws | 222 |
|      | C. Open Manifolds of Negative Curvature—Weak Laws  | 228 |
| VI.  | Bibliographical Remarks                            | 234 |
|      | References                                         | 234 |

## I. Introduction

Imagine a particle moving at random on a curved surface or higher dimensional manifold. The (stochastic) law of motion is dictated by the geometry of the manifold; locally the motion is isotropic and moves at constant speed, on the average.

The mathematical structure used to describe this stochastic process requires that the manifold admit a local measure of angle and distance, i.e., a Riemannian metric. Once the motion has been defined locally, one may study the global interaction between the geometry and the stochastic process. In particular the influence of curvature is of central importance.

The modern period of stochastic differential geometry began with the work of Itô and was continued by McKean. In these works the stochastic process is determined by a second-order differential operator on the manifold, with no explicit mention of a Riemannian metric. More recently Malliavin and his co-workers began a series of inquiries dealing with the specific relations between the Riemannian metric and the canonical stochastic process, referred to as the *Brownian motion*. These papers form the point of departure of our interest in this area.

In Section II we review the necessary formulas from Riemannian geometry. The Brownian motion is defined in terms of the Laplace–Beltrami operator and is shown to exist by Itô's method of stochastic differential equations. Although analytically powerful, this method does not provide any geometric insight. For this purpose we introduce a Poisson-type stochastic process, the *isotropic transport process*. This latter motion takes place along geodesics, changing directions at random at the jump times of a Poisson process. It is proved that under an appropriate scaling the Brownian motion is the (weak) limit of a sequence of isotropic transport processes.

Section III contains results on "semilocal" stochastic Riemannian geometry. In a geodesic ball we compare the Brownian motion with a motion on a manifold of constant curvature. The mean exit time and principal eigenvalue of the Laplacian are both monotone functions of the curvature, in a certain sense. We also give explicit formulas for the infinitesimal variations of these quantities due to an infinitesimal change in the metric.

In Section IV we survey the results of Molchanov concerning the transition density for small time. If $x$ and $y$ are sufficiently close, this probability density behaves almost as if the space were Euclidean when $t \to 0$. But if $x$ and $y$ are sufficiently separated (for instance if there are several shortest geodesics joining $x$ and $y$), then the transition density has a radically different behavior when $t \to 0$.

In Section V we examine the asymptotic behavior of the Brownian motion for large time. If the manifold is compact, the motion is recurrent; the Riemannian volume element serves as an invariant measure, which is approached exponentially fast when $t \to \infty$. From the realm of possibilities in the noncompact case, we choose the case of simply connected manifolds of negative curvature. From classical theorems in geometry the Brownian motion is equivalent to a diffusion process on Euclidean space. The negative curvature implies that the motion is transient; the explicit a.s. rate of approach to infinity can be calculated in terms of the curvature in a neighborhood of infinity. The rate of decay of the transition density can be estimated in terms of the global behavior of the curvature. This is shown to be related to some new estimates on the spectrum of the Laplace–Beltrami operator.

## II. Brownian Motion

### A. Laplacian of a Riemannian Manifold

A Riemannian manifold is a pair $(M, g)$, where $M$ is a $C^\infty$ manifold and $g$ is a Riemannian metric, i.e., a positive definite symmetric bilinear form which depends smoothly on $x \in M$. Following established convention, we write

$$ds^2 = g_{ij}(x)\, dx^i\, dx^j, \qquad (2.1)$$

where $(x^1, \ldots, x^n)$ are the local coordinates of a patch, and where the summation convention is used.

$M$ becomes a metric space when we impose the distance function

$$d(A, B) = \inf_\gamma \int_A^B |ds|, \qquad (2.2)$$

where the infimum is taken over all continuous, piecewise differentiable paths $t \to \gamma(t)$, $0 \leq t \leq 1$, such that $\gamma(0) = A$, $\gamma(1) = B$. If the infimum in (2.2) is achieved by a curve $\gamma = (x^1(t), \ldots, x^n(t))$, then $\gamma$ must satisfy the geodesic equations

$$\frac{d^2 x^i}{dt^2} + \Gamma^i_{jk} \frac{dx^j}{dt} \frac{dx^k}{dt} = 0, \qquad (2.3)$$

where $\Gamma^i_{jk}$ are the Christoffel coefficients corresponding to $g_{ij}$.

We assume that the metric $d$ is *complete*, i.e., that every Cauchy sequence has a limit. By the theorem of Hopf–Rinow [5a; p. 11] this is equivalent to the existence of global solutions of (2.3), valid for every initial position and direction.

The Riemannian manifold has a natural volume element given by the local coordinate formula

$$dV = G\, dx^1\, dx^2 \cdots dx^n, \qquad (2.4)$$

where $G = [\det(g_{ij})]^{\frac{1}{2}}$. This choice of volume element determines a natural inner product on smooth functions:

$$\langle f_1, f_2 \rangle = \int_M f_1(x) f_2(x)\, dV(x), \qquad (2.5)$$

where $f_1, f_2 \in C_0^\infty(M)$, the space of $C^\infty$ functions with compact support.

A Dirichlet inner product is defined by the formula

$$\langle\!\langle f_1, f_2 \rangle\!\rangle = \int_M (df_1 | df_2) \, dV(x), \tag{2.6}$$

where $(\,|\,)$ indicates the dual inner product of covectors; in local coordinates this takes the form

$$(df_1 | df_2) = g^{ij} \frac{\partial f_1}{\partial x^i} \frac{\partial f_2}{\partial x^j},$$

where $(g^{ij})$ is the matrix inverse of $(g_{ij})$.

Finally, we can define the Laplacian uniquely by the formula

$$\langle \Delta f_1, f_2 \rangle = \langle\!\langle f_1, f_2 \rangle\!\rangle, \tag{2.7}$$

which is valid for all $f_1, f_2 \in C_0^\infty(M)$. The following properties are immediate:

$$\langle \Delta f_1, f_2 \rangle = \langle f_1, \Delta f_2 \rangle, \quad f_1, f_2 \in C_0^\infty(M), \tag{2.8}$$

$$\langle \Delta f, f \rangle \geq 0, \quad f \in C_0^\infty(M), \tag{2.9}$$

$$\langle \Delta f, f \rangle = 0 \quad \text{iff} \quad f \text{ is constant}, \tag{2.10}$$

$$-\Delta f = G^{-1} \frac{\partial}{\partial x^i} \left( G g^{ij} \frac{\partial f}{\partial x^j} \right), \quad f \in C_0^\infty(M), \tag{2.11}$$

and

$$-\Delta f = g^{jk} \left( \frac{\partial^2 f}{\partial x^j \partial x^k} - \Gamma^i_{jk} \frac{\partial f}{\partial x^i} \right), \quad f \in C_0^\infty(M). \tag{2.12}$$

Indeed, (2.8)–(2.10) are a consequence of the symmetry and nonnegativity properties of the Dirichlet inner product (2.6). (2.11) results from performing an integration by parts in the right-hand side of (2.7). To prove (2.12), notice that the right-hand side is independent of local coordinates. If we take Riemann normal coordinates at $x \in M$, then $\Gamma^i_{jk} = 0 = (\partial g^{ij}/\partial x^k)$. Thus the expressions (2.11) and (2.12) are identical in this case, which proves (2.12) in general.

## B. Brownian Motion on M

From the many possible descriptions of Brownian motion, we choose the familiar martingale formulation: $\{X_t^x, t \geq 0\}$ is a stochastic process with continuous sample paths $X_0^x = x$ such that

$$f(X_t^x) - f(x) + \int_0^t (\Delta f)(X_s^x) \, ds \tag{2.13}$$

is a martingale for every $f \in C_0^\infty(M)$. From this it is immediate that we have Dynkin's identity,

$$E\{f(X_t^x)\} = f(x) - E\left\{\int_0^t (\Delta f)(X_s^x)\,ds\right\}, \qquad (2.14)$$

for bounded stopping times $t$. Furthermore the left-hand side of (2.14) satisfies the heat equation $(\partial u/\partial t) + \Delta u = 0$ with $u(0^+, x) = f(x)$.

To prove the existence of $\{X_t^x, t \geq 0\}$ we may employ the method of Itô stochastic differential equations [31; pp. 90–96]. In a coordinate patch $U$, we consider the equation

$$x^i(t) = x^i + \sqrt{2}\int_0^t \sigma^{ij}(x^1(s),\ldots,x^n(s))\,dw_j(s) - \int_0^t (g^{jk}\Gamma_{jk}^i)(x^1(s),\ldots,x^n(s))\,ds, \qquad (2.15)$$

where $(\sigma^{ij})$ is the positive definite symmetric square root of the matrix $(g^{ij})$ and $(w_1(t), \ldots, w_n(t))$ is an $n$-dimensional Wiener process.

**Examples** (1) Let

$$M = S^n(R) = \{(x_1,\ldots,x_{n+1}): x_1^2 + \cdots + x_{n+1}^2 = R^2\}$$

with the imbedded metric. We use the spherical coordinate parametrization $x_1 = R\cos\theta_1$, $x_2 = R\sin\theta_1\cos\theta_2$, ..., $x_{n+1} = R\sin\theta_1\sin\theta_2\cdots\sin\theta_n$, where $0 < \theta_1 \leq \pi$, ..., $0 < \theta_{n-1} < \pi$, $0 < \theta_n \leq 2\pi$. From the general formulas

$$g_{ij} = \frac{\partial \mathbf{x}}{\partial \theta_i} \cdot \frac{\partial \mathbf{x}}{\partial \theta_j},$$

we obtain the metric in the form

$$ds^2 = R^2\,d\theta_1^2 + R^2\sin^2\theta_1\,d\theta_2^2 + \cdots + R^2\sin^2\theta_1 + \cdots + \sin^2\theta_{n-1}\,d\theta_n^2.$$

Therefore

$$G = R^n \sin^{n-1}\theta_1 \sin^{n-2}\theta_2 \cdots \sin\theta_{n-1},$$

and thus from (2.11) we have

$$-\Delta f = R^{-2}\left[\frac{\partial^2 f}{\partial\theta_1^2} + (n-1)\cot\theta_1\,\frac{\partial f}{\partial\theta_1}\right]$$

$$+ (R\sin\theta_1)^{-2}\left[\frac{\partial^2 f}{\partial\theta_2^2} + (n-2)\cot\theta_2\,\frac{\partial f}{\partial\theta_2}\right]$$

$$+ (R\sin\theta_1\sin\theta_2)^{-2}\left[\frac{\partial^2 f}{\partial\theta_3^2} + (n-3)\cot\theta_3\,\frac{\partial f}{\partial\theta_3}\right]$$

$$+ \cdots + (R\sin\theta_1\cdots\sin\theta_{n-1})^{-2}\frac{\partial^2 f}{\partial\theta_n^2}.$$

Thus the stochastic differential equations (2.15) take the form

$$d\theta_1 = R^{-1}\sqrt{2}\, dw_1 + (n-1)R^{-2}\cot\theta_1\, dt,$$
$$d\theta_2 = (R\sin\theta_1)^{-1}\sqrt{2}\, dw_2 + (n-2)(R\sin\theta_1)^{-2}\cot\theta_2\, dt,$$
$$\vdots$$
$$d\theta_{n-1} = (R\sin\theta_1 \sin\theta_2 \cdots \sin\theta_{n-1})^{-1}\sqrt{2}\, dw_{n-1}$$
$$+ (R\sin\theta_1 \cdots \sin\theta_{n-2})^{-2}\cot\theta_{n-1}\, dt,$$
$$d\theta_n = (R\sin\theta_1 \cdots \sin\theta_{n-1})^{-1}\sqrt{2}\, dw_n.$$

From the above form of the equations it is not difficult to show that $\{\theta_1(t), t \geq 0\}$ is a Markov process; given $\{\theta_1(t), \ldots, \theta_{j-1}(t), t \geq 0\}$, the process $\{\theta_j(t), t \geq 0\}$ is a temporarily inhomogeneous Markov process.

(2) Let $M = H^n(\rho) = \{(x_1, \ldots, x_n): x_1^2 + \cdots + x_n^2 < \rho^2\}$ with the metric

$$ds^2 = \frac{dx_1^2 + \cdots + dx_n^2}{[\rho^2 - x_1^2 \cdots - x_n^2]^2}.$$

$M$ is the standard $n$-dimensional "hyperbolic space," the canonical $n$-dimensional simply connected manifold of constant negative sectional curvature. A convenient parametrization of $M$ is afforded by the spherical polar coordinates $(r, \theta_1, \ldots, \theta_{n-1})$, where

$$x_1 = r\cos\theta_1,$$
$$x_2 = r\sin\theta_1 \cos\theta_2, \ldots, x_n = r\sin\theta_1 \sin\theta_2 \cdots \sin\theta_{n-1}$$

In this coordinate system we have

$$ds^2 = (\rho^2 - r^2)^{-2}[dr^2 + r^2\, d\theta_1^2 + (r\sin\theta_1)^2\, d\theta_2^2$$
$$+ \cdots + (r\sin\theta_1 \cdots \sin\theta_{n-2})^2\, d\theta_{n-1}^2].$$

Thus,

$$G = (\rho^2 - r^2)^{-n} r^{n-1} \sin^{n-2}\theta_1 \cdots \sin\theta_{n-2}$$

$$-\Delta f = r^{1-n}(\rho^2 - r^2)^n \frac{\partial}{\partial r}\left[(\rho^2 - r^2)^{2-n} r^{n-1} \frac{\partial f}{\partial r}\right]$$

$$+ r^{-2}\sin^{2-n}\theta_1 (\rho^2 - r^2)^2 \frac{\partial}{\partial \theta_1}\left(\sin^{n-2}\theta_1 \frac{\partial f}{\partial \theta_1}\right)$$

$$+ \cdots + r^{-2}\sin^{-1}\theta_{n-2}(\rho^2 - r^2)^2 \frac{\partial}{\partial \theta_{n-2}}\left(\sin\theta_{n-2} \frac{\partial f}{\partial \theta_{n-2}}\right)$$

$$+ r^{-2}(\rho^2 - r^2)^2 \frac{\partial^2 f}{\partial \theta_{n-1}^2}.$$

In this case the stochastic equations take the form

$$dr = (\rho^2-r^2)\sqrt{2}\, dw_1 + (\rho^2-r^2)^{-n-1}r^{n-2}[n(\rho^2+r^2) + 5r^2 - \rho^2]\, dt$$
$$d\theta_1 = r^{-1}(\rho^2-r^2)\sqrt{2}\, dw_2 + r^{-2}(\rho^2-r^2)^2(n-2)\cot\theta_1\, dt$$
$$\vdots$$
$$d\theta_{n-1} = r^{-1}(\rho^2-r^2)\sqrt{2}\, dw_n.$$

As in Example 1 we see that $\{r(t), t \geq 0\}$ is a Markov process. Given the distributions of $\{r(t), \theta_1(t), \ldots, \theta_{j-1}(t)\}$, the process $\{\theta_j(t), t \geq 0\}$ is a temporarily inhomogeneous Markov process.

(3)  Let $M$ be the surface of revolution in $R^3$ defined by the equations $x = \Phi(u)\cos\theta$, $y = \Phi(u)\sin\theta$, $z = u$, where $0 \leq u \leq L$, $0 \leq \theta \leq 2\pi$. This leads to the metric

$$ds^2 = (1+\Phi'^2)\, du^2 + \Phi^2\, d\theta^2,$$

with

$$G = \Phi(1+\Phi'^2)^{\frac{1}{2}}$$

and

$$-\Delta f = (1+\Phi'^2)^{-1}\frac{\partial^2 f}{\partial u^2} + \Phi^{-1}(1+\Phi'^2)^{-\frac{3}{2}}\Phi'(1+\Phi'^2 - \Phi\Phi'')\frac{\partial f}{\partial u} + \Phi^{-2}\frac{\partial^2 f}{\partial \theta^2}.$$

The stochastic equations take the form

$$du = (1+\Phi'^2)^{-\frac{1}{2}}\sqrt{2}\, dw_1 + \Phi^{-1}(1+\Phi'^2)^{-\frac{3}{2}}\Phi'(1+\Phi'^2-\Phi\Phi'')\, dt,$$
$$d\theta = \Phi^{-1}\sqrt{2}\, dw_2.$$

In this case we see that $\{u(t), t \geq 0\}$ is a temporarily inhomogeneous Markov process. Conditional on $\{u(t), t \geq 0\}$, the process $\{\theta(t), t \geq 0\}$ is a temporarily inhomogeneous Markov process.

C.  *Isotropic Transport Process on M*

Brownian motion on $M$ has been defined by means of the martingale property (2.13). While analytically tractable, this definition lacks a clear geometrical meaning. It is the purpose of this section to define a new stochastic process—the isotropic transport process, whose geometric meaning is clear. In Section I.D it will be proved that the Brownian motion may be viewed as a limit of the isotropic transport process when a certain parameter coverges to zero.

To define this new process, we first introduce the *geodesic flow*. This is a deterministic motion on $M$, obtained by solving the second-order differential equations

$$\frac{d^2 x^i}{dt^2} + \Gamma^i_{jk}\frac{dx^k}{dt}\frac{dx^j}{dt} = 0, \qquad (2.16)$$

with the initial conditions $x^i(0) = x^i$, $(dx^i/dt)(0) = \xi^i$. The solution is denoted by $Y_0^{(x,\xi)}(t)$ and is a flow on the tangent bundle of $M$. In fact, if we define

$$(T_t^0 f)(x, \xi) = f(Y_0^{(x,\xi)}(t), \dot{Y}_0^{(x,\xi)}(t)), \tag{2.17}$$

then it can be shown that $T_{t_1+t_2}^0 = T_{t_1}^0 T_{t_2}^0$ and (2.17) satisfies the partial differential equation

$$\frac{\partial u}{\partial t} = \xi^i \frac{\partial u}{\partial x^i} - \Gamma_{jk}^i \xi^j \xi^k \frac{\partial u}{\partial \xi^i}. \tag{2.18}$$

In other words $\{T_t^0, t \geq 0\}$ is a semigroup of operators whose infinitesimal operator is given by (2.18). This first-order differential operator is also called the *geodesic flow field*.

To define the isotropic transport process we introduce a probability space $(\Omega, \mathcal{B}, P)$ on which is defined a sequence $\{e_n\}, n \geq 1$ of independent random variables with the exponential distribution

$$P\{e_n > t\} = e^{-t}, \quad (n \geq 1, t > 0). \tag{2.19}$$

Let $X_0 = x$, $\xi_0 = \xi$; given $\{X_0, \xi_0, \ldots, X_n, \xi_n\}$, we let

$$X_{n+1} = Y_0^{(X_n, \xi_n)}(e_{n+1}). \tag{2.20}$$

We then require that the conditional distribution of $\xi_{n+1}$ is a uniform distribution on the unit sphere of the tangent space at the point $X_{n+1} \in M$. Finally, we set

$$Y^{(x,\xi)}(t) = Y_0^{(X_n, \xi_n)}(t - e_1 - \cdots - e_n) \tag{2.21}$$

where $e_1 + \cdots + e_n \leq t < e_1 + \cdots + e_{n+1}$. $\{Y^{(x,\xi)}(t), t \geq 0\}$ is the *isotropic transport process*; the curve $t \to Y^{(x,\xi)}(t)$ is continuous and piecewise geodesic. To compute the infinitesimal operator of this process, we introduce the following operators:

$$(T_t^0 f)(x, \xi) = f(Y_0^{(x,\xi)}(t), \dot{Y}_0^{(x,\xi)}(t)), \tag{2.22}$$

$$(R_\lambda^0 f)(x, \xi) = \int_0^\infty e^{-\lambda t} (T_t^0 f)(x, \xi) \, dt, \tag{2.23}$$

$$(T_t f)(x, \xi) = E\{f(Y^{(x,\xi)}(t), \dot{Y}^{(x,\xi)}(t))\}, \tag{2.24}$$

$$(R_\lambda f)(x, \xi) = \int_0^\infty e^{-\lambda t} (T_t f)(x, \xi) \, dt, \tag{2.25}$$

$$(Pf)(x, \xi) = \int_{T_x(M)} f(x, \xi) \mu_x(d\xi) \tag{2.26}$$

$$(D_Z f)(x, \xi) = \xi^i \frac{\partial f}{\partial x^i} - \Gamma_{jk}^i \xi^j \xi^k \frac{\partial f}{\partial \xi^i}. \tag{2.27}$$

# STOCHASTIC RIEMANNIAN GEOMETRY

In the above formulas $f$ is a function on $T(M)$, the tangent bundle of the manifold $M$; $\mu_x$ is the uniform distribution on the unit sphere of $T_x(M)$, the tangent space at $x$. Then we have the following result.

**Theorem** Let $f \in C_0^\infty [T(M)]$, $u = T_t f$. Then $\dot{u} = u(t, x, \xi)$ satisfies the integrodifferential equation

$$\frac{\partial u}{\partial t}(t, x, \xi) = \xi^i \frac{\partial u}{\partial x^i}(t, x, \xi) - \Gamma^i_{jk} \xi^j \xi^k \frac{\partial u}{\partial \xi^i}(t, x, \xi)$$

$$+ \int_{T_x(M)} [u(t, x, \eta) - u(t, x, \xi)] \mu_x(d\eta). \qquad (2.28)$$

To prove this result, we require a series of results pertaining to the resolvent operators $R_\lambda$, $R_\lambda^0$. In these lemmas $C(T(M))$ (resp $C^1(T(M))$) is the space of continuous (resp differentiable) functions which vanish at infinity.

**Lemma 1** $R_\lambda^0$ maps $C^1(T(M))$ into $C^1(T(M))$ and satisfies

$$(\lambda - D_Z) R_\lambda^0 f = f, \qquad f \in C^1(T(M)). \qquad (2.29)$$

**Lemma 2** $R_\lambda$ maps $C(T(M))$ into $C(T(M))$ and satisfies

$$R_\lambda f = R_{1+\lambda}^0 f + R_{1+\lambda}^0 P R_\lambda f, \qquad f \in C(T(M)). \qquad (2.30)$$

**Lemma 3** $R_\lambda$ maps $C^1(T(M))$ into $C^1(T(M))$ and satisfies

$$(\lambda - D_Z - P + I) R_\lambda f = f, \qquad f \in C^1(T(M)). \qquad (2.31)$$

*Proof of Lemma 1* Let $f \in C^1(T(M))$. By the smooth dependence on initial conditions $(x, \xi) \to Y_0^{(x, \xi)}(t)$ is a $C^\infty$ mapping for each $t > 0$. Therefore $T_t^0 f$ is a differentiable function. The dominated convergence theorem implies that $R_\lambda^0 f \in C^1(T(M))$. To show that $R_\lambda^0 f$ vanishes at infinity, let $\varepsilon > 0$. Choose $T > 0$ such that $e^{-\lambda T} \|f\| \leq \varepsilon \lambda / 2$. Since $f \in C(T(M))$, we can find $R > 0$ such that $|f(x, \xi)| \leq \varepsilon \lambda / 2$ if $d((x, \xi); (x_0, \xi_0)) \geq R$. Now let $d((x, \xi); (x_0, \xi_0)) \geq R + T$; then $d(Y_0^{(x, \xi)}(t); (x_0, \xi_0)) \geq R$ for $t \leq T$. Hence,

$$|(R_\lambda^0 f)(x, \xi)| \leq \left| \int_0^T e^{-\lambda t} f(Y_0^{(x, \xi)}(t)) dt \right| + \left| \int_T^\infty e^{-\lambda t} f[Y_0^{(x, \xi)}(t)] dt \right|$$

$$\leq \frac{\varepsilon \lambda}{2} \left( \frac{1 - e^{-\lambda T}}{\lambda} \right) + \frac{\varepsilon}{2} \leq \varepsilon,$$

for $d((x, \xi); (x_0, \xi_0)) \geq R + T$. Thus $R_\lambda^0 f \in C(T(M))$.

To verify (2.28) we write, for $h > 0$,

$$T_h^0 R_\lambda^0 f = \int_0^\infty e^{-\lambda t} T_{t+h}^0 f \, dt = e^{\lambda h} \int_h^\infty e^{-\lambda u} T_u^0 f \, du$$

$$= e^{\lambda h} \left[ R_\lambda^0 f - \int_0^h e^{-\lambda u} (T_u^0 f) \, du \right].$$

But $(T_u^0 g)(x,\xi) = g(Y_0^{(x,\xi)}(u), \dot{Y}_0^{(x,\xi)}(u))$. Therefore,

$$D_Z(R_\lambda^0 f) = \lim_{h\to 0} [T_h^0(R_\lambda^0 f) - (R_\lambda^0 f)]/h$$

$$= \lim_{h\to 0} \left[ h^{-1}(e^{\lambda h}-1) R_\lambda^0 f - h^{-1} \int_0^h e^{-\lambda u}(T_u^0 f)\, du \right]$$

$$= \lambda R_\lambda^0 f - f,$$

where we have used the fact that $T_t^0$ satisfies the differential equation (2.18).

*Proof of Lemma 2* To prove Lemma 2, we follow a similar argument to prove that $R_\lambda f \in C(T(M))$ whenever $f \in C(T(M))$; the details are omitted. To prove Eq. (2.30), we write

$$(R_\lambda f)(x,\xi) = E\left\{ \int_0^{e_1} + \int_{e_1}^\infty \right\} e^{-\lambda t} f(Y^{(x,\xi)}(t), \dot{Y}^{(x,\xi)}(t))\, dt$$

$$= E\left\{ \int_0^\infty I_{(t<e_1)} e^{-\lambda t} f(Y^{(x,\xi)}(t), \dot{Y}^{(x,\xi)}(t))\, dt \right\}$$

$$+ E\left\{ e^{-\lambda e_1} \int_0^\infty e^{-\lambda s} f(Y^{(x,\xi)}(e_1+s), \dot{Y}^{(x,\xi)}(e_1+s))\, ds \right\}.$$

The first integral is equal to $(R_{1+\lambda}^0 f)(x,\xi)$. For the second, note that $Y^{(x,\xi)}(e_1+s) = Y_0^{(Y_1,\xi_1)}(s)$. Taking the conditional expectation with respect to $(Y_1, \xi_1)$, we have

$$E\left\{ e^{-\lambda e_1} \int_0^\infty e^{-\lambda s} f(Y^{(x,\xi)}(e_1+s)\, ds) \right\}$$

$$= E\{e^{-\lambda e_1}(R_\lambda f)(Y_1, \xi_1)\}$$

$$= E\{e^{-\lambda e_1}(PR_\lambda f)(Y_1)\}$$

$$= \int_0^\infty e^{-\lambda s}(PR_\lambda f)(Y_0^{(x,\xi)}(s), \dot{Y}_0^{(x,\xi)}(s)) e^{-s}\, ds$$

$$= (R_{1+\lambda}^0 PR_\lambda f)(x,\xi)$$

*Proof of Lemma 3* Using Lemma 2, we iterate (2.30), obtaining

$$R_\lambda f = R_{1+\lambda}^0 f + \sum_{n=1}^\infty (R_{1+\lambda}^0 P)^n (R_{1+\lambda}^0 f).$$

This series converges uniformly, due to the estimation $\|R_{1+\lambda}^0 P\| \leq (1+\lambda)^{-1}$. Thus, $R_\lambda f \in C(T(M))$. For $f \in C^1(T(M))$, this series may be differentiated term-by-term and the differentiated series also converges uniformly. Hence, $R_\lambda f \in C^1(T(M))$.

To prove (2.31) we apply the operator $(I+\lambda-D_Z)$ to both sides of (2.30) and use (2.29). Thus $(I+\lambda-D_Z)R_\lambda f = f + PR_\lambda f$, which was to be proved.

## D. Convergence to Brownian Motion

We will show that the Brownian motion on $M$ may be approximated, in the sense of weak convergence, by a sequence of isotropic transport processes. The method depends on general theorems from semigroup theory.

For this purpose, let $\varepsilon > 0$ and repeat the construction of Section II.C with $\{\xi_n\}_{n \geq 1}$ replaced by $\{\varepsilon\xi_n\}_{n \geq 1}$, thus obtaining a stochastic process $\{Y_\varepsilon^{(x,\xi)}(t), t \geq 0\}$. Let $u(t,x,\xi) = E\{f(Y_\varepsilon^{(x,\xi)}(t/\varepsilon^2), \dot{Y}_\varepsilon^{(x,\xi)}(t/\varepsilon^2))\}$. Then $u$ satisfies the integrodifferential equation

$$\frac{\partial u}{\partial t} = \varepsilon^{-1}\left[\xi^i \frac{\partial u}{\partial x^i} - \Gamma^i_{jk}\xi^j\xi^k \frac{\partial u}{\partial \xi^i}\right]$$

$$+ \varepsilon^{-2} \int_{T_x(M)} [u(t,x,\eta) - u(t,x,\xi)]\mu_x(d\eta). \qquad (2.32)$$

We shall use the following theorem from semigroup theory.

**Theorem** (T. G. Kurtz, *J. Functional Analysis*, **12** (1973), 55–67.) *Let $A$, $B$ generate strongly continuous contraction semigroups $e^{tA}$, $e^{tB}$ on a Banach space $L$. Assume the existence of $P = s-\lim_{t\to\infty} e^{tB}$, a projection operator. Suppose that $B+\varepsilon A$ generates a strongly continuous contraction semigroup on $L$ for every $\varepsilon > 0$. Furthermore, assume that there exists a set $D$ such that $PAf = 0$ for all $f \in D$ and that the equation $Bh = -Af$ has a solution whenever $f \in D$, which we denote by $h = V(Af)$. Define $Cf = PAVAf$ for $f \in D$, and assume that the range of $\lambda - C$ contains $\bar{D}$ for some $\lambda > 0$. Then the closure of $C$ restricted so that $Cf \in D$ is the generator of a strongly continuous contraction semigroup and $\lim_{\varepsilon \to 0} \exp[t(A/\varepsilon + B/\varepsilon^2)]f = e^{tC}f$ for all $f \in L$.*

To apply the theorem to Eq. (2.31), we first take $L = C(T(M))$, $A = D_z$, $B = P - I$; it remains to compute the operator $C$. By the above formula,

$$C(x,\xi) = \int_{T_x(M)} (D_z D_z f)(x,\xi)\mu_x(d\xi)$$

$$= \int_{T_x(M)} \xi^j\xi^k\left\{\frac{\partial^2 f}{\partial x^j \partial x^k} - \Gamma^i_{jk}\frac{\partial f}{\partial x^i}\right\}\mu_x(d\xi)$$

$$= \frac{1}{n}g^{ij}(x)\left\{\frac{\partial^2 f}{\partial x^j \partial x^k} - \Gamma^i_{jk}\frac{\partial f}{\partial x^i}\right\}$$

$$= (-1/n)(\Delta f).$$

From this we have the following result on the approximation of Brownian motion.

**Proposition** *Assume that the Brownian motion semigroup preserves the class*

$C(M)$. *Then for every $f \in C(M)$*

$$\lim_{\varepsilon \to 0} E\{f(Y_\varepsilon^{(x,\xi)}(t/\varepsilon^2))\} = E\{f(X_{t/n}^x)\}.$$

In order to apply the above criterion in concrete cases, it is necessary to show that the Brownian motion does not move "too fast" in a neighborhood of infinity. This can be ensured by appropriate bounds on the curvature. For instance, we have the following theorem of S. T. Yau [50].

**Theorem** *Let $M$ be a complete Riemannian manifold with Ricci curvature bounded from below by a constant. Then the Brownian motion semigroup preserves the class of continuous functions which vanish at infinity.*

### III. Semilocal Properties

#### A. Geodesic Polar Coordinates

In order to analyze the Brownian motion in the neighborhood of a point, we introduce a canonical coordinate system in the neighborhood of $x_0 \in M$. Let $\xi = (\theta_1, \ldots, \theta_{n-1})$ be spherical polar coordinates on the unit sphere in the tangent space at $x_0$. The mapping

$$(t, \xi) \to \exp_{x_0}(t\xi) \tag{3.1}$$

carries $[0, \infty) \times S^{n-1}$ into $M$, where $\exp_{x_0}(t\xi)$ is the value at instant $t$ of the geodesic with initial velocity $\xi$. From Riemannian geometry, we recall the normal neighborhood theorem: for some $\delta > 0$ the mapping (3.1) is a diffeomorphism from $[0, \delta) \times S^{n-1}$ onto a neighborhood $U$ of $x_0$. In this neighborhood we use the coordinates $(r, \theta_1, \ldots, \theta_{n-1})$, where $r = d(x, x_0)$. The geodesics emanating from $x_0$ intersect the geodesic spheres orthogonally. In these coordinates the metric takes the form

$$ds^2 = dr^2 + \bar{g}_{ij} \, d\theta_i \, d\theta_j \tag{3.2}$$

where $1 \leq i, j \leq n-1$. Let $G = (\det \bar{g}_{ij})^{\frac{1}{2}}$. Substituting in (2.11) we see that for $r > 0$

$$\begin{aligned}
-\Delta f &= G^{-1} \frac{\partial}{\partial r}\left(G \frac{\partial f}{\partial r}\right) + G^{-1} \frac{\partial}{\partial \theta_i}\left(G \bar{g}_{ij} \frac{\partial f}{\partial \theta_j}\right) \\
&= \frac{\partial^2 f}{\partial r^2} + \left(G^{-1} \frac{\partial G}{\partial r}\right) \frac{\partial f}{\partial r} + a_{ij} \frac{\partial^2 f}{\partial \theta_i \, \partial \theta_j} + b_i \frac{\partial f}{\partial \theta_i},
\end{aligned} \tag{3.3}$$

where $a_{ij}$, $b_i$ are $C^\infty$ functions of $(r, \theta_1, \ldots, \theta_{n-1})$. From (2.15) we have the fundamental stochastic integral equation for the radial motion:

$$r(t) = r_0 + \sqrt{2}\, w(t) + \int_0^t [G^{-1}(\partial G/\partial r)][r(s), \theta_1(s), \ldots, \theta_{n-1}(s)] \, ds, \tag{3.4}$$

where $\{w(t), t \geqslant 0\}$ is a standard Wiener process. The precise statement is that the solution of (3.4) is identical in law to the radial component of the Brownian motion until the exit time from $U$.

Let $T_\rho^x = \inf\{t > 0 : r(X_t^x) = \rho\}$, the exit time from the sphere of radius $\rho$. Let $u(x) = E\{T_\rho^x\}$. By the general theory of Markov processes [11, Vol 2, p. 51] $u$ is the solution of

$$\Delta u = 1, \quad \lim_{r(x) \to \rho} u(x) = 0. \tag{3.5}$$

Another quantity of interest is $\lambda(\rho)$, defined by

$$\lambda(\rho) = \inf\{\lambda : \exists\, f \neq 0,\, \ni \Delta f = \lambda f,\, f = 0 \text{ on } \partial B(x_0, \rho)\}, \tag{3.6}$$

where $\lambda(\rho)$ can be calculated by the formula

$$-\lambda(\rho) = \limsup_{t \to \infty} t^{-1} \log P\{T_\rho^x > t\}.$$

**Examples** Explicit calculations of $G'/G$, $E(T_\rho^x)$, $\lambda(\rho)$ are available in the special case of three-dimensional manifolds of constant curvature. These include the Euclidean ball of arbitrary radius, a geodesic ball in a hyperbolic space of arbitrary radius, and the spherical cap in $S^3(R)$, of radius less than $\pi R$. The results are compiled in Table I.

The value of $\Delta f$ is computed for a radial function $f = f(r)$. Because of dimension three, the values of $E(T_\rho^0)$ and $\lambda(\rho)$ can be computed from familiar ordinary differential equations with explicit solutions. As a consequence we note the following facts.

(1) When $K$ increases from negative to positive values, the value of $E(T_\rho^0)$ increases from below the Euclidean value to above the Euclidean value.

(2) When $K$ increases from negative to positive values, the value of $\lambda(\rho)$ decreases from above the Euclidean value to below the Euclidean value.

**B. Monotone Dependence on Curvature**

On the basis of these examples we have the following theorem.

**Theorem** Let the sectional curvatures be bounded in the form

$$K(x, P) \leqslant k_0, \quad x \in B(x_0, \rho), \tag{3.7}$$

where $k_0$ is a constant and $P$ is a 2-plane in the tangent space at $x$. Let $E^K$, $E^{k_0}$ denote the expectation for the given Riemannian diffusion and for the corresponding canonical manifold of constant curvature. Similarly, let $\lambda^K$, $\lambda^{k_0}$ denote the corresponding eigenvalues. Then

$$E^K(T^{x_0}) \leqslant E^{k_0}(T^{x_0}), \quad \lambda^K(\rho) \geqslant \lambda^{k_0}(\rho).$$

TABLE I

| Geometric quantity | $K$ | $K = 0(R^3)$ | $K = a^2 > 0(S^3)$ | $K = -k^2 < 0(H^3)$ |
|---|---|---|---|---|
| $G'/G$ | | $2/r$ | $2a \cot ar$ | $2k \coth kr$ |
| $-(\Delta f)(r)$ | | $r^{-1}(rf)_{rr}$ | $(\sin ra)^{-1}[((\sin ra)f)_{rr} + a^2(\sin ra)f]$ | $(\sinh kr)^{-1}[((\sinh kr)f)_{rr} - k^2(\sinh kr)f]$ |
| $E(T_\rho^0)$ | | $\rho^2/6$ | $1/(2a^2) - (\rho/2a)\cot \rho a$ $= \rho^2/6 + (7a^2/180)\rho^4 + 0(\rho^6), \quad \rho \to 0$ | $(\rho/2k)\coth \rho k - 1/(2k^2)$ $= \rho^2/6 - (7k^2/180)\rho^4 + 0(\rho^6), \quad \rho \to 0$ |
| $\lambda(\rho)$ | | $\pi^2/\rho^2$ | $(\pi^2/\rho^2) - a^2$ | $(\pi^2/\rho^2) + k^2$ |

The proof of this theorem is an immediate consequence of the following inequality on the distribution function of the variable $T$:

$$P^K(T^{x_0} > t) \geq P^{k_0}(T^{x_0} > t). \tag{3.8}$$

Indeed $E(T_\rho^{x_0}) = \int_0^\infty P(T_\rho^{x_0} > t)\, dt$ and

$$-\lambda = \limsup_{t \to \infty} t^{-1} \log P\{T_\rho^{x_0} > t\}.$$

To prove the inequality (3.8), we make the preliminary observation that (3.7) implies that $(G'/G) \geq \phi(r)$, where

$$\phi(r) = \begin{cases} (n-1)k^{\frac{1}{2}} \cot rk^{\frac{1}{2}} & k > 0, \\ (n-1)/r & k = 0, \\ (n-1)(-k)^{\frac{1}{2}} \coth r(-k)^{\frac{1}{2}} & k < 0. \end{cases}$$

We then use the following comparison theorem.

**Theorem** *Let $r_t$, $\tilde{r}_t$ be solutions of the following stochastic differential equations*

$$r_t = r_0 + \sqrt{2}\, W_t + \int_0^t (G^{-1}(\partial G/\partial r))(X_s^x)\, dx,$$

$$\tilde{r}_t = r_0 + \sqrt{2}\, \tilde{W}_t + \int_0^t \phi(\tilde{r}_s)\, ds,$$

*where $W_t$, $\tilde{W}_t$ are Wiener processes. Then*

$$P\{r_t > R\} \leq P\{\tilde{r}_t > R\}, \quad 0 < R < \infty.$$

The proof appears in [16] and [25].

C. *Small Perturbations of the Metric*

The results of Section III.B show that $E(T_\rho)$ is an increasing functional of the curvature, whereas $\lambda(\rho)$ is a decreasing functional of the curvature. To obtain a more precise functional dependence we shall compute the first-order change in $E(T_\rho)$ and $\lambda(\rho)$ when we make a small change in the metric. For this purpose we assume that the volume element has the form

$$G_\varepsilon = G_0(r) \exp(\varepsilon h(r)), \tag{3.9}$$

where $G_0$ is a fixed volume element and $h$ is a $C^\infty$ function with $h(0) = 0$. This is satisfied, in particular, if $M$ is a strongly harmonic Riemannian manifold [33]. We then have the following results.

**Theorem** *With the metric (3.9) we have for $\varepsilon \to 0$*

$$E^\varepsilon(T_\rho) = E^0(T_\rho) - \varepsilon \int_0^\rho H_1(\rho, r) h'(r)\, dr + 0(\varepsilon^2), \tag{3.10}$$

and
$$\lambda^\varepsilon(\rho) = \lambda^0(\rho) + \varepsilon \int_0^\rho H_2(\rho, r) h'(r)\, dr + 0(\varepsilon^2), \tag{3.11}$$

where
$$H_1(\rho, r) = \int_0^r G_0(z_2)\, dz_2 \int_r^\rho G_0^{-1}(z_1)\, dz_1,$$

$$H_2(\rho, r) = \frac{f_0(r) f_0'(r) G_0(r)}{\int_0^\rho f_0(r)^2 G_0(r)\, dr},$$

and $f_0$ is the (unique up to a constant multiple) solution of $\Delta f_0 + \lambda^0(\rho) f_0 = 0$, $f_0(\rho) = 0$.

*Proof* $u(x) = E(T_\rho^x)$ is the solution of $\Delta u = 1$, $u(\rho) = 0$ and therefore

$$u(0) = \int_0^\rho G_\varepsilon^{-1}(z_1)\, dz_1 \int_0^{z_1} G_\varepsilon(z_2)\, dz_2$$

$$= \int_0^\rho G_0^{-1}(z_1)\, dz_1 \int_0^{z_1} G_0(z_2) \exp(\varepsilon(h(z_2) - h(z_1)))\, dz_2.$$

Thus
$$E^\varepsilon(T_\rho) - E^0(T_\rho) = \varepsilon \int_0^\rho G_0^{-1}(z_1) \int_0^{z_1} G_0(z_2) [h(z_2) - h(z_1)]\, dz_2 + 0(\varepsilon^2)$$

$$= -\varepsilon \int_0^\rho G_0^{-1}(z_1) \int_0^{z_1} G_0(z_2)\, dz_2 \int_{z_2}^{z_1} h'(z_3)\, dz_3 + 0(\varepsilon^2),$$

from which the result follows by changing the order of integration.

To prove (3.9) we consider the following perturbation problem:
$$f'' + [(G_0^{-1} G_0') + \varepsilon h'] f' = \lambda f.$$

When $\varepsilon = 0$ we have a simple eigenvalue $\lambda_0$ with eigenfunction $f_0$. Assuming an expansion, $\lambda = \lambda_0 + \varepsilon \lambda_1 + \cdots$, we have

$$\lambda_1 = \frac{\int_0^\rho h' f_0' f_0 G_0\, dr}{\int_0^\rho (f_0)^2 G_0\, dr},$$

which is equivalent to (3.9).

*Special Case of Surfaces* In the case where $n = 2$ the above formulas can be expressed directly in terms of the Gaussian curvature. Indeed, we have the "Jacobi equation,"
$$G'' + KG = 0,$$

where the curvature $K$ may be supposed to have the expression
$$K = K_0 + \varepsilon K_1 + \cdots.$$

Substituting (3.9) and equating powers of $\varepsilon$, we have

$$h'' + 2(G_0^{-1}G_0')h' = -K_1,$$

which can be solved explicitly for $h$, subject to the conditions that $h(0) = 0$, $h'(0)$ be finite. This leads to

$$h'(r) = -G_0^{-2}(r)\int_0^r K_1(z)G_0^2(z)\,dz,$$

and thus

$$E^\varepsilon(T_\rho) = E^0(T_\rho) + \varepsilon\int_0^\rho K_1(r)\tilde{H}_1(\rho,r)\,dr + 0(\varepsilon^2), \tag{3.12}$$

$$\lambda^\varepsilon(\rho) = \lambda^0(\rho) - \varepsilon\int_0^\rho K_1(r)\tilde{H}_2(\rho,r)\,dr + 0(\varepsilon^2), \tag{3.13}$$

where

$$\tilde{H}_1(\rho,r) = \int_r^\rho G_0^{-2}(z_3)\,dz_3 \int_0^{z_3} G_0(z_2)\,dz_2 \int_{z_2}^r G_0^{-1}(z_1)\,dz_1,$$

$$\tilde{H}_2(\rho,r) = \frac{G_0^2(r)\int_r^\rho G_0^{-1}(z)f_0(z)f_0'(z)\,dz}{\int_0^\rho f_0(z)^2 G_0(z)\,dz}.$$

**Example** As a concrete illustration of the above perturbation problem, let us consider a three-dimensional Riemannian manifold with metric

$$ds^2 = dr^2 + \left(\frac{\sin kr}{k}\right)^2[d\theta^2 + \sin^2\theta\,d\phi^2], \quad 0 \leqslant r < r_1$$

$$= dr^2 + r^2[d\theta^2 + \sin^2\theta\,d\phi^2], \quad r_1 \leqslant r < R.$$

The solution of the equation $\Delta f = \lambda f$ with $f(R) = 0$ has the form

$$f(r) = \frac{\sin r(\lambda+k)^{\frac{1}{2}}}{\sin k r}, \quad 0 \leqslant r < r_1,$$

$$= [C\sin(R-r)\lambda^{\frac{1}{2}}/r], \quad r_1 \leqslant r < R.$$

When we impose the condition that $f$ and $f'$ be continuous, we find that $C$ is determined and that $\lambda$ must satisfy the transcendental equation

$$(\lambda+k)^{\frac{1}{2}}\cot[r_1(\lambda+k)^{\frac{1}{2}}] + \lambda^{\frac{1}{2}}\cot[(R-r_1)\lambda^{\frac{1}{2}}] = k^{\frac{1}{2}}\cot r_1 k^{\frac{1}{2}} - r_1^{-1}.$$

By the implicit function theorem this equation has a unique solution defined in a neighborhood of $k=0$, with $\lambda(0)=(\pi^2/R^2)$. Differentiating with respect to $k$ and setting $k=0$, we have

$$\lambda'(0) = [-\tfrac{1}{3} - \Psi'(\alpha_0)]\bigg/\left[\left(\frac{R-r}{r}\right)\Psi'(\tilde{\alpha}_0) + \Psi'(\alpha_0)\right],$$

where $\Psi(x) = x^{\frac{1}{2}} \cot(x^{\frac{1}{2}})$, $\alpha_0 = (\pi^2 r^2/R^2)$, $\bar{\alpha}_0 = [\pi^2(R-r)^2/R^2]$. Clearly, $\lambda'(0) < 0$, in accordance with the above theory.

The computation of $\lambda''(0)$ is unwieldy in general. In the special case $R = 2r_1$ the calculations become simpler. Indeed, in this case $\alpha_0 = \bar{\alpha}_0 = (\pi^2/4)$. Differentiating the basic relation twice and setting $k = 0$, we find that

$$\lambda''(0) = \left(\frac{2}{45} - \frac{26}{36\pi^2}\right) r_1^2 \cong (-0.028) r_1^2.$$

## IV. Asymptotic Properties, $t \to 0$

In this section we review the important results of Molchanov [35] on the behavior of the transition density of the Brownian motion where $t \to 0$. The reader is referred to the original article for a treatment in the case of a diffusion corresponding to an additional first-order term.

### A. Near Points

Let $M$ be a complete Riemannian manifold with a global radius of injectivity $\rho > 0$. This means that for any $x \in M$, the exponential map is a diffeomorphism of some open neighborhood of $0 \in R^n$ onto the geodesic ball of radius $\rho$, centered at $x$.

The fundamental solution of the heat equation is denoted by $p(t, x, y)$. By definition $p$ is a nonnegative $C^\infty$ function on $(0, \infty)$, which satisfies $(\partial/\partial t + \Delta_x) p = 0$ for each $y$ and satisfies $\lim_{t \to 0} \int_M p(t, x, y) f(y) \, dV(y) = f(x)$ for any $C^\infty$ function $f$. We shall assume in addition that $p$ is *unique*; this is automatically satisfies in case $M$ is compact [5, pp. 205–207].

**Theorem** *For any compact set $D \subseteq M$, there is a constant $\delta > 0$ such that whenever $d(x, y) < \delta$, we have uniformly for $(x, y) \in D \times D$*

$$p(t, x, y) \sim \frac{\exp[-d^2(x,y)/4t]}{(4\pi t)^{n/2}} H(x, y), \tag{4.1}$$

where $H(x, y) = [G^{-\frac{1}{2}}/d(x,y)^{(n-1)/2}]$ and $G$ is the density at $y$ in geodesic polar coordinates centered at $x$.

*Example 4.1* In the case where $M = R^n$, $H = 1$ and (4.1) is an equality.

*Example 4.2* In the case where $M = S^n(R)$ we have $G = [R \sin(r/R)]^{n-1}$; therefore in this case we have

$$p(t, x, y) \sim \frac{\exp(-r^2/4t)}{(4\pi t)^{n/2}} \left(\frac{r/R}{\sin(r/R)}\right)^{(n-1)/2}, \qquad r = d(x, y) < \pi R.$$

**Example 4.3** In the case where $M = H^n(R)$, we have $\rho = +\infty$, $G = [R \sinh(r/R)]^{n-1}$. Therefore in this case we have

$$p(t,x,y) \sim \frac{\exp(-r^2/4t)}{(4\pi t)^{n/2}} \left(\frac{r/R}{\sinh(r/R)}\right)^{(n-1)/2},$$

*Proof* We follow the analytical method of [5; pp. 207–215]. Consider a function of the form

$$p_k(t,x,y) = \frac{\exp(-r^2/4t)}{(4\pi t)^{n/2}} [u_0(x,y) + tu_1(x,y) + \cdots + u_k(x,y)], \quad (4.2)$$

where $r = d(x,y)$ is the geodesic distance and $u_0, \ldots, u_k$ are functions to be defined. If we apply the operator $(\partial/\partial t) + \Delta_x$ to (4.2) and equate powers of $t$, we find that

$$\left(\frac{\partial}{\partial t} + \Delta_x\right) p_k = \left(\sum_{j=0}^{k+1} t^{j-1} q_j\right) \frac{\exp(-r^2/4t)}{(4\pi t)^{n/2}}, \quad (4.3)$$

where

$$q_0 = ru_0' + \tfrac{1}{2}r(G'/G)u_0,$$

$$q_j = ru_j' + [(r/2)(G'/G) + j]u_j + \Delta_x u_{j-1}, \quad 1 \leq j \leq k.$$

We let $u_0$ be a $C^\infty$ function that equals $G^{-\frac{1}{2}}(x,y)$ for $d(x,y) < \delta/2$ and is zero for $d(x,y) \geq \delta$. Then, assuming that $(u_0, \ldots, u_{j-1})$ have been defined, let $u_j$ be a $C^\infty$ function which is zero for $d(x,y) \geq \delta$ and which is the solution of $q_j = 0$ for $d(x,y) < \delta/2$. Finally, let

$$\Phi(t,x,y) = [(\partial/\partial t) + \Delta] p_k(t,x,y),$$

when $k > n/2$. Then $\Phi$ has the following properties:

$$\Phi \in C^\infty([0,\infty) \times M \times M) \quad (4.4a)$$

$$\lim_{t \to 0} \Phi(t,x,y) = 0, \quad (x,y) \in M \times M \quad (4.4b)$$

$$|\Phi(t,x,y)| \leq (\text{const}) t^{k-n/2}, \quad (x,y) \in M \times M. \quad (4.4c)$$

Now let

$$\Psi(t,x,y) = E\left\{\int_0^t \Phi(t-s, X_s^x, y)\, ds\right\},$$

where $\Psi$ is the solution of $[(\partial/\partial t) + \Delta] \Psi = \Phi$ [11] with the initial condition $\Psi(0^+, x, y) = 0$. Now we claim that

$$p = p_k + \Psi, \quad (4.5)$$

is the fundamental solution of the heat equation. Indeed $[(\partial/\partial t) + \Delta] p = \Psi - \Psi = 0$ and clearly $\lim_{t \to 0} p(t,x,y) = \delta_x(y)$.

Therefore (4.5) is the fundamental solution of the heat equation. From (4.2) and (4.4a), we see that $p(t, x, y)$ is asymptotic to the first term of (4.2), whence the result.

### B. Distant Points

If $d(x, y) \geq \delta$, then (4.1) will not be true, in general. To analyze the general case, Molchanov has proved the following preliminary result, which generalizes Varadhan's theorem [42, 43].

**Proposition** *For any* $(x, y) \in M \times M$

$$\lim_{t \to 0} 4t \log p(t, x, y) = -d^2(x, y).$$

On the basis of this result and the result of Section IV.A, we can analyze the general case. We first require a preliminary estimate which allows us to focus attention on the set of minimal geodesics from $x$ to $y$.

Given $(x, y) \in M \times M$, choose an integer $l$ such that $d(x, y) < (l+1)\delta$. Given $\varepsilon > 0$, let $\delta_1(\varepsilon) = \varepsilon + (l+1)^{-1} d(x, y)$. Let $y_0 = x$, $y_{l+1} = y$ and let

$$\overline{M} = \Big\{(y_1, \ldots, y_l): d(y_j, y_{j+1}) < \delta_1 \quad \text{for} \quad 0 \leq j \leq l,$$

$$\sum_{j=0}^{l} d(y_j, y_{j+1}) < d(x, y) + \varepsilon\Big\}.$$

**Proposition** $P\{\overline{M}\} \leq (\text{const}) \exp\{-[d(x, y) + \varepsilon]^2/4t\}$, $t \to 0$.

*Proof* We write $\overline{M} = \bigcup_{j=0}^{l} M_j$, where

$$M_0 = \Big\{(y_1, \ldots, y_l): d(y_j, y_{j+1}) < \delta_1 \quad \text{for} \quad 0 \leq j \leq l,$$

$$\sum_{j=0}^{l} d(y_j, y_{j+1}) \geq d(x, y) + \varepsilon\Big\},$$

$$M_j = \Big\{(y_1, \ldots, y_l): d(y_j, y_{j+1}) > \delta_1,$$

$$\sum_{j=0}^{l} d(y_j, y_{j+1}) < d(x, y) + \varepsilon\Big\}.$$

To estimate $P(M_0)$, we write

$$P(M_0) \leq \int_{M_0} \cdots \int \prod_{j=0}^{l} p(t/(l+1), y_j, y_{j+1}) \, dV(y_1) \cdots dV(y_l)$$

$$\times (\text{const}) t^{-l} \int_{M_0} \cdots \int \exp[-(l+1) d^2(y_j, y_{j+1})/t] \, dV(y_1) \cdots dV(y_l),$$

$$\times (\text{const}) t^{-l} \exp[-(d(x, y) + \varepsilon)^2/t],$$

where we have used (4.1) and Schwarz's inequality, in the form

$$(l+1)(d_1^2 + \cdots + d_{l+1}^2) \geq (d_1 + \cdots + d_{l+1}).$$

To estimate $P(M_j)$, $(j \geq 1)$, we first prove an auxiliary result.

**Lemma** *Let $d_1, \ldots, d_n$ be positive real numbers with $d_1 + \cdots + d_n = D$, $d_1 \geq \delta > D/n$. Then $n(d_1^2 + \cdots + d_n^2) \geq D^2[1 + (1 - n\delta/D)^2/(n-1)]$.*

*Proof* For a fixed value of $d_1$, the minimum of $d_1^2 + \cdots + d_n^2$ is achieved when $d_2 = d_3 = \cdots = d_n = (D - d_1)/(n-1)$. Substituting this into $d_2^2 + \cdots + d_n^2$ we see that $d_1^2 + \cdots + d_n^2 \geq d_1^2 + (D - d_1)^2/(n-1)$. For $d_1 \in [\delta, D]$, the minimum of this expression occurs when $d_1 = \delta$. Finally, a lengthy algebraic manipulation shows that

$$n(\delta^2 + (D-\delta)^2/(n-1)) = D^2[1 + (1 - n\delta/D)^2/(n-1)],$$

the stated result.

*Application to the estimation of $P(M_i)$* We have

$$P(M_i) = \int_{M_i} \cdots \int \prod_{j=0}^{l} p[t/(l+1), y_j, y_{j+1}] \, dV(y_1) \cdots dV(y_l)$$

$$\leq \int_{M_i} \cdots \int \prod_{j=0}^{l} \exp[-(l+1)d^2(y_j, y_{j+1})/t] \, dV(y_1) \cdots dV(y_l)$$

$$\leq \int_{M_i} \cdots \int \exp \frac{-D^2}{t} \left[1 + \left(1 - \frac{(l+1)\delta}{D}\right)^2\right] dV(y_1) \cdots dV(y_l).$$

Here we have used the estimate of Varadhan, together with the above lemma. Thus we see that $P(M_i) \leq (\text{const}) \exp[-(d^2(x, y) + \varepsilon)/4t]$ when $t \to 0$.

Following Molchanov, we now proceed to give the explicit asymptotic behavior of $p(t, x, y)$ in some typical cases. Let $\Omega_{xy}$ be the set of minimal geodesics from $x$ to $y$. By general theorems [5a]; $\Omega_{xy}$ is nonempty. We consider various hypotheses on the structure of $\Omega_{xy}$.

*Case 1* $\Omega_{xy}$ consists of a finite number of geodesics $\{\gamma_1, \ldots, \gamma_m\}$ and $x$ and $y$ are nonconjugate along each of them. Then

$$p(t, x, y) \sim [\exp(-r^2/4t)/(4\pi t)^{n/2}] H(x, y), \qquad r = d(x, y), \quad t \to 0.$$

*Case 2* $\Omega_{xy}$ consists of a finite number of geodesics $\{\gamma_1, \ldots, \gamma_m\}$ and $x$ and $y$ are conjugate along each of them. In suitable coordinates $(z^i)$ near $\gamma(r/2)$, the action has the form

$$E(\gamma) = d^2(x, y) + (z^1)^{2p} + \sum_{i=2}^{n} (z^i)^2,$$

where $\rho > 1$. Then

$$p(t, x, y) \sim [\exp(-r^2/4t)/t^{(n+1-p^{-1})/2}] H(x, y),$$

where $H(x, y)$ is an expression that depends on the Hessian of the action.

Case 3  $\Omega_{xy}$ has the structure of a smooth manifold of dimension $d$. Then

$$p(t, x, y) \sim [\exp(-r^2/4t)/t^{(n+d)/2}] H(x, y),$$

where $H(x, y)$ is an expression that depends on the Hessian of the action.

These results, although not exhaustive of all cases, indicate the effect of many "paths of least action." In each case the proof is carried out by means of Laplace's asymptotic method [35; pp. 20–23].

Case 1 includes the case of a compact manifold of negative curvature. In particular, for a symmetric space of negative curvature we have

$$p(t, x, y) \sim \frac{\exp(-r^2/4t)}{(4\pi t)^{n/2}} \prod_{i=1}^{n} \left( \frac{r(-\lambda_i^{\frac{1}{2}})}{\sinh r(-\lambda_i^{\frac{1}{2}})} \right)^{\frac{1}{2}},$$

where $(\lambda_i)$ are the principal curvatures of any geodesic hypersurface orthogonal to $\gamma$.

Case 3 includes the case of a surface of revolution in $R^3$, with the metric

$$ds^2 = [1 + \Phi'(u)^2] du^2 + \Phi(u)^2 d\theta^2, \quad 0 \leq u \leq L, \quad 0 \leq \theta \leq 2\pi,$$

with $\Phi(0) = 0$, $\Phi(L) = 0$ and $\Phi(u) > 0$ for $0 < u < L$. Let $N = (0, 0)$, $S = (L, 0)$. We first write the Chapman–Kolmogorov equation in the form

$$p(t, N, S) = \int_M p(t/2, N, x) p(t/2, x, S) \, dV(x)$$

and use the result of (IV.A) in the form

$$p(t, N, x) \sim [\exp(-d^2/4t)/4\pi t] (d(u)/\Phi(u))^{\frac{1}{2}},$$

where $d = d(u) = \int_0^u (1 + \Phi'^2)^{\frac{1}{2}} ds$. Making the similar estimate for $p(t/2, x, S)$, we have to examine the exponent

$$F(u) = \left( \int_0^u (1 + \Phi'^2)^{\frac{1}{2}} ds \right)^2 + \left( \int_u^L (1 + \Phi'^2)^{\frac{1}{2}} ds \right)^2.$$

$F(u)$ is positive on $(0, L)$ and has a single interior minimum at $u_0$, where

$$\int_0^{u_0} (1 + \Phi'^2)^{\frac{1}{2}} ds = \int_{u_0}^L (1 + \Phi'^2)^{\frac{1}{2}} ds.$$

Applying the method of Laplace yields the result

$$p(t, N, S) \sim [\exp(-r^2/4t)/(4\pi t)^{\frac{3}{2}}] \pi r, \quad r = d(N, S),$$

in accordance with Case 3.

# STOCHASTIC RIEMANNIAN GEOMETRY

## V. Asymptotic Properties, $t \to \infty$

In this section we will survey the asymptotic behavior of the Brownian motion for large times. In case of a compact manifold the Brownian motion is recurrent and the Riemannian volume element is an invariant measure which is approached exponentially fast when $t \to \infty$. In the case of a complete simply connected manifold of negative curvature, the Brownian motion is transient, with a limiting direction when $t \to \infty$. In this case we can study the rates of approach to infinity a.s. and in the sense of probability law. In the first case the estimates depend on the behavior of the curvature at infinity; in the second case the estimates depend on the global behavior of the curvature.

### A. Compact Manifolds

We borrow the following analytical facts from the theory of partial differential equations [44, pp. 254ff]

(i) zero is a simple eigenvalue of $\Delta$;
(ii) the remainder of the spectrum is discrete and consists of positive eigenvalues $0 < \lambda_1 \leq \lambda_2 \leq \cdots$ with $\lim_{n\to\infty} \lambda_n = +\infty$.

From these facts it follows that we have an orthogonal decomposition of $L^2(M)$; if $f = \sum_{n=1}^{\infty} \langle f, \phi_n \rangle \phi_n$, then

$$e^{-t\Delta} f = \sum_{n=0}^{\infty} e^{-\lambda_n t} \langle f, \phi_n \rangle \phi_n, \tag{5.1}$$

where $\phi_0 = (\text{vol } M)^{-\frac{1}{2}}$. In particular, when $t \to \infty$ we have

$$e^{-t\Delta} f = (\text{vol } M)^{-1} \int_M f \, dV + O(e^{-\lambda_1 t}). \tag{5.2}$$

Let us use these analytical facts to prove the following asymptotic behavior of $X_t$ in the compact case.

**Theorem** *For any open set $U \subset M$ and any $x \in M$, we have $P\{X_t^x \in U$ infinitely often when $t \to \infty\} = 1$.*

**Proof** Let $T_U^x = \inf\{t > 0: X_t^x \in U\}$, $f(x) = P\{T_U^x < \infty\}$. From the general theory of elliptic operators [31; pp. 98–99] $f$ is a $C^\infty$ function on the exterior of $U$ and $f(x) \to 1$ when $x$ approaches $\partial U$. By the Markov property, we have $f(x) - T_t f(x) = P\{T_U^x < t, T_U^x \circ \theta_t = \infty\}$. Thus $T_t f \leq f$. On the other hand, by the compactness of $M$, we know from (5.2) that

$$\lim_{t \to \infty} T_t f = (\text{vol } M)^{-1} \int_M f \, dV = c,$$

a positive constant. Thus $0 < c \leq f(x)$ and $\int_M [f(x) - c] \, dV = 0$. Since the integrand is nonnegative and continuous, we see that $f(x) - c = 0$. Letting $x \to \partial U$ we see that $c = 1$. To complete the proof, we note that $1 = \lim_{t \to \infty} T_t f = \lim_{t \to \infty} P\{T_U^x \circ \theta_t < \infty\} = P\{X_t^x \in U$ infinitely often when $t \to \infty\}$.

### B. Open Manifolds of Negative Curvature—Strong Laws

In this section we assume that $M$ is a simply-connected complete Riemannian manifold of nonpositive sectional curvature. By the classical theorem of Cartan–Hadamard [5a; p. 36], the exponential mapping at $x_0$ is a diffeomorphism for any base point $x_0 \in M$. Taking geodesic polar coordinates in $T_{x_0}(M)$, we can write

$$r_t = r_0 + \sqrt{2} W_t + \int_0^t \left( G^{-1} \frac{\partial G}{\partial r} \right)(r_s, \theta_s) \, ds, \tag{5.3}$$

where $\{W_t, t \geq 0\}$ is a Wiener process. In order to estimate the integral term we need the following fact, a proof of which does not seem easily available in the literature.

**Lemma** *Assume that the Ricci curvature tensor satisfies the bound*

$$\mathrm{Ric}(X, X) \leq (n-1)k, \quad k < 0 \tag{5.4}$$

*for all unit vectors $X \in T_x(M)$. Then*

$$G^{-1} \frac{\partial G}{\partial r} \geq (n-1)(-k)^{\frac{1}{2}} \coth r(-k)^{\frac{1}{2}}. \tag{5.5}$$

*Proof* Let $x \in M$ and let $\gamma$ be a minimal geodesic joining $x_0$ to $x$. By the hypothesis of nonpositive curvature, $\gamma$ can have no conjugate points. Let $y_1, \ldots, y_n$ be an orthonormal basis in $T_x(M)$, where $y_1 = \gamma'$. From the formula of second variation of arc length [5; p. 135], we have

$$G^{-1} \frac{\partial G}{\partial r} = \sum_{i=2}^n \langle Y_i', Y_i \rangle,$$

where $\{Y_i\}_{i=2}^n$ are Jacobi fields along $\gamma$ with $Y_i(0) = 0$, $Y_i(r) = y_i$. Let

$$I(V, W) = \int_0^r [\langle V_i', W' \rangle + \langle R(W, \gamma') V, \gamma' \rangle] \, ds,$$

where $V$ and $W$ are vector fields along $\gamma$. $I(V, W)$ is a symmetric bilinear form on the space of vector fields along $\gamma$, which are orthogonal to $\gamma$. We use the index lemma of Morse theory [5a; p. 24].

**Index Lemma** *Let $W$ be a vector field along $\gamma$ and let $V$ be the unique Jacobi* field such that $V(0) = 0 = W(0)$, $V(x) = W(x)$. Then $I(V,V) \leqslant I(W,W)$ with equality iff $V = W$.

We will now compare $M$ with $M_0$, a manifold of constant sectional curvature equal to $k$, where we have identified the tangent spaces $T_{x_0}(M)$ and $T_{x_0}(M_0)$ by means of a fixed isomorphism. Let $\hat{Y}_i$ be the vector field in $M_0$, which is obtained by parallel transport of $Y_i$ backward along $\gamma$ in $M$ and then forward along the corresponding geodesic $\hat{\gamma}$ in $M$. Thus,

$$\begin{aligned}
G^{-1}\frac{\partial G}{\partial r} &= \sum_{i=2}^{n} \langle Y'_i, Y_i \rangle \\
&= \sum_{i=2}^{n} \int_0^r [\langle Y'_i, Y'_i \rangle - \langle R(Y_i, \gamma')\gamma', Y_i \rangle ]\, ds \\
&= \int_0^r \left[ \sum_{i=2}^{n} \langle Y'_i, Y'_i \rangle - \mathrm{Ric}(\gamma', \gamma') \right] ds \\
&\geqslant \int_0^r \left[ \sum_{i=2}^{n} \langle Y'_i, Y'_i \rangle - (n-1)k(\gamma', \gamma') \right] ds \\
&= \int_0^r \sum_{i=2}^{n} \langle \hat{Y}'_i, \hat{Y}'_i \rangle - (n-1)k \langle \hat{\gamma}', \hat{\gamma}' \rangle\, ds \\
&= \sum_{i=2}^{n} I(\hat{Y}_i, \hat{Y}_i).
\end{aligned}$$

Now let $\{_0Y_i\}_{i=2}^n$ be the Jacobi fields in $M_0$, which satisfy $_0Y_i(0) = 0$, $_0Y_i(r) = y_i$. Then by the index lemma, we have $I(_0Y_i, _0Y_i) \leqslant I(Y_i, Y_i)$. Thus,

$$G^{-1}\frac{\partial G}{\partial r} \geqslant \sum_{i=2}^{n} I(_0Y_i, _0Y_i).$$

But the final expression is the value of $G^{-1}(\partial G/\partial r)$ computed in the manifold of constant sectional curvature. By a standard calculation (see Example 2, Section II.B), we have that this quantity equal to

$$(n-1)(d/dr)\log(\sinh r(-k)^{\frac{1}{2}}/(-k)^{\frac{1}{2}}) = (n-1)(-k)^{\frac{1}{2}} \coth r(-k)^{\frac{1}{2}}.$$

With the above preliminary information, we can give the fundamental asymptotic result.

**Theorem** *If the sectional curvature $K$ satisfies the bounds $\overline{K} \leqslant K \leqslant k < 0$, then*

$$\liminf_{t \to \infty}(t_t/t) \geqslant (n-1)(-k)^{\frac{1}{2}} \quad \text{a.s.} \tag{5.6}$$

$$\lim_{t \to \infty} \theta(t) \quad \text{exists a.s.} \tag{5.7}$$

*Proof* From (5.5) we have $G^{-1}(\partial G/\partial r) \geq (n-1)(-k)^{\frac{1}{2}} \coth(r(-k)^{\frac{1}{2}}) \geq (n-1)(-k)^{\frac{1}{2}}$. Therefore, $r_t \geq r_0 + W_t + (n-1)(-k)^{\frac{1}{2}}t$. But we know that $W_t = O(t^{\frac{1}{2}} \log t)$ when $t \to \infty$. Therefore result (5.6) follows immediately.

To prove result (5.7) we prove a series of results in which we write the argument $t$ rather than a subscript.

*Step 1* Consider a geodesic triangle connecting the points $O, A, B$; let $x = d(O, A)$, $y = d(O, B)$, $d = d(A, B)$, $\phi = \cos^{-1}(OA, OB)$. Then

$$1 - \cos \phi \leq [\cosh d(-k)^{\frac{1}{2}}/\sinh x(-k)^{\frac{1}{2}} \sinh y(-k)^{\frac{1}{2}}]. \qquad (5.8)$$

*Proof* By the Rauch comparison theorem [5a; p. 29] we have $d \geq \bar{d}$ where $\bar{d}$ is the distance from $A$ to $B$ in the space form of constant curvature $k$. But for a space form we have the "law of cosines"

$$\cosh d(-k)^{\frac{1}{2}} = \cosh x(-k)^{\frac{1}{2}} \cosh y(-k)^{\frac{1}{2}} - \cos \phi \sinh x(-k)^{\frac{1}{2}} \sinh y(-k)^{\frac{1}{2}}.$$

If we now replace $\cosh x(-k)^{\frac{1}{2}} \cosh y(-k)^{\frac{1}{2}}$ by $\sinh x(-k)^{\frac{1}{2}} \sinh y(-k)^{\frac{1}{2}}$ we make the right-hand member smaller. Solving the resulting inequality for $1 - \cos \phi$ gives the stated result.

*Step 2* $\sum_{j=1}^{\infty} d_0[\theta(j), \theta(j+1)] < \infty$ a.s. where $d_0$ is the metric in $T_0(M)$.

*Proof* Let $A = X(j)$, $B = X(j+1)$ and substitute in (5.8). From (5.6) we see that $x \geq Cj$, $y \geq Cj$ for a positive constant $C$. To estimate $d$, we note that

$$d(X(j+1), X(j)) = W_{j+1} - W_j + \int_j^{j+1} (\Delta r)(X_s^x) \, ds.$$

When $j \to \infty$ the integral term is bounded by a constant. The Brownian term is bounded a.s. by $(\log j)^{\frac{1}{2}}$. Substituting this into (5.8), we have

$$d_0(\theta(j), \theta(j+1))^2 \leq (\text{const})(\log j)^{\frac{1}{2}} \exp(-2Cj).$$

The proof is complete.

*Step 3* $\sup_{j \leq t < j+1} d_0(\theta(t), \theta(j)) \to 0$ a.s. when $j \to \infty$.

*Proof* Repeating the above reasoning with $A = X(j)$ and $B = X(t)$, we see that

$$\sum_{j=1}^{\infty} P\{\sup_{j \leq t < j+1} d[X(t), X(j)] \geq (\text{const})(\log j)^{\frac{1}{2}}\} < \infty.$$

The conclusion follows from the Borel–Cantelli lemma.

*Special Case of Surfaces* In case $n = 2$, the preceding results can be given a more precise form, in particular (5.6) can be restated as an ergodic theorem.

To state this result, recall that the metric can be written in the form
$$ds^2 = dr^2 + G(r,\theta)^2 \, d\theta^2,$$
where $G(0,\theta) = 0$, $G'(0,\theta) = 1$ and
$$G''_{rr}(r,\theta) + K(r,\theta) G(r,\theta) = 0.$$

**Theorem** *We have, almost surely, when $t \to \infty$,*
$$\lim_{t \to \infty} \frac{r_t}{t} = \left[ \lim_{r \to \infty} \frac{\int_0^r -K(r,\theta) G(r,\theta) \, dr}{\int_0^r G(r,\theta) \, dr} \right]^{\frac{1}{2}}.$$

*In particular, if $\lim_{r \to \infty} K(r,\theta) = k(\theta)$ exists, then $\lim_{t \to \infty} r_t/t = (-k(\theta))^{\frac{1}{2}}$.*

**Step 1** $G_r/G \geq 1/r$ for $r > 0$.

*Proof* Let $\psi = G/G_r$. Then $\psi(0) = 0$ and $\psi_r = (G_r^2 - GG_{rr})/G_r^2 \leq 1$. Thus $\psi(r) \leq r$.

In particular, $G_r/G$ is bounded away from zero on any compact subset of $[0, \infty)$.

In the following two steps we will show that
$$\liminf_{r \to \infty} (-K)^{\frac{1}{2}} \leq \liminf_{r \to \infty} G_r/G \leq \limsup_{r \to \infty} G_r/G \leq \limsup_{r \to \infty} (-K)^{\frac{1}{2}}.$$

**Step 2** If $\liminf_{r \to \infty} G_{rr}/G \geq k^2 > 0$, then $\liminf_{r \to \infty} G_r/G \geq |k|$.

*Proof* We may suppose, without loss of generality, that $k = 1$. Indeed, it suffices to define $\tilde{G}(r,\theta) = kG(r/k, \theta)$. Then $\tilde{G}_{rr}/\tilde{G} = k^{-2} G_{rr}/G$ and $\tilde{G}_r/\tilde{G} = k^{-1} G_r/G$.

For the proof, we let $h = G_r/G$; we must show that $\liminf_{r \to \infty} h \geq 1$. We then consider two cases.

If $\limsup_{r \to \infty} h(r) = \liminf_{r \to \infty} h(r)$, let $m$ be the common value. If $m < 1$, let $\varepsilon < (1-m)/2$. For all sufficiently large $r$, we have $G_{rr}/G \geq (1-\varepsilon)^2$. Now, from the differential equation for $h$ we have $h_r = G_{rr}/G - h^2 \geq (1-\varepsilon)^2 - (m+\varepsilon)^2 > 0$ for all sufficiently large $r$. Thus, $h(r) > c_1 + c_2 r$ for some positive $c_2$ and thus $\liminf_{r \to \infty} h(r) = \infty$, a contradiction: thus $m \geq 1$.

In the second case we may assume that $\limsup_{r \to \infty} h(r) > \liminf_{r \to \infty} h(r)$. If the latter is strictly less than 1, we may construct an infinite sequence $r_k \to \infty$ with $h(r_k) = A$, $h_r(r_k) \leq 0$ with $A < 1$. Choose $\varepsilon < 1-A$; for all sufficiently large $r$ we have $G_{rr}/G \geq (1-\varepsilon)^2$. Now, from the differential equation we have $h_r(r_k) = G_{rr}/G - h^2(r_k) \geq (1-\varepsilon)^2 - A^2 > 0$, a contradiction. Therefore, $\liminf_{r \to \infty} h(r) \geq 1$, which was to be proved.

**Step 3** If $\limsup_{r \to \infty} G_{rr}/G \leq k^2$, then $\limsup_{r \to \infty} G_r/G \leq |k|$.

*Proof* As before, we may assume that $k = 1$. Defining $h = G_r/G$, we must show that $\limsup_{r\to\infty} h(r) \leq 1$.

As before we consider two cases.

If $\lim\sup_{r\to\infty} h(r) = \liminf_{r\to\infty} h(r)$, let $m$ be the common value. If $m > 1$, let $\varepsilon < (m-1)/2$. For all sufficiently large $r$, we have $G_{rr}/G \leq (1+\varepsilon)^2$. Now from the differential equation for $h$, we have

$$h_r = G_{rr}/G - h^2 \leq (1+\varepsilon)^2 - (m-\varepsilon)^2 < 0$$

for all sufficiently large $r$. Thus, $\limsup_{r\to\infty} h(r) = -\infty$, a contradiction, and thus $m \leq 1$.

In the second case we may assume that $\liminf_{r\to\infty} h(r) < \limsup_{r\to\infty} h(r)$. If the latter is strictly greater than 1, we may construct an infinite sequence $r_k \to \infty$ with $h(r_k) = A$, $h_r(r_k) \geq 0$, where $A > 1$. Choose $\varepsilon < A - 1$; for all sufficiently large $r$ we have $G_{rr}/G \leq (1+\varepsilon)^2$. Now, from the differential equation we have $h_r(r_k) = G_{rr}/G - h^2(r_k) \leq (1+\varepsilon)^2 - A^2 < 0$, a contradiction. Therefore $\limsup_{r\to\infty} h(r) \leq 1$, which was to be proved.

*Remark 1* Steps 2 and 3 show, in particular, that if $\lim_{r\to\infty} K(r,\theta) = -k^2$ then $\lim_{r\to\infty}(G_r/G)(r,\theta) = |k|$. The proof shows further that if the first limit is uniform in $\theta \in [0, 2\pi]$ then the second one is also uniform.

*Remark 2* The converse is not true, in general. Indeed, suppose that $G$ is of the form

$$G(r) = r \exp\left(\int_0^r p(s)\, ds\right),$$

where $p$ is a $C^1$ function with $p(r) = 0$ for $0 \leq r \leq r_0$ and $p(r) = \sin^2(r^2)/4r$ for $r \geq r_1$. Then we have

$$G_r/G = 1 + \sin^2(r^2)/4r, \quad r > r_1$$
$$G_{rr}/G = 1 + \sin(r^2)\cos(r^2) + O(1/r), \quad r \to \infty.$$

Then $\lim_{r\to\infty} G_r/G = 1$, but

$$\limsup_{r\to\infty} G_{rr}/G = \tfrac{3}{2}, \quad \liminf_{r\to\infty} G_{rr}/G = \tfrac{1}{2}.$$

*Remark 3* The proof of Steps 2 and 3 shows that if $K(r,\theta)$ oscillates sufficiently slowly when $r \to \infty$, we may have

$$\liminf_{r\to\infty}(-K)^{\frac{1}{2}} = \liminf_{r\to\infty} G_r/G < \limsup_{r\to\infty} G_r/G = \limsup_{r\to\infty}(-K)^{\frac{1}{2}}.$$

We consider various conditions on the curvature at infinity:

(H$_1$)   $\liminf_{r\to\infty}(-K) \geq k_-^2 > 0$,    uniformly, $0 \leq \theta \leq 2\pi$,

(H$_2$)   $\limsup_{r\to\infty}(-K) \leq k_+^2 < \infty$,    uniformly, $0 \leq \theta \leq 2\pi$,

(H$_3$)  $\lim_{r \to \infty} (-K) = k^2 > 0$,  uniformly, $0 \leq \theta \leq 2\pi$,

(H$_4$)  $\lim_{r \to \infty} \dfrac{\int_0^r -KG\, dr}{\int_0^r G\, dr} = k^2 > 0$,  uniformly, $0 \leq \theta \leq 2\pi$.

We formulate the different a.s. asymptotic behaviors of $r(t), t \to \infty$:

(C$_1$)  $\liminf_{t \to \infty} r(t)/t \geq k_-$,

(C$_2$)  $\limsup_{t \to \infty} r(t)/t \leq k_+$,

(C$_3$)  $\lim_{t \to \infty} r(t)/t = k$.

**Proposition** (H$_1$) *implies* (C$_1$); (H$_1$) *and* (H$_2$) *imply* (C$_2$); *either* (H$_2$) *or* (H$_4$) *implies* (C$_3$).

*Proof* To prove the first statement, we must first show that $r(t) \to \infty$. To see this, note that Steps 1 and 2 together imply that $G_r/G \geq \delta$ ($0 < r < \infty$) for some positive constant $\delta$. Thus, from (5.3), we see that

$$r(t) \geq r(O) + 2\sqrt{W(t)} + \delta t.$$

Hence, $r(t) \to \infty$. Now return to (3.1a) and use (H$_1$) and Step 2 again on the integral term:

$$(G_r/G)(r(s), \theta(s)) \geq k_- - \varepsilon \quad \text{for} \quad s \geq T.$$

Hence, $\liminf_{t \to \infty} r(t)/t \geq k_- - \varepsilon$ for any $\varepsilon > 0$.

To prove the second statement, apply Step 3 to the integral term in (5.3). Since we already know that $r(t) \to \infty$, we see that $\limsup_{t \to \infty} r(t)/t \leq k_+$, as was to be proved.

To prove the final statement, note that (H$_3$) implies (H$_4$). Therefore we need only prove that (H$_4$) implies (C$_3$). To do this, define $H = G/\int_0^r G\, dr$. Then

$$H_r = G_r \bigg/ \left(\int_0^r G\, dr\right) - H^2.$$

By the differential equation $G_{rr} + KG = 0$, we see that

$$\int_0^r KG\, dr = G_r(0, \theta) - G_r(r, \theta).$$

Thus (H$_4$) states that $G_r/(\int_0^r G\, dr \to k^2)$. Therefore $H_r = \overline{K} - H^2$, where $\lim_{r \to \infty} \overline{K} = k^2$. Using the proof of Steps 2 and 3, it follows that $\lim_{r \to \infty} H = k > 0$. Finally, we have $G_r/G = (H^{-1}G_r)/(\int_0^r G\, dr \to k^{-1}k^2) = k$. Substituting this into (5.3), we see that $r(t)/t \to k$.

## C. Open Manifolds of Negative Curvature—Weak Laws

In this section we will prove two theorems on the distribution of $r_t$ when $t \to \infty$. In both cases $M$ is a two-dimensional manifold of nonpositive curvature. The first theorem is a "central limit theorem" for $r_t$. The second theorem is the analog of the "large deviation estimate"

**Theorem 1** *Assume that the metric density $G$ satisfies*

$$G'/G \geq k, \quad (r, \theta) \in (0, \infty) \times S^1 \tag{5.9}$$

$$G'/G = k + o(1/r), \quad r \to \infty, \quad \theta \in S^1 \tag{5.10}$$

*uniformly for $\theta \in S^1$. Then $(r_t - kt)/t^{\frac{1}{2}}$ has a limiting normal distribution with mean 0 and variance 1.*

The proof will be broken into several steps.

*Step 1* $\lim_{t \to \infty} E(r_t/t) = k$.

*Proof* Recall the basic stochastic equation

$$r_t = r_0 + \sqrt{2}\, W(t) + \int_0^t (G'/G)(r_s, \theta_s)\, ds. \tag{5.11}$$

Thus,

$$r_0 \leq E(r_t - kt) \leq \int_0^t E[(G'/G) - k](r_s, \theta_s)\, ds + r_0.$$

The integral is to be written in the form

$$\int_0^t E(A)\, ds = \int_0^t E(A;\, r_s > 1)\, ds + \int_0^t E(A;\, r_s \leq 1)\, ds.$$

In the first term the integrand is bounded by a constant and because of (5.10) converges to zero when $s \to \infty$. Therefore the expectation converges to zero and thus the integral $= o(t)$, $t \to \infty$. To analyze the second term, use the fact [6; Corollary 1] that $K \leq 0$ implies

$$p(t, x, y) \leq (4\pi t)^{-1} \exp(-d^2(x, y)/4t).$$

Noting that $G'/G = O(1/r)$, $r \to 0$ we have

$$\int_0^t E((G'/G) - k)((r_s, \theta_s);\, r_s \leq 1)\, ds \leq \text{const} \int_0^t \int_0^1 \exp(-r^2/4s)\, dr\, ds,$$

which is $o(t)$, $t \to \infty$. This proves that $E(r_t/t) = k + o(1)$, $t \to \infty$.

*Step 2* $E(r_t - kt)^2 \leq 2t + o(t)$, $t \to \infty$.

*Proof* By Itô's formula, we have

$$(r_t - kt)^2 = r_0^2 + (2^{\frac{3}{2}}) \int_0^t (r_s - ks)\, dW_s + \int_0^t (r_s - ks)((G'/G) - k)(r_s, \theta_s)\, ds + 2t.$$

Therefore,
$$E((r_t-kt)^2) = r_0^2 + 2t + \int_0^t E((r_s-ks)((G'/G)-k)(r_s,\theta_s))\,ds.$$

By (5.9) the final integral can be bounded by
$$\int_0^t E(r_s((G'/G)-k)(r_s,\theta_s))\,ds.$$

Now the function $r \to r((G'/G)-k)$ is bounded for $0 < r < \infty$ and converges to zero when $r \to \infty$. Therefore, the expectation converges to zero when $s \to \infty$, by the dominated convergence theorem.

Step 3  $E((r_t-kt)(W_t)) = t\sqrt{2}$.

*Proof* Apply Itô's lemma to the product and take expectations.

Step 4  $E((r_t-(kt+2^{\frac{1}{2}}W_t))^2) = o(t)$, $t \to \infty$.

*Proof* The above expectation can be written as
$$E((r_t-kt)^2) - 2^{\frac{3}{2}}E((r_t-kt)W_t) + 2t.$$

Using Steps 2 and 3, we have the bound $2t+o(t)-4t+2t = o(t)$.

*Proof of the Theorem* We have proved that the random variable
$$\frac{r_t-kt}{t^{\frac{1}{2}}} - \frac{W_t}{t^{\frac{1}{2}}}$$
converges to zero in $L^2$ when $t \to \infty$. But $W_t/t^{\frac{1}{2}}$ has a standard normal distribution. Since convergence in $L^2$ implies convergence in distribution, the theorem is proved.

We now consider estimates of the form
$$P\{r_t < R\} \leq (\text{const})\,e^{-\lambda t}.$$

For this purpose, let
$$-\lambda_1(f) = \limsup_{t \to \infty} t^{-1} \log E\{f(X_t^x)\}, \qquad \lambda_1 = \sup_f \lambda_1(f),$$
where the supremum is taken over all continuous functions with compact support. We propose to estimate $\lambda_1$ in terms of the global behavior of the metric. Recalling the spectral decomposition of the Laplace operator,
$$e^{-t\Delta}f = \int_0^\infty e^{-\lambda t}\,dE_\lambda f,$$
we see that $\lambda_1$ is the infimum of the spectrum of $\Delta$, acting on $L^2(M)$.

We assume explicitly that $M$ is a simply connected, complete, two-dimensional Riemannian manifold of nonpositive curvature that satisfies the technical condition

$$|(G_{rr}/G_r)_\theta| \leq \text{const}, \quad r \geq r_0, \quad 0 \leq \theta \leq 2\pi.$$

**Theorem 2**

$$\inf_M (G_r/G) \leq (4\lambda_1)^{\frac{1}{2}} \leq \inf_{0 \leq \theta \leq 2\pi} \limsup_{r \to \infty} (G_{rr}/G_r)$$

*As a corollary, we will show that if $G_r/G$ is nondecreasing along each ray and the curvature has a limit equal to $-k^2(\theta_0)$ along the ray $\theta = \theta_0$, then $\lambda_1 = \frac{1}{4} \inf_{0 \leq \theta \leq 2\pi} k^2(\theta)$.*

This result shows, for instance, that when the curvature is constant and is equal to $-k^2$. then $\lambda_1 = k^2/4$; no explicit calculations with special functions are needed in our approach.

To prove the lower half of Theorem 2, we modify the methods used in [32]. To obtain the upper bound we first obtain a comparison function and apply the variational characterization of $\lambda_1$. It is shown that if $G_{rr}/G_r$ satisfies an upper bound on a sufficiently long sector, then a corresponding upper bound can be obtained.

*Step 1 The lower bound* To prove the lower bound, we recall the variational characterization of $\lambda_1$:

$$\lambda_1 = \inf_{f \neq 0} \frac{\int_0^{2\pi} \int_0^\infty (f_r^2 + f_\theta^2/G^2) G \, dr \, d\theta}{\int_0^{2\pi} \int_0^\infty f^2 G \, dr \, d\theta}, \quad (5.12)$$

where the infimum is taken over continuous, piecewise $C^1$ functions $f$ with compact support. We assume that $G_r/G \geq \delta > 0$ (if $\delta = 0$ there is nothing to prove). Following [32] we have

$$\int_0^\infty f^2 G \, dr = \delta^{-1} \int_0^\infty f^2 G_r \, dr$$

$$= -2\delta^{-1} \int_0^\infty f f_r G \, dr.$$

Therefore, by Schwarz's inequality

$$\left( \int_0^\infty f^2 G \, dr \right)^2 \leq 4\delta^{-2} \left( \int_0^\infty f^2 G \, dr \right) \left( \int_0^\infty f_r^2 G \, dr \right), \quad (5.13)$$

with the conclusion that

$$\int_0^\infty f_r^2 G \, dr \geq (\delta^2/4) \int_0^\infty f^2 G \, dr. \quad (5.14)$$

When we add in the angular term, do the angular integration, and divide by the denominator of (5.12), we see that for any $f \neq 0$ this quotient is bounded below by $\delta^2/4$, which was to be proved.

*Step 2* Assume that $G_{rr}/G_r \leq m$ for $R_0 \leq r \leq R_1$, $\alpha \leq \theta \leq \beta$. Then
$$\lambda_1 \leq m^2/4 + \pi^2/(R_1 - R_0)^2 + \pi^2/[(\beta - \alpha)^2 g(R_0)^2].$$

*Proof* Let
$$f(r, \theta) = \exp(-mr/2) \sin[\pi(r - R_0)/(R_1 - R_0)] \sin[\pi(\theta - \alpha)/(\beta - \alpha)]$$

in the indicated region and let $f = 0$ elsewhere. By direct computation $f$ is a solution of the differential equation
$$f_{rr} + mf_r + [m^2/4 + \pi^2/(R_1 - R_0)^2] f = 0, \qquad (5.15)$$

with the end condition $f(R_0, \theta) = 0$, $f(R_1, \theta) = 0$. Thus,
$$f_{rr} + (G_r/G) f_r + [m^2/4 + \pi^2/(R_1 - R_0)^2] f = (G_r/G - m) f_r. \qquad (5.16)$$

Multiply (5.16) by $fG$ and integrate on $(R_0, R_1)$; thus,
$$-\int_{R_0}^{R_1} f_r^2 G \, dr + [(m^2/4 + \pi^2/(R_1 - R_0)^2] \int_{R_0}^{R_1} f^2 G \, dr = \int_{R_0}^{R_1} (G_r - mG) f f_r \, dr. \qquad (5.17)$$

We now integrate the right-hand member by parts. The boundary term is zero and the new integrand has the same sign as $mG_r - G_{rr}$ which is non-negative by assumption. Therefore,
$$\int_{R_0}^{R_1} f_r^2 G \, dr \leq [m^2/4 + \pi^2/(R_1 - R_0)^2] \int_{R_0}^{R_1} f^2 G \, dr. \qquad (5.18)$$

To treat the $\theta$ terms in (5.12) we note that $f$ also satisfies
$$f_{\theta\theta} + [\pi^2/(\beta - \alpha)^2] f = 0$$

with $f(r, \alpha) = 0 = f(r, \beta)$. Multiplying this equation by $f$ and integrating on $(0, 2\pi)$ we have
$$\int_0^{2\pi} f_\theta^2 \, d\theta = \pi^2/(\beta - \alpha)^2 \int_0^{2\pi} f^2 \, d\theta.$$

By hypothesis we can make the following estimations:
$$\int_{R_0}^{R_1} \int_\alpha^\beta (f_\theta^2/G) \, d\theta \, dr \leq g(R_0)^{-1} \int_{R_0}^{R_1} \int_\alpha^\beta f_\theta^2 \, d\theta \, dr$$
$$= \pi^2 (\beta - \alpha)^{-2} g(R_0)^{-1} \int_{R_0}^{R_1} \int_\alpha^\beta f^2 \, d\theta \, dr$$
$$\leq \pi^2 (\beta - \alpha)^{-2} g(R_0)^{-2} \int_{R_0}^{R_1} \int_\alpha^\beta f^2 G \, d\theta \, dr.$$

Combining this with (5.18) we have that

$$\int_\alpha^\beta \int_{R_0}^{R_1} (f_r^2 + f_\theta^2/G^2) G\, dr\, d\theta$$

$$\leq [m^2/4 + \pi^2/(R_1-R_0)^2 + \pi^2/(\beta-\alpha)^2 (g(R_0)^2)] \int_\alpha^\beta \int_{R_0}^{R_1} f^2 G\, dr\, d\theta.$$

Inserting the above $f$ into the variational characterization we have proved the lemma.

*Proof of the Theorem* For this purpose, let $\bar{m} = \inf_\theta \overline{\lim}_{r\to\infty} (G_{rr}/G_r)$. Given $\varepsilon > 0$ we can find $R_0' > 0$ and $(\alpha, \beta)$ such that $G_{rr}/G_r \leq \bar{m} + \varepsilon$ when $r \geq R_0'$ and $\alpha \leq \theta \leq \beta$. Let $R_0$ be such that $g(R_0) > \pi/[\varepsilon(\beta-\alpha)]$ and $R_0 \geq R_0'$. Therefore for any $R_1 > R_0$ we have by Lemma 3.1

$$\lambda_1 \leq (\tfrac{1}{4})(\bar{m}+\varepsilon)^2 + \pi^2/(R_1-R_0)^2 + \varepsilon^2.$$

In this inequality we let $R_1 \to \infty$. Since the resulting inequality is valid for every $\varepsilon > 0$, we have proved that $\lambda_1 \leq \bar{m}^2/4$.

*Discussion of the Result—Applications and Examples* In previous work upper and lower bounds for $\lambda_1$ were obtained in terms of the curvature. Recall that in a system of geodesic polar coordinates we have

$$-K = G_{rr}/G, \tag{5.19}$$

where $K$ is the Gaussian curvature. Thus we obtain the two-dimensional case of reference 32.

**Corollary 1** *Suppose that $K < -k^2 < 0$ on $M$. Then $\lambda_1 \geq k^2/4$.*

*Proof* $G$ satisfies the inequality $G_{rr} \geq k^2 G$ with $G(0^+, \theta) = 0$, $G_r(0^+, \theta) = 1$. Thus $h = G_r/G$ satisfies the inequality $h_r + h^2 \geq k^2$, $\lim_{r\to 0} rh(r,\theta) = 1$. Therefore $h(r,\theta) \geq k \coth kr$ with the conclusion that $(G_r/G) \geq k$ everywhere. Applying the lower bound, we obtain the stated result.

In reference 5b it was proved that $K \geq -k^2$ implies $\lambda_1 \leq k^2/4$. Although our method does not yield this result, we do obtain a related localized result.

**Corollary 2** *Suppose that $k^2(\theta) = \lim_{r\to\infty} [-K(r,\theta)]$ exists for $\theta \in (\alpha, \beta)$ and satisfies $k^2(\theta) \leq k^2$. Then $\lambda_1 \leq k^2/4$.*

*Proof* We have $G_{rr}/G_r = (-K)/h$, where $h = G_r/G$ is a solution of the equation $h_r + h^2 = -K$. Using a comparison estimate when $r \to \infty$ we have $\lim_{r\to\infty} h(r,\theta) = k(\theta)$. (This also follows by an appropriate use of the Rauch comparison theorem.) Therefore $\lim_{r\to\infty} (G_{rr}/G_r) = k_y\theta)$. Applying the upper half of Theorem 2, we see that $\lambda_1 \leq k^2/4$, which was to be proved.

Using the same idea, we can obtain another variation of Cheng's result.

**Corollary 3** *Suppose that* $K(r,\theta) \geq -k^2$ *and* $(G_r/G)(r,\theta) \geq k_1$ *for* $\theta \in (\alpha, \beta)$ *and* $r \geq R_1$. *Then* $\lambda_1 \leq k^2/4k_1$.

*Proof* In this case we have $G_{rr}/G_r \leq k^2/k_1$ for $\theta \in (\alpha, \beta)$ and $r \geq R_1$, whence the result follows.

From Theorem 2, we see that the upper bound depends only on the details of the metric in a neighborhood of infinity. It is natural to ask whether a lower bound can be obtained which only depends on the metric in a neighborhood of infinity. The following example shows, in particular, that in the lower bound part of Theorem 2, the infimum cannot be replaced by lim inf when $r \to \infty$.

*Example* Let $K(r)$ be a $C^\infty$ function such that

$$K(r) = -1, \quad 0 \leq r \leq R_1,$$
$$K(r) = -4, \quad R_1 + 1 \leq r < \infty, \quad (5.20)$$

where $R_1 > \pi\sqrt{4/3}$; let $G(r)$ be the solution of $G_{rr} + KG = 0$, $G(0^+) = 0$, $G_r(0) = 1$. Thus $G(r) = \sinh r$ and $G_{rr}/G_r = \tanh r$ for $0 \leq r \leq R_1$. Following the proof of Step 2, we let $f(r) = \exp(-mr/2)\sin(\pi r/R_1)$ for $0 \leq r \leq R_1$ and $f = 0$ for $r > R_1$, where $m = \tanh R_1$. Substituting in the variational characterization (5.12) we have

$$\lambda_1 \leq \tfrac{1}{4}(\tanh R_1)^2 + \pi^2/R_1^2$$
$$< \tfrac{1}{4} \quad\quad\quad + \tfrac{3}{4}$$

and thus $\lambda_1 < 1$. On the other hand, it is clear from (5.20) that $\lim_{r\to\infty}(G_r/G)$ exists and is equal to 2. Hence, $\liminf_{r\to\infty}(G_r/G) = 2 > (4\lambda_1)^{1/2}$.

By modifying the constants in this example and taking $R_1$ sufficiently large, we obtain the following proposition: *Given $0 < a < b$, there exists a rotationally invariant metric $G(r)$ with curvature function $K(r)$ such that $\lim_{r\to\infty}[-K(r)] = b^2$ and $\lambda_1 < a^2/4$.*

It is natural to ask if, under suitable regularity conditions, $\lambda_1$ depends only on the details of the metric in a neighborhood of infinity. We have the following result.

**Corollary 4** *Assume that $G_r/G$ is nonincreasing along each ray and that $\lim_{r\to\infty} K(r,\theta) = -k^2(\theta)$ exists for each $\theta \in [0, 2\pi]$. Then*

$$(4\lambda_1)^{1/2} = \inf_{0 \leq \theta \leq 2\pi} k(\theta).$$

*Proof* From the proof of Corollary 2, we have $\lim_{r\to\infty}(G_{rr}/G_r) = k(\theta)$. On the other hand, the same argument shows that $\lim_{r\to\infty}(G_r/G) = k(\theta) = \inf_{r>0}(G_r/G)$. Hence the left- and righthand members of Theorem 2 are equal in this case, and the proof is complete.

## VI. Bibliographical Remarks

Brownian motion on a homogeneous Riemannian space was studied in the works of Yosida [46-48] using semigroup methods. Meanwhile, Itô [17-19] gave a construction of Brownian motion on a differentiable manifold using the method of stochastic differential equations. This method was further developed by Gangolli [15] and McKean [31], where the Riemannian point of view plays a less important role. From another direction, several discrete approximations of Brownian motion were developed; Malliavin [27] generalized Poincaré's "methode de balyage" to Riemannian manifolds. Jorgensen [22] constructed a random walk approximation to Brownian motion; Pinsky [36] gave a transport approximation of Brownian motion, described in Section II.D.

The method of comparison estimates was emphasized in the works of Malliavin and his coworkers [25, 6-9]. An independent proof of the comparison theorem in Section III.B was given by Ikeda and Watanabe [16]. Analytic proofs of the eigenvalue comparison theorems were given by Cheng [5b]. The perturbation calculation in Section III.C appears in [37].

The asymptotic behavior of the fundamental solution when $t \to 0$ was studied by Varadhan [42, 43] and Molchanov [35]. An analytical proof of the localized estimate with a full asymptotic expansion appears in [5]. See also [22a].

The asymptotic ($t \to \infty$) behavior of the Brownian motion of a manifold of negative curvature was first studied in [27, 40, 41, 45]. The two-dimensional case of Section V.B appears in [38]; the estimate for $\lambda_1$ in Section V.C appears in [39]. Further results appear in [22b].

### REFERENCES

1. Airault, H., Subordination du processus dans le fibre tanqent et formes harmoniques *C. R. Acad. Sci. Ser. A* **282** (1976), 1311-1314.
2. Azencott, R., Behavior of Diffusion Semigroups at infinity, *Bull. Soc. Math. (France)* **102** (1974), 193-240.
3. Azencott, R., Diffusions sur les variétés differentiables, *C. R. Acad. Sci. Ser. A* **274** (1972), 651-654.
4. Azencott, R., Methods of localization and diffusion on manifolds, Univ. Paris **7**, preprint.
5. Berger, M., Gauduchon, P., and Mazet, E., "Le Spectre d'une Variété Riemannienne," Vol. 194 Springer-Verlag Lecture Notes in Mathematics, New York, 1971.
5a. J. Cheeger and D. G. Ebin, "Comparison Theorems in Riemannian Geometry." North Holland, Amsterdam.
5b. Cheng, S. Y., Eigenvalue Comparison theorems and its goemetric applications, *Math. Z.* **143** (1975), 289-297.
6. Debiard, A., Gaveau, B., and Mazet, E., Theorems de comparison en géometrie riemannienne, *Publ. Res. Inst. Math. Sci., Kyoto Univ.* **12** (1976), 391-425.

7. Debiard, A., Gaveau, B., and Mazet, E., Temps d'arrêt des diffusions riemanniennes *C. R. Acad. Sci.* **278** (1974), 723–725.
8. Debiard, A., Gaveau, B., and Mazet, E., Temps de sortie des boules normales et minoration locale de $\lambda_1$, *C. R. Acad. Sci.* **278** (1974), 795–798.
9. Debiard, A., Gaveau, B., and Mazet, E. Theorèmes de comparison en geometrie riemannienne, *C. R. Acad. Sci.* **281** (1975), 455–458.
10. Duncan, T. E., Stochastic systems on Riemannian manifolds, Univ. of Kansas, preprint.
11. Dynkin, E. B., "Markov Processes," (2 Vols.) Springer-Verlag, Berlin and New York, 1965.
12. Dynkin, E. B., Diffusion of tensors, *Dokl Akad. Nauk. SSSR* **141** (1961), 288–291.
13. Eells, J., Random walk on the fundamental group, *in AMS Symp. in Differential Geometry*, Vol. 27, Pt. 2, Stanford University, 1973, pp. 211–217.
14. Gangolli, R., On the construction of certain diffusions on a differentiable manifold *Z. Wahrscheinlichkeitstheorie* **2** (1964), 406–419.
15. Hunt, G., Some theorems concerning Brownian motion., *Trans. Amer. Math. Soc.* **81** (1956), 264–293.
16. Ikeda, N. and Watanabe, S., A comparison theorem for solutions of stochastic differential equations and applications, Univ. of Kyoto, preprint.
17. Itô, K., On stochastic differential equations on a differentiable manifold 1, *Nagoya Math. J.* **1** (1950), 35–47.
18. Itô, K., Brownian motions on a Lie group, *Proc. Japan Acad.* **26** (1950), 1–10.
19. Itô, K., On stochastic differential equations on a differentiable manifold 2, *Mem. Coll. Sci., Kyoto Imp. Univ.* **28** (1953), 82–85.
20. Itô, K., The Brownian motion and tensor fields on Riemannian manifolds, *Proc. Int. Cong. Math.* Stockholm, 1963, pp. 536–539.
21. Itô, K., "Stochastic Parallel Displacement." Vol. 451, Springer-Verlag Lecture Notes in Mathematics, 1975, pp. 1–7.
22. Jorgensen, E., A central limit theorem for geodesic random walks, *Z. Wahrscheinlichkeitstheorie und Verw. Gebiete.* **32** (1975), 1–6.
22a. Kannai, Y., Off diagonal short time asymptotics for fundamental solutions of diffusion equations, Weitzmann Institute, Rehovot, preprint.
22b. Kifer, Yu. I., Brownian motion and harmonic functions on manifolds of negative curvature, *Theory Prob. Appl.* **21** (1976), 81–95.
23. Kobyashi, S., and Nomizu, K., "Foundations of differential Geometry." (2 Vols.), Wiley (Interscience), New York, 1963.
24. Malliavin, P., Formules de la moyenne, calcul de perturbations, et annulation des formes harmoniques, *J. Functional Anal.* **17** (1974), 274–291.
25. Malliavin, P., Asymptotics of the Green's function of a Riemannian manifold and Itô's stochastic integrals, *Proc. Natl. Acad. Sci. USA* **71** (1974), 381–383.
26. Malliavin, P., Integration stochastique des systèmes semi-elliptiques diagonaux dans leur symbole principale, *C. R. Acad. Sci.* **280** (1975), 941–944.
27. Malliavin, P., Diffusions et Géometrie Differentielle Globale, Lecture Notes, Estivio 1975. Edizione Cremonese, 1976.
28. Malliavin, P., Resolution stochastiques de certaines problèmes de derivée oblique dans le demi-espace, *C. R. Acad. Sci.* **283** (1976).
29. Malliavin, P., Annulation de cohomologie et calcul des perturbations, *C. R. Acad. Sci.* **283** (1976).
30. McKean, H. P., Brownian motions on the 3-dimensional rotation group, *Mem. Coll. Sci. Kyoto Imp. Univ.* **33** (1960), 25–38.
31. McKean, H. P., "Stochastic Integrals." Academic Press, New York, 1969.

32. McKean, H. P., An upper bound to the spectum of the Laplacian on a manifold of negative curvature, *J. Diff. Geom.* **4** (1970), 359–366.
33. Michel, D., Comparison des notions de varietes riemanniennes globalement harmoniques et fortement harmoniques, *C. R. Acad. Sci.* **282** (1976), 1007–1010.
34. Molchanov, S. A., Strong Feller property of diffusion processes on smooth manifolds, *Theory Prob. Appl.* **13** (1968), 471–475.
35. Molchanov, S. A., Diffusion processes and Riemannian geometry. *Usp. Math. Nauk.* **30** (1975), 3–59, [translated in *Russian Math. Surveys* **30** (1975), 1–63.]
36. Pinsky, M., Isotropic transport process on a Riemannian manifold, *Trans. Amer. Math. Soc.* **218** (1976), 353–360.
37. Pinsky, M., Temps de sortie des boules normale en fonction de la courbure moyene, *C. R. Acad. Sci.* **282** (1976), 57–58.
38. Pinsky, M., An Individual Ergodic Theorem for the Diffusion on a Manifold of Negative Curvature, *Proc. Conf. Stochastic Differential Equations*, pp. 231–240. Academic Press, New York, 1977.
39. Pinsky, M., The spectrum of the Laplacian on a manifold of negative curvature, *J. Differential Geometry*, 1978.
40. Pratt, J. J., Etude asymptotique du mouvement brownien sur une variété riemannienne à courbure négative, *C. R. Acad. Sci.* **272** (1971), 1586–1589.
41. Pratt, J. J., Etude asymptotiques et convergence angulaire du mouvement brownien sur une variété à courbure négative, *C. R. Acad. Sci.* **280** (1975), 1539–1542.
42. Varadhan, S., On the behavior of the fundamental solution of the heat equation with variable coefficients, *Commun. Pure Appl. Math.* **20** (1967), 431–455.
43. Varadhan, S., Diffusion processes in a small time interval, *Comm. Pure Appl. Math.* **20** (1967), 659–685.
44. Warner, F., "Foundations of Differentiable Manifolds and Lie Groups." Scott Foreseman, Glenview, Illinois, 1971.
45. Vauthier, J., Diffusion sur une variété riemannienne complete à courbure négative, *C. R. Acad. Sci.* **275** (1972), 925–926.
46. Yosida, K., Brownian motion in a homogeneous Riemannian space, *Pacific J. Math.* **2** (1952), 263–270.
47. Yosida, K., On the integration of diffusion equations in Riemannian spaces, *Amer. Math. Soc.* **3** (1952), 864–873.
48. Yosida, K., Brownian motion on the surface of the 3-sphere, *Ann. Math. Stat.* **20** (1949), 292–296.
49. Yosida, K., "Functional Analysis." Springer-Verlag, Berlin and New York, 1965.
50. Yau, S. T., On the heat kernel of a complete Riemannian manifold, UCLA, California, preprint.

# Index

## A

Absolutely continuous measure, 170
Abstract Cauchy problem, 10
Abstract Wiener space, 5, 8
Admissibility, 117
Approximation
   Cauchy polygonal, 113
   closure, 27
   parabolic, 24

## B

Bounded variation, 95
Brownian motion process, 5, 99, 199
   multiparameter, 178
Burgers' equation, 26, 39

## C

Cameron–Yeh process, 178, 186
Canonical measure, 100
Cauchy polygonal approximation, 113
Characteristic functional, 4
Characteristics, method of, 22
Closure error, 27
Closure approximation, 27
Conditional backward derivative, 97
Conditional expectation functional, 13
Conditional forward derivative, 96
Continuously sample differentiable, 95
Controllable dynamical system(s), 49
   mild solution of, 50
Covariance function, 175
Covariance operator, 4, 53
   incremental, 57

## D

Delay equation(s), 48, 62
   stochastic, 52, 72
   filtering for, 68

Delayed observation(s), 80
Dirichlet inner product, 202
Dynamical system(s)
   autonomous, 46
   controllable, 49
   linear, 46
   semigroup property of, 46

## E

Equation(s)
   Burgers', 26, 39
   delay, 48, 52, 62, 68, 72
   heat, 49, 52, 61, 67, 72
   Hopf, 15, 19, 34
   Itô, 93, 119, 140
   Jacobi, 214
   Kolmogorov, 14, 34
   Langevin, 92, 109
      generalized, 90, 109
   Navier–Stokes, 13, 18, 19, 39
   random Helmholtz, 23
   random hyperbolic, 9, 22
   random parabolic, 9, 23
Equilibrium measure, 21
Equilibrium solution, 21
Equivalent measures, 170
   Wiener, 9
Evolution operator(s)
   almost strong, 75
   mild, 17, 74
   quasi, 75

## F

Feynman–Kac formula, 25
Filtering problem, 64
Fréchet derivative, 7
Functional(s)
   characteristic, 4
   conditional expectation, 13

Functional(s) (*continued*)
  moment-generating, 25
  multiplicative, 29
  nonanticipating, 100
Functional integration, 25

## G

Gaussian measure, 5, 179, 188
Gaussian process(es), 177
  generalized, 179, 189
  stationary, 179, 187
  with independent increments, 186
Gaussian random variable, 54
Gaussian white noise, 6, 19, 179, 189
Gauss–Markov measure, 21
Geodesic flow, 205
Geodesic flow field, 206

## H

Heat equation, 49, 61
  noisy, 52, 72
    filtering for, 67
Hilbert–Schmidt norm, 5
Hilbert–Schmidt operator, 5
Hille–Yoshida theorem, 47
Hopf equation, 15, 19, 34

## I

Independent random variables, 54
Index lemma, 222
Innovations process(es), 66
Invariant measure, 155
Isotropic transport process(es), 200
  on Riemannian manifold, 205
Itô equation(s)
  integrodifferential, 119, 140
  second-order, 93, 119
Itô's formula, 101
Itô's lemma, 7

## J

Jacobi equation, 214

## K

Kakutani's theorem, 173
Kolmogorov equation(s), 14, 34
Kraft's theorem, 173

## L

$L_2$ - solution, 104
Langevin equation, 92, 109
  generalized, 80, 110, 128, 136, 139
  nonlinear, 109, 131
Limit set, 154

## M

Mean-square continuity, 54
Mean-square convergence, 54
Mean vector, 4
Measurable norm, 5
Measurable process, 94
Measure(s)
  absolutely continuous, 170
  canonical, 100
  equilibrium, 21
  equivalent, 9, 170
  Gauss–Markov, 21
  Gaussian, 5, 179, 188
    product, 174
  invariant, 155
  singular, 170
  weak convergence of, 54
  Wiener, 8, 9
Mild solution, 50, 59, 106
Moment-generating functional, 25
Multiparameter Brownian motion, 178
Multiplicative functional, 29

## N

Navier–Stokes equations, 13, 18
  linearized, 19
  with random forcing, 19, 39
Nonanticipating functional, 100
Normal neighborhood theorem, 210

## O

Observation process, 80
Operator(s)
  covariance, 4, 53, 57
  evolution, 17, 74, 75
  Hilbert–Schmidt, 5
Ornstein–Uhlenbeck process, 22
  generalized, 21

# INDEX

## P

Parabolic approximation, 24
Prediction problem, 64
Product Gaussian measures, 174

## R

Radon–Nikodym derivative, 9, 171
Random evolution, 29
Random Helmholtz equation, 23
Random hyperbolic equation(s), 9, 22
Random matrix, 22
Random parabolic equation(s), 9, 23
Random transport equation, 29
Random variables, Hilbert space-valued, 53
  expectation of, 53
  Gaussian, 53
  independent, 53
Regulator problem, 61
Reproducing kernel Hilbert space, 176
Riemannian manifold, 201
  Brownian motion on, 202
  isotropic transport process on, 205

## S

Sample continuity, 54, 95
Sample solution, 102
Semigroups of operators
  equicontinuous, 11
  infinitesimal generator of, 47
  stable, 50
  strongly continuous, 47
Separation principle, 71
Sequential Wiener integral, 25
Signal process, 80
Singular measure, 170
Smoothing problem, 64
Sobolev space, 10, 34, 179

Solution process, 11, 12
  nonanticipating, 13
Stabilizibility, 50
State estimation problem, 64
Stationary Gaussian process(es), 179, 187
Stochastic control problem, general, 73
Stochastic initial boundary value problem, 9
  hyperbolic, 9
  parabolic, 9
Stochastic integral(s), 5, 6
  central, 6, 8
Stochastic process(es)
  Hilbert space-valued, 54
  nonanticipating, 5
  simple, 5
  sample continuous, 54, 95
  strongly continuous, 95
Strong derivative, 95
Strong solution, 59, 104

## T

Tracking problem, 73
Transition probability, 13
Turbulent diffusion approximation, 28

## V

Vibrating string, 157

## W

Weak convergence, 54
Wiener measure, 8
  equivalent, 9
  linear transformation of, 9
Wiener process, 5
Wiener–Lévy Brownian motion process, 178, 186

RAYMOND H. FOGLER LIBRARY
DATE DUE